Handbook of Spinal Cord Monitoring

Handbook of Spinal Cord Monitoring

Proceedings of the Fifth International Symposium on Spinal Cord Monitoring, London, UK, June 2–5, 1992

Edited by

S. J. Jones
Department of Clinical Neurophysiology
The National Hospital for Neurology and Neurosurgery
Queen Square
London WC1N 3BG
UK

S. Boyd
Department of Clinical Neurophysiology
The Hospital for Sick Children
Great Ormond Street
London WC1N 3JH
UK

M. Hetreed
Royal National Orthopaedic Hospital
Brockley Hill
Stanmore
Middlesex HA7 4LP
UK

N. J. Smith
Clinical Neurophysiology Department
Queen's Medical Centre
University Hospital
Nottingham NG7 2UH
UK

Sponsored by the Dunhill Medical Trust

KLUWER ACADEMIC PUBLISHERS
DORDRECHT / BOSTON / LONDON

Distributors

for the United States and Canada: Kluwer Academic Publishers, PO Box 358, Accord Station, Hingham, MA 02018-0358, USA
for all other countries: Kluwer Academic Publishers Group, Distribution Center, PO Box 322, 3300 AH Dordrecht, The Netherlands

A catalogue record for this book is available from the British Library.

ISBN 0-7923-8833-X

Library of Congress Cataloging in Publication Data

International Symposium on Spinal Cord Monitoring (5th : 1992 : London, England)
Handbook of spinal cord monitoring : proceedings of the Fifth International Symposium on Spinal Cord Monitoring, London, UK, June 2–5, 1992 / edited by S.J. Jones ... [et al.].
p. cm.
Includes bibliographical references and index.
ISBN 0-7923-8833-X (case bound)
1. Spinal cord—Surgery—Congresses. 2. Intraoperative monitoring—Congresses. 3. Evoked potentials (Electrophysiology)—Congresses. 4. Spinal cord—Diseases—Diagnosis—Congresses. 5. Spinal cord—Physiology—Congresses I. Jones, S. J. (Stephen J.) II. Title.
[DNLM: 1. Spinal Cord—physiology—congresses. 2. Spinal Cord—surgery—congresses. 3. Monitoring, Physiologic—congresses. 4. Evoked Potentials—congresses. WL 400 I615h 1992]
RD594.3.I58 1992
617.4′82—dc20
DNLM/DLC
for Library of Congress 93-31307
CIP

Published in the United Kingdom by Kluwer Academic Publishers, PO Box 55, Lancaster, UK.

Kluwer Academic Publishers BV incorporates the publishing programmes of D. Reidel, Martinus Nijhoff, Dr W. Junk and MTP Press.

Typeset by Lasertext Limited, Stretford, Manchester M32 0JT.

Printed in Great Britain by Hartnolls, Bodmin, Cornwall.

Contents

CONTENTS

Animal studies

List of Principal Authors

H. Baba
Department of Orthopaedic Surgery
Fukui Medical School
Shimoaizuki 23
Matsuoka, Fukui 910-11
Japan

S. M. Bowles
Emory University School of Medicine
Atlanta
Georgia
USA

F. Bracchi
Institute of Human Physiology
University of Milan
Via Mangiagalli 32
20133 Milan
Italy

C. Callegarini
Neurological Department
S. Maria delle Croci Hospital
Ravenna
Italy

D. Claus
Department of Neurology
University of Erlangen-Nürnberg
Schwabachanlage 6
W-8520 Erlangen
Germany

V. Deletis
New York University Medical Center
New York
USA

M. E. Fennelly
The Royal National Orthopaedic Hospital Trust
Stanmore
UK

J. E. Fitzgerald
Regional Medical Physics
Freeman Hospital and Royal Victoria Infirmary
Newcastle upon Tyne
UK

M. H. Fraser
Regional Spinal Injuries Centre
Southport District General Hospital
Town Lane
Southport, Merseyside
UK

W. T. Frazier
Department of Anesthesiology
Emory University School of Medicine
Atlanta
Georgia
USA

A. J. Gale
Wessex Neurological Centre
Southampton General Hospital
Tremona Road, Shirley
Southampton SO9 4XY
UK

C. J. Griffiths
Freeman Hospital and Royal Victoria Infirmary
Newcastle upon Tyne
UK

J. M. Guérit
Cliniques Universitaires St-Luc
Unité d'Explorations Electrophysiologiques du Système Nerveux
Avenue Hippocrate 10
B-1200 Brussels
Belgium

J. A. Halter
Division of Restorative Neurology and Human Neurobiology
Baylor College of Medicine
One Baylor Plaza
Houston Texas 77030
USA

P. D. Heath
Department of Orthopaedics
North Staffordshire Royal Infirmary
Stoke-on-Trent
UK

L. Henderson
University Hospital
Nottingham
UK

J. Herdmann
Department of Neurosurgery
Heinrich-Heine-University
Moorenstr. 5
4000 Düsseldorf 1
Germany

M. A. Hetreed
Royal National Orthopaedic Hospital
Brockley Hill
Stanmore
Middx HA7 4LP
UK

Y. Hiraizumi
Department of Orthopaedic Surgery
Showa University School of Medicine
1-5-8 Hatanodai
Shinagawa-ku, Tokyo 142
Japan

T. Iizuka
Department of Orthopaedic Surgery
Saitama Medical School
38 Norohongo, Moroyama, Iruma
Saitama 350-04
Japan

M. Imai
Department of Orthopaedic Surgery
Asahikawa Medical College
Nishikagura 4-5
078 Asahikawa
Japan

T. Iwahara
Department of Orthopaedic Surgery
Asahikawa Medical College
Nishikagura 4-5
078 Asahikawa
Japan

T. Izumi
Department of Orthopaedic Rehabilitation
Kanazawa Medical University
1-1 Daigaku, Uchinada-machi
Kahoku-gun, Ishikawa 920-02
Japan

S. J. Jones
Department of Clinical Neurophysiology
The National Hospital for Neurology and Neurosurgery
Queen Square
London WC1N 3BG
UK

H. Kajihara
Department of Orthopaedic Surgery
Juntendo University
School of Medicine
Tokyo
Japan

R. Kakigi
Department of Internal Medicine
Saga Medical School
Nabeshima
Saga City 849
Japan

Y. Katayama
Department of Neurological Surgery
Nihon University School of Medicine
Tokyo 173
Japan

N. Kawahara
Department of Orthopaedic Surgery
Kanazawa City Hospital
3-7-3 Heiwamachi Kanazawa
Ishikawa 921
Japan

H. Komori
Department of Orthopaedic Surgery
Toride Kyodo General Hospital
5901-1 Terada Toride-City
Ibaragi Pref. 302
Japan

K. Kothbauer
Department of Neurosurgery
Inselspital
University of Berne
Berne
Switzerland

LIST OF PRINCIPAL AUTHORS

B. Meyer
Neurochirurgische Klinik
Klinikum Kalkweg
Zu den Rehwiesen 9
4100 Duisburg
Germany

M. Murakami
Department of Orthopaedic Surgery
School of Medicine Chiba University
1-8-1 Inohana, Chuô-ku, Chiba-shi
Chiba 260
Japan

Y. Node
Department of Neurosurgery
Nippon Medical School
Tokyo
Japan

M. Nozawa
Department of Orthopaedic Surgery
Juntendo University School of Medicine
2-1-1 Hongo, Bunkyo-ku
Tokyo
Japan

M. R. Nuwer
Department Head Clinical Neurophysiology
Reed Neurological Research Center
710 Westwood Plaza Room 1-194
Los Angeles, CA 90024-6987
USA

S. L. Papastefanou
Department of Orthopaedic and Accident Surgery
Queen's Medical Centre
University Hospital
Nottingham
UK

U. Pechstein
Neurosurgical Clinical
University of Bonn
Sigmund-freud-Str. 25
W-5300 Bonn 1
Germany

S. Sato
Department of Orthopaedic Surgery
University of the Ryukus
207 Uehara, Nishihara
Okinawa 903-01
Japan

J. Schramm
Department of Neurosurgery
University Hospital
Sigmund-Freud-Str. 25
W-5300 Bonn 1
Germany

E. M. Sedgwick
Clinical Neurological Sciences
Southampton General Hospital
Tremona Road
Southampton SO9 4XY
UK

K. Shimoji
Department of Anesthesiology
Niigata University School of Medicine
#1-757 Asahi-Machi
Niigata 951
Japan

T. Sloan
The University of Texas
Health Science Center
San Antonio
Texas
USA

N. J. Smith
Clinical Neurophysiology Department
Queen's Medical Centre
University Hospital
Nottingham NG7 2UH
UK

T. Tamaki
Department of Orthopedic Surgery
Wakayama Medical College
Wakayama
Japan

K. T. Tamura
Department of Neurosurgery
Hiratsuka City Hospital Minamihara 1-19-1
Hiratsuka
Kanagawa 254
Japan

T. Tani
Kochi Medical School
Kochi
Japan

B. A. Taylor
The Royal National Orthopaedic Hospital Trust
Stanmore
UK

Y. Toda
Department of Orthopaedic Surgery
Otokoyama Hospital
Kansai Medical University
Japan

W. Wang
Department of Neurosurgery
Mount Sinai Medical Center
New York, NY
USA

T. Yamamoto
Department of Neurological Surgery
Nihon University School of Medicine
Tokyo 173
Japan

M. Zileli
Ege University Faculty of Medicine
Department of Neurosurgery
Bornova
Izmir
Turkey

Preface

S. J. JONES, S. BOYD, M. HETREED and N. J. SMITH

The fifth in the series of International Symposia on Spinal Cord Monitoring, held in the fashionable heart of London, was an occasion notable both for the warmth of friendship amongst colleagues from all over the globe and for the exciting developments which had taken place in the three years since the previous meeting in Japan. The consolidated ground was represented by the invited views of surgeons, anaesthetists and neurophysiologists who had been pre-eminent in the field for many years. Most notably, results were presented of the first large-scale survey of the use of sensory spinal cord monitoring techniques in almost 60 000 operations. The direction of future development, on the other hand, was clearly indicated by the large proportion of research papers devoted to techniques for monitoring the motor pathways of the cord. This has been a major objective ever since surgeons first began to perceive the limitations of the 'wake-up' test.

Our aim in planning the Symposium and Proceedings was to extend their interest beyond those actively engaged in research, to provide an introduction and source of reference for surgeons and others who perceive the desirability of spinal cord monitoring but have been deterred from implementing it by a lack of specialist knowledge or assistance. Such readers are directed mainly to the first eight chapters of this book. The research papers which form the bulk of the text are largely of specialist interest, although many are written from a practical perspective and should not deter the less experienced reader.

Our gratitude is due to many companies and individuals who have supported us, particularly the British Scoliosis Research Foundation, the British Orthopaedic Association and the Scoliosis Research Trust ARISE. Not least, grateful thanks are extended to the Dunhill Medical Trust for making possible the publication of this volume.

Invited Chapters

1
Somatosensory evoked potentials and their use for spinal cord monitoring

S. J. JONES

INTRODUCTION

Spinal cord monitoring is now used in many centres specializing in different kinds of surgery, and those who are responsible for implementing it come from a wide range of fields. It is understandable, therefore, that sometimes the individuals concerned have little grounding in human neurophysiology and no previous experience of recording evoked potentials. The object of this chapter is to provide a primer in the neurophysiology of the sensory pathways and the use of SEPs for spinal cord monitoring.

EVOKED POTENTIALS OF THE NERVOUS SYSTEM

The sole function of the peripheral nerve fibres (axons) and a major function of the spinal cord (largely consisting of bundles of axons) is to feed sensory information from sense organs to the brain and motor commands from the brain to the musculature. Nerve fibres operate by the flow of electricity, but the analogy with a simple electrical lead is not a close one. For one thing, whereas conduction along a lead is effectively instantaneous, the velocity of impulses along a nerve fibre is actually quite slow, ranging from about 1 m/s to 80 m/s according to the diameter (larger meaning faster) and construction of the axon. The other important difference is that the electricity of nerve fibres and brain cells 'leaks', such that a small amount of current flows in the surrounding tissue and causes differences in electrical potential to be detectable on the surface of the skin. Although the potential difference between any two points on the skin is seldom more than a few microvolts (μV, volts $\times 10^{-6}$), it can be amplified and its change over the period during which the nerve is active displayed as a waveform on an oscilloscope screen.

Most nerve trunks contain a mixture of sensory and motor fibres. The

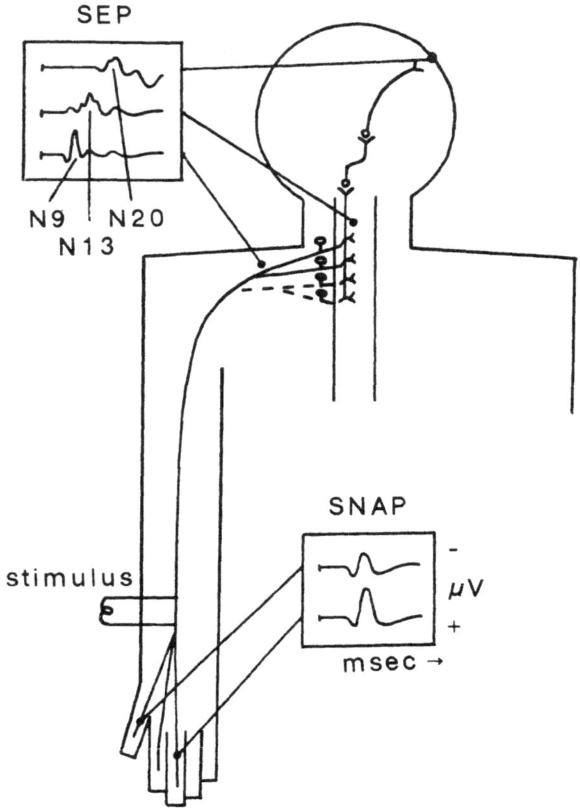

Figure 1 Somatosensory evoked potentials and antidromic sensory nerve action potentials after median nerve stimulation at the wrist

former can be activated by stimulating their associated sense organs with an appropriate 'natural' stimulus. Alternatively, both sorts of nerve fibre can be made active by applying a brief electrical shock directly over the nerve with a pair of stimulating electrodes. Electrodes are placed on the skin over the nerve at a point where it runs relatively close to the surface, for example the median nerve at the wrist. The applied electric shock initiates an impulse (the nerve action potential, NAP) in each stimulated fibre, starting at the point of stimulation and travelling along the nerve in both directions at a certain fixed velocity. The 'compound NAP' is the sum of all the individual NAPs when many fibres are activated. Action potentials in a purely sensory nerve are SNAPs; SNAPs can be recorded from appropriate fingers of the hand when the mixed (sensory and motor) median nerve is stimulated at the wrist (Figure 1), since the digital nerves are purely sensory. In this instance the SNAPs are being conducted the wrong or 'antidromic' way, away from the head. The currents flowing in the tissue surrounding the nerve cause the electrical potential, where an adjacent recording electrode is attached, first to become positive, reach a peak, turn negative, reach a second peak and finally become positive again before diminishing to zero. This characteristic triple-peaked waveform of voltage against time may be displayed on the screen with positive up or down, as indicated by polarity symbols on the voltage calibration bar (Figure 1).

A brief (less than 1 ms) electrical shock delivered to a nerve trunk is capable of activating a large proportion of the large diameter fibres present, without causing significant pain. Each fibre is activated in an 'all-or-nothing' fashion (i.e. on increasing the stimulus intensity the fibre will abruptly commence firing and the discharge will be the same at all higher intensities) but the number of fibres which are activated by a given stimulus, and thus the size of the compound NAP, depends on the intensity of the shock and the range of diameters of the fibres present at the site of stimulation. The largest diameter fibres are activated at the lowest intensities; these are sensory axons concerned with tactile and position sense, and motor fibres responsible for movement. Large fibres are all ensheathed in layers of the fatty substance known as myelin, which prevents electrical current from leaking into or out of the fibre for most of its length and increases the speed of conduction by causing the NAP to progress along the nerve in a series of leaps from one break in the myelin to the next.

Electrical potentials can only be measured at one site relative to another, and when the reference electrode is relatively remote from the nerve such that it may be assumed to be 'indifferent' to the generated activity, the recording is said to be 'monopolar'. A second electrode a few cm further along the nerve will record a similar sequence of potential changes as the first, but a short time later. In a 'bipolar' recording the signal from the first electrode is displayed relative to that at the second, by feeding them to the two inputs of a differential amplifier. The amplifier compares the two signals and amplifies the difference between them. Bipolar recordings tend to be more complex than monopolar, but have the advantage that any spurious signals picked up by the electrodes (usually due to electrical apparatus or metal objects in the vicinity) are of similar size at each, and hence tend to be cancelled out by the amplifier which is sensitive only to the difference in potential. Even after amplification the NAP may be contaminated by electrical 'noise', hence it is usual to accumulate the responses to a number (sometimes several hundreds) of stimuli and compute their average in a purpose-built computer (see below).

In addition to the antidromic SNAPs in the fingers, stimulating the median nerve at the wrist induces a compound NAP in sensory and motor fibres which travels up the arm to the spinal cord. Here and in the brainstem sensory fibres originating in the periphery connect with neurones of the central nervous system. The potentials recorded from the head and neck in association with the transmission and processing of sensory information in the spinal cord and the brain are the 'somatosensory evoked potentials' (SEP, Figure 1). After the commencement of activity in the sensory cortex, cortical SEPs can be recorded from quite wide areas of the scalp. Electrodes on the head can also detect auditory evoked potentials generated in the

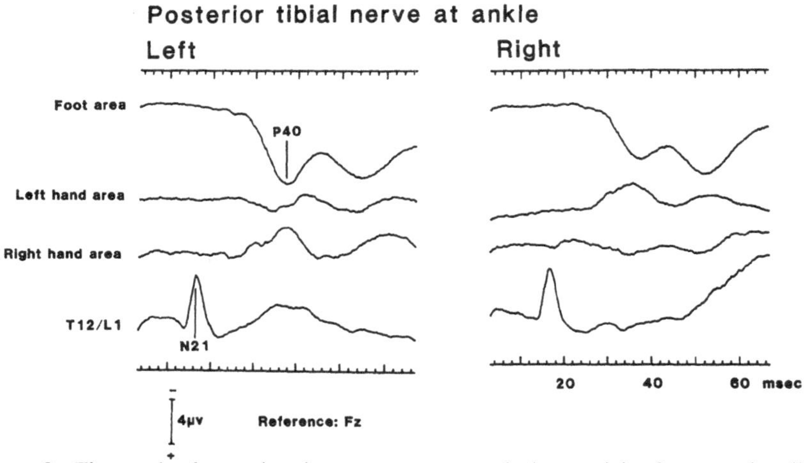

Figure 2 Thoraco-lumbar and scalp somatosensory evoked potentials after posterior tibial nerve stimulation at the ankle

acoustic nerve, brainstem and auditory cortex, and visual evoked potentials generated in the occipital cortex.

The conventional way of identifying particular waves or components of the evoked potential is by the polarity of the peak, positive (P) or negative (N), and its latency in milliseconds (i.e. the time elapsed since the nerve or sense organ was stimulated). So the main SEP components to median nerve stimulation at the wrist are an N9 over the clavicle, an N13 over the neck and an N20 on the scalp, followed by other waves of positive and negative polarity. When the posterior tibial nerve is stimulated at the ankle an N21 can usually be recorded on the back within one or two segments of the thoraco-lumbar junction, while the first cortical SEP component is a P40 (Figure 2). The latency nomenclature represents the notional mean latency across the normal adult population. In a given subject, however, the precise latency depends on a number of factors including height, arm length and age. The largest peripheral nerve fibres tend to degenerate during adulthood, causing SEPs to become progressively dominated by the contribution of smaller fibres which have a slower velocity of conduction. SEP latencies may also be prolonged when the nerves are diseased or partially injured.

The other useful measure of an SEP component is its amplitude, i.e. the highest or lowest voltage attained relative to the zero-potential baseline or the preceding peak of the opposite polarity. It is noteworthy that, whereas latency is the more useful parameter when SEPs are recorded for the investigation of chronic neurological disease (its variability across the normal population is proportionately much less than that of amplitude), for monitoring purposes amplitude changes are of greater significance since the problems which occur are usually due to acute blockage of conduction. Normal variability is also much lower between SEPs recorded sequentially from the same individual.

RECORDING APPARATUS

The most basic form of EP recorder, sufficient for many surgical monitoring applications, consists of an electrical nerve stimulator, an amplifier, a signal averaging computer, an oscilloscope display and a plotter or printer for hard copy. Most current machines are capable of recording from two or more electrodes simultaneously, and possess a floppy disc drive for storage of the waveforms. The amplification is in fact performed in a small preamplifier or headbox, connected to the main machine by a lead 2 m or more in length. Signals from the recording electrodes are fed to the headbox via relatively short (c. 1 m) leads. The reason for this is that the leads from the patient carrying very small signals are liable to pick up interference from other electrical apparatus; the amount of such interference is proportional to the length of the lead, hence it is important that the amplification be made as close as possible to the patient. The headbox should be compact enough to be hung underneath the operating table, out of the way of the surgeon, while the main recording machine can be anywhere in the theatre.

Intraoperative monitoring machines are required to conform to the same safety standards as other electrical apparatus in contact with the patient. It is particularly important that currents generated by the diathermy or bipolar coagulator cannot escape to earth via the recording or stimulating electrodes. If this happens, the area of skin, dura or nerve tissue in contact with the electrode may carry a high current and suffer burning. It is also essential that the amplifier be protected against such currents, usually by some sort of automatic circuit breaker. Monitoring is never possible while diathermy is in use but the system should be capable of recovery in a very few seconds after diathermy ceases. The accumulation of responses by the signal averaging computer should cease automatically when the amplifier is ‘blocked’ (i.e. the input circuit is overloaded) and when occasional signals of large amplitude occur, since these are sure to be artifacts.

With up-to-date machines offering multiple (usually four) channel operation it is possible to monitor SEPs from several sites simultaneously; at the same time as recording the crucial signals which depend on conduction through the segments at risk, potentials can be recorded from caudal sites in order to establish that, for whatever reason, there has been no interruption of input to the cord from the stimulated limb. Some machines offer the possibility of alternating the stimuli rapidly between two sites and monitoring the SEPs concurrently in separate signal averaging stores. This is particularly appropriate when monitoring lower limb SEPs, since the same electrodes are used to record spinal and cortical SEPs to stimulation on both sides. It is desirable to have a further semi-permanent store which holds a ‘baseline’ recording for quick comparison with subsequent SEPs. Some machines offer an automatic mode in which new recordings are made at predetermined intervals and plotted or stored to disc with a note of the time of recording. On storing individual records, it should be possible for the operator to add some comment about the surgical and anaesthetic conditions at the time. Table 1 lists the main features of an ideal machine for spinal cord monitoring, roughly in order of priority.

Table 1 Desirable features of the recording machine used for spinal cord monitoring

Full electrical isolation
Input stage protection circuitry
Fast recovery from block
Compact size
Amplification in remote head box
Resolution >500 points/channel
High contrast display
High quality hard copy
Multiple memory stores
Flexible artefact rejection
Trigger synchronization with external stimulator
On-board, fast access disc storage
Fine control of stimulus intensity for direct nerve stimulation
Stackable display
Four channels
Automatic monitoring mode
Automatic timekeeping
Full keyboard
Measurements and comments stored with trace
Superimposition, addition and subtraction of traces
Double pulse stimulator
Dual alternating stimulators with separate averager stores
Adaptable program, parameters stored on switch-off
PC-compatible discs
Serial interface
Dual timebase
Video connection for screen display
Auditory stimulator/earpiece for AEP monitoring
Visual stimulator/LED goggles for VEP monitoring

NERVE STIMULATION

The electrical stimulus is usually a square pulse which can be adjusted in terms of its duration (typically 0.1–1.0 msec) and intensity (up to about 300 volts for a 'constant voltage' stimulator or about 50 milliamps for a 'constant current' type). The difference between constant voltage and constant current concerns how the stimulator reacts to any change in the resistance of the contact between the stimulating electrode and the skin, but for most purposes can be neglected. Increasing the duration of the pulse affects the nerve in a similar way as increasing the intensity, that is more fibres of progressively smaller diameter will begin to be activated. Using a standard stimulator it is not possible to activate in any significant numbers the smallest diameter myelinated and unmyelinated fibres which are responsible for sensation of heat, cold and pain.

The pair of electrodes placed on the skin for nerve stimulation usually has a separation of at least 3 cm, although closer pairing may be necessary in very young children. The electrode which becomes negatively charged (the cathode, connected to the black socket of the stimulator) is the one actually responsible for activating the nerve and needs to be in as close proximity as possible. The position of the positive electrode (the anode) is less critical.

The surface which contacts the skin may be of metal, although frequently this is covered in lint or felt soaked in saline. The skin resistance may be reduced slightly by rubbing with an alcohol-impregnated wipe, and a conducting gel is applied to the electrode to prevent the contact from drying too quickly. Subdermal needle electrodes can be used, but are only of any real advantage when it is necessary to stimulate a nerve which runs very deep. Self-adhesive ECG electrodes make highly satisfactory stimulating electrodes for intraoperative spinal cord monitoring, since they are secure and place no pressure on the stimulation site. Their large size may make it convenient to place the anode on the opposite side of the limb from the cathode, taking care to avoid proximity with other nerve trunks (it might be thought advantageous to stimulate more than one nerve simultaneously, but this is generally to be avoided since the associated SEPs may have slightly different latencies such that overlapping positive and negative potentials at the recording electrodes tend to cancel one another out). If the electrodes are slightly misplaced, or the nerve trunk runs very deep, some degree of compensation can be achieved by increasing the stimulus intensity, but it is better to find the optimum location. When a mixed (motor and sensory) nerve is stimulated, a good indicator of an adequate stimulus intensity is a brisk twitch of the innervated muscles, but this may be difficult to obtain in surgical patients who have had muscle relaxants administered.

Each nerve fibre requires a millisecond or two in which to recover before it can be stimulated again. Cortical SEPs, however, take very much longer. Particularly in patients under anaesthesia, cortical SEPs are reduced in size when the stimuli are delivered at a rate of more than two or three per second, and may be abolished completely at rates exceeding about 10/s. The optimal rate of stimulation is one which causes a minimal degree of attenuation, but yet is fast enough to be able to average a reasonable number of responses in a given time. The best compromise will usually be between 3 and 5 stimuli/s, although lower rates may be necessary under deep anaesthesia. Spinal SEPs are considerably more resistant to attenuation, both by anaesthetic agents and high rates of stimulation, and the latter can be as high as 10 or 20/s without causing any significant degradation.

In addition to the electrodes used for stimulation, a third electrode is usually attached to the limb proximal to the stimulation site and connected to earth (ground, often a green socket) via the headbox. Only one such connection is necessary, even if the stimulus is to be given to more than one limb. An electrode on another limb or the head will usually serve equally well. Although the location of the earth electrode is unimportant, it is vital that the resistance of the contact be low (less than 5 kΩ) or the recordings will be contaminated by noise of various sorts. It is often a good idea to change the earth connection if a recording is unacceptably noisy.

RECORDING ELECTRODES

Because the electrical potentials recorded from the body are so much smaller than those used to stimulate a nerve, more care needs to be taken with the

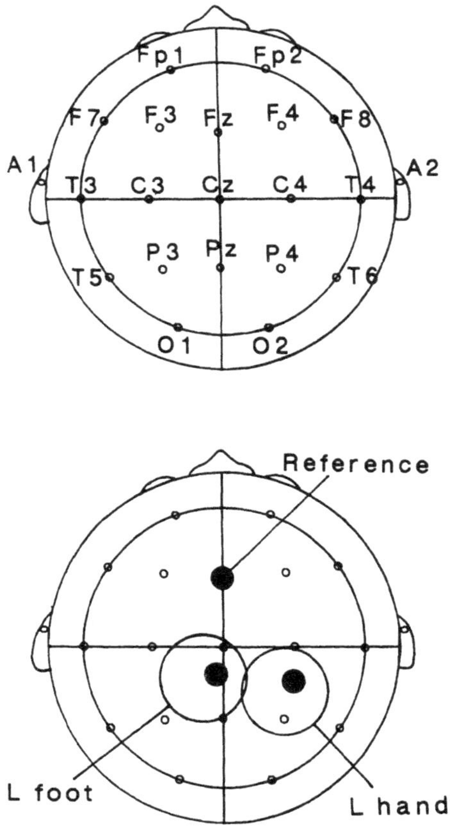

Figure 3 Upper: principal scalp recording sites of the International 10–20 System. Lower: optimal (filled circle) and acceptable (open circle) recording locations for left upper and lower limb SEPs

attachment of recording electrodes. Conventional silver or gold EEG electrodes are attached to the skin with fast-setting glue (collodion), tape or conductive paste. It is critical that the contact resistance be reduced below 5 kΩ, either by rubbing the area with a mildly abrasive agent or by subsequently scratching the surface layer of skin with a serrated-tip needle inserted through the central hole in the electrode. The same needle is used to fill the electrode cup with conducting gel. Subdermal needle electrodes have the advantage that they are quicker to attach, but they are also more easily dislodged and are prone to artifacts when the patient is moved on the operating table.

The optimal location for recording cortical SEPs to stimulation of nerve trunks in either leg, the 'sensory foot area', is on the midline of the scalp about 2 cm posterior to the vertex (Cz, the mid-point between the nasion and the inion, Figure 3). The response can be recorded within a few cm all round, but tends to fall off more rapidly in a frontal direction. This 'active' electrode is connected to the black input (Grid 1) of the headbox. The reference electrode, connected to the red input or Grid 2, is usually located in the frontal region at Fz. For recording cortical SEPs from the arms the active electrode location (the 'sensory hand area') is found by marking the foot area and measuring about 7 cm towards the ear on the opposite side from the stimulus. The reference electrode is again situated in the frontal region.

When more than one recording channel is available, SEPs can also be recorded at levels below the segments at risk so as to be sure that, in the event of the loss of cortical SEPs, the input from the periphery to the spinal cord is still present. Unfortunately it is often difficult to record from the skin potentials generated in the lumbar nerve roots or the cauda equina, but the response is more consistent where sensory nerves make their first synapse in the cord at about T12 vertebral level. With upper limb stimulation, the peripheral NAP can be monitored with an electrode over the brachial plexus at Erb's point and spinal cord potentials can be recorded from the cervical region. However, if a scalp reference electrode is used the spinal SEPs are mixed with potentials of about the same latency generated further rostrally and picked up by the reference. Such potentials, generated at about the level of the foramen magnum, can be recorded from widespread scalp locations following upper or lower limb stimulation, as can be seen when the reference electrode is at some more neutral site off the head.

Whereas on the skin spinal SEPs can only be reliably recorded adjacent to the root entry level of the nerve fibres concerned, using electrodes attached to the spinal column or inserted into the epidural space they can be recorded at all active levels of the cord. Various possibilities have been explored for locating invasive spinal cord monitoring electrodes during correction of spinal deformities; these include steel screws in the dorsal processes and needles in the interspinous ligaments. More satisfactory than either, from the point of view of reliability and minimal interference with surgery, are leads about 1 mm in diameter threaded into the dorsal epidural space, either directly when the dura has been exposed or percutaneously through a Tuohy needle (Figure 4). When the lead is introduced it naturally enters the epidural space at an oblique angle and can be advanced so that the active recording tip (about 3 mm of stainless steel) is well above the operating zone. As a reference electrode a needle can be inserted into the paraspinal muscles, or a second steel contact can be mounted on the epidural lead 1–2 cm caudal to the active electrode. The latter is probably the most satisfactory arrangement from the point of view of minimizing the electrical 'noise', since when the active and reference electrodes are close together their noise signals are similar and are mostly cancelled out by the differential amplifier. Spinal cord potentials, resembling SNAPs but with two, three or more negative peaks, can be recorded from all levels of the spine (Figure 5). They possess several advantages over cortical SEPs for intraoperative monitoring, the most important being their greater resistance to anaesthetic agents and to high rates of stimulation. Their only significant disadvantage is the invasive

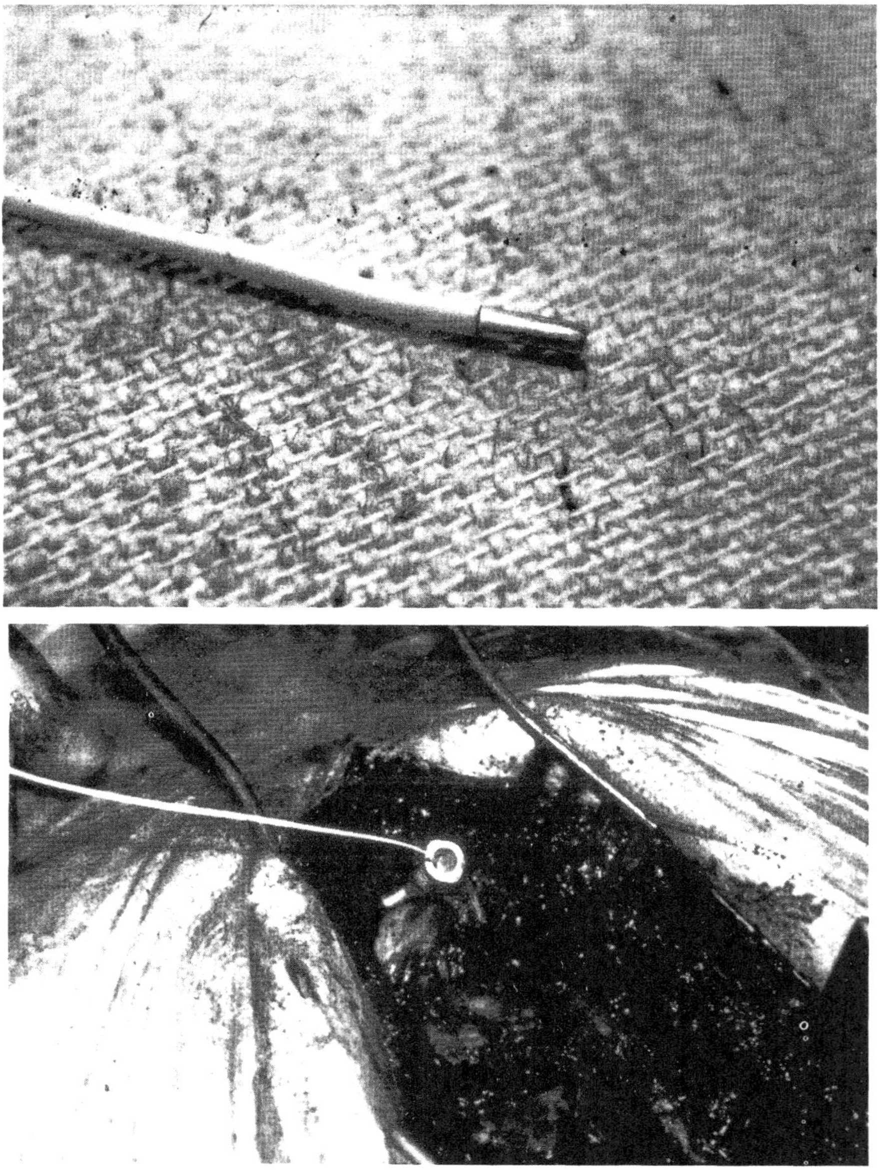

Figure 4 Upper: epidural electrode used for spinal cord monitoring (tip dimensions 3 × 1 mm). Lower: Insertion of epidural recording electrode via 16-gauge catheter

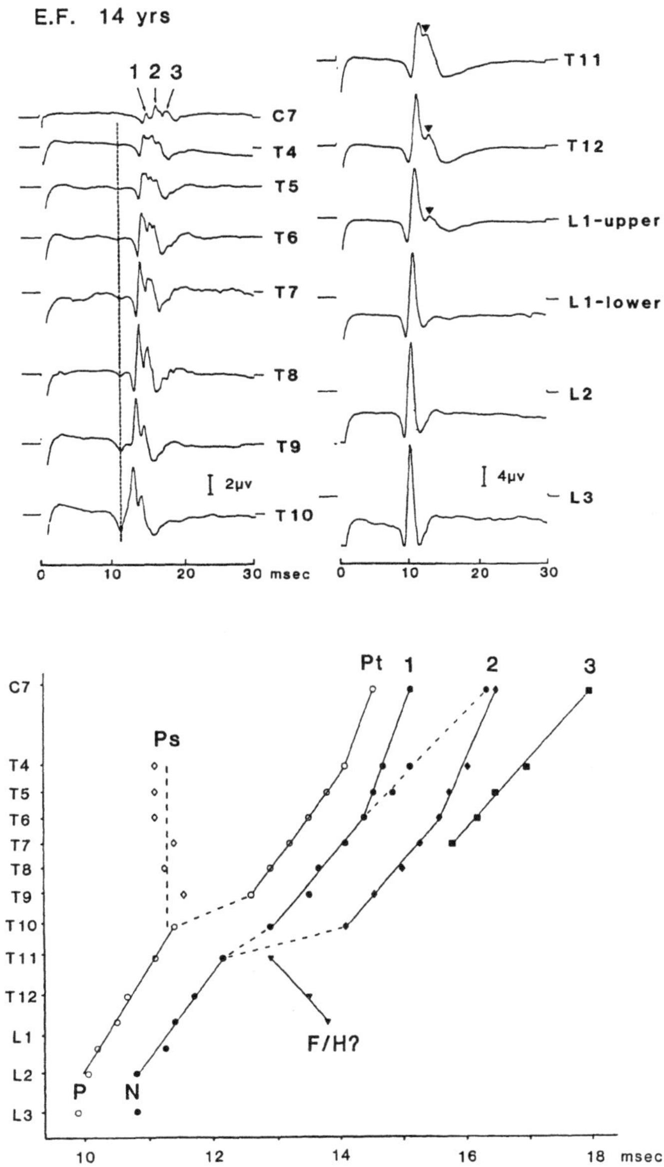

Figure 5 Multi-level recording of epidural SEPs to tibial nerve stimulation at the knee, peak latency plotted against length and level of spinal cord[1]

placement, making their use undesirable in patients with an obstructed spinal canal and impossible for operations on the high cervical region.

AMPLIFICATION

On the recording machine it is necessary to select the limits of the frequency range of the signals which are to be amplified, the gain factor by which they are to be raised and the length of time after each stimulus which is to be displayed on the screen. The frequency range of SEP activity extends from a few cycles/sec (Hz) for the late cortical waves to more than 1000 Hz for peripheral nerve and spinal cord potentials. When spinal SEPs are recorded from the epidural space, low frequency activity is largely irrelevant and the 'high pass' amplifier filter setting can be set to exclude frequencies below, say, 100 Hz. This has the advantage of excluding much of the noise due to other electrical apparatus, based on the mains frequency of 50 Hz or 60 Hz. In cortical SEPs a large proportion of the electrical energy is at a much lower frequency, so the high pass filter will usually be set at 10–30 Hz. Unless conditions are unusually noisy, the upper frequency limit ('low pass' filter) can be set at 2000 Hz for both spinal and cortical SEPs, but for the latter a lower setting of 1000 Hz or 500 Hz may occasionally give cleaner-looking recordings.

Neurophysiological amplifiers have a wide range of gain settings, but the most appropriate for spinal cord monitoring is usually one which magnifies signals of 20 microvolts to a height of about 1 cm on the screen. Further amplification will have the effect of sending the trace off the screen for long periods, causing the amplifier to 'block'. At 20 μV/cm it is still usually impossible to distinguish the SEP signals from the background noise, so further enhancement is obtained by averaging the responses to a number of stimuli.

The principle of electrophysiological averaging is that the nerve signals of interest occur with a fixed latency after each stimulus and are of the same polarity (positive or negative) and shape each time. Background noise, on the other hand, is assumed to be entirely random. Consequently, when the signals recorded after each stimulus are held in a buffer memory, aligned with one another using the moment of stimulus delivery as the marker and then averaged, the SEP peaks and troughs which occur with consistent polarity and latency will tend to remain in the average but the noise will be diminished. Usually the computation is performed anew after each stimulus and the 'running average' displayed on the screen underneath (or optionally to) the raw input signal. If both the signal and the mean noise level are constant, the signal/noise ratio will improve by a factor equal to the square root of the number of responses averaged. For spinal cord monitoring, depending on the size of the signal and the level of the noise, it may be necessary to average as many as 1000 or as few as 10 or 20 responses. It is unrealistic, however, to expect the averaging process to cope with very high noise levels associated, for example, with use of the diathermy, and it may also be defeated when the noise occurs in sudden bursts or consists of large, abrupt spikes.

It takes approximately 20 ms for nerve signals to arrive at the sensory cortex from the wrist and about 40 ms from the ankles. In conscious subjects cortically generated SEPs can be recorded for at least a further 100 ms, but these late waves are usually abolished by anaesthesia. Consequently the epoch over which signals are averaged and displayed should be no longer than 100 ms after each stimulus pulse, possibly only 50 ms for the upper limb SEPs. An even shorter epoch of 30 ms may be chosen if it is only spinal cord potentials which are being monitored.

PROBLEMS

Much has been made of the difficulty of recording small bioelectrical signals in the operating theatre. In fact, with today's recording machines it is often easier to record SEPs from anaesthetized patients than from the same individuals while awake, when the same degree of relaxation (muscle-generated electrical activity being the most potent source of interference in tense individuals) and cooperation cannot be achieved. Fundamental precautions for intraoperative monitoring are to ensure that the electrode contacts are of low resistance and that leads carrying unamplified signals do not pass too close to other electrical apparatus. At worst there may be sources of interference (X-ray lights, electrically heated blankets) which need to be switched off before satisfactory recordings can be obtained. Interference is likely to increase when surgical instruments are in close proximity to the recording electrodes, since they tend to act as antennae. Practical experience is by far the most effective way of discovering the source of recording difficulties, and how to overcome them.

GENERATORS OF HUMAN SEPS

Peripheral nerve

After median nerve stimulation at the wrist, the compound NAP travelling towards the head can be recorded from various sites where the nerve runs close to the surface. At the elbow the main negative peak has a latency of about 4 ms and an amplitude usually more than 20 μV. Over the brachial plexus the latency is about 9 ms and the amplitude usually less than 10 μV (sometimes as little as 1–2 μV). The clavicular potential is usually given the same nomenclature as later SEP components, termed N9 according to its negative polarity and mean latency in ms. Some of this activity is conducted to the head, and is recorded by scalp electrodes as a positive peak of similar latency (this requires the reference electrode to be located on a relatively uninfluenced region such as the opposite shoulder). Following lower limb stimulation it is not possible to record an equivalent of the median nerve N9, since there is no point where the nerve roots of the lumbosacral plexus run close enough to the skin, but as a monitor of the peripheral nerve volley

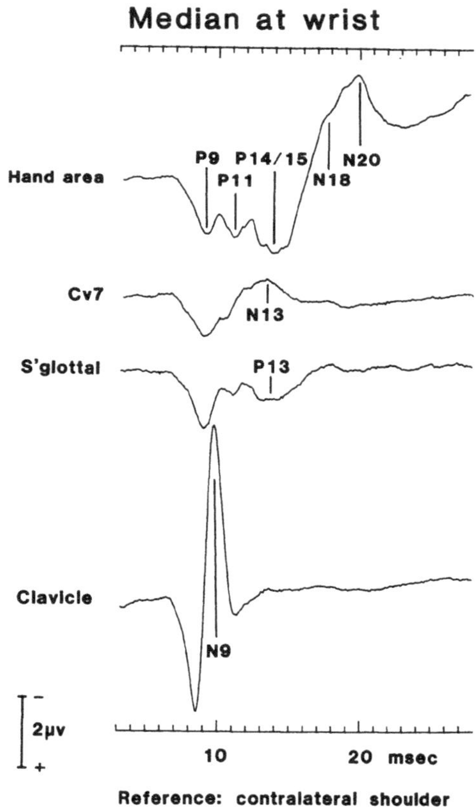

Figure 6 Main short-latency components of SEP to median nerve stimulation at the wrist: scalp, neck (posterior, anterior) and clavicle electrodes with shoulder reference. Scalp reference is usually preferable for monitoring

one can record the NAP at the knee with a pair of electrodes over the popliteal fossa.

Spinal cord

About 2 ms after the N9, potentials generated in the proximal dorsal root segments and the spinal cord at the root entry level for median nerve sensory fibres (mainly C6 and C7) are recorded as a negative peak on the dorsum of the neck (N11) and a positive peak of similar latency on the scalp. Following N11 at the neck electrode there is a larger peak, N13, generated by interneurones in the spinal cord. N13 spreads for 2–3 segments rostral and caudal to the root entry level on the posterior side. On the anterior side activity in the same interneurones gives rise to a P13 (Figure 6). One can therefore optimize the recorded potential due to spinal cord interneurones by feeding signals from the posterior electrode to one input of the differential

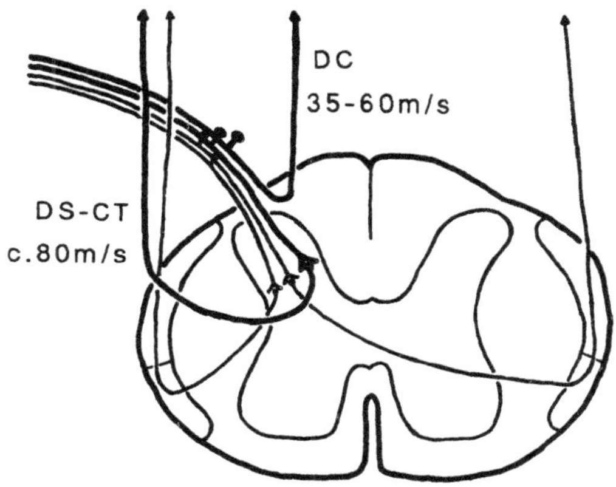

Figure 7 Afferent spinal cord tracts (dorsal spinocerebellar tract, dorsal columns) responsible for main components of spinal SEP to tibial nerve stimulation at the knee

amplifier and the anterior electrode to the other. Following tibial nerve stimulation at the ankle, the negative peak equivalent to N13 occurs at a latency of 20–25 ms and is maximal at T12 or L1. Unfortunately this potential is impossible to record in some neurologically normal subjects.

After averaging a large number of responses to tibial nerve stimulation, a very small 'travelling' wave can sometimes be recorded from thoracic and cervical levels. This represents the envelope of activity in sensory pathways of the spinal cord, seen in greater detail in epidural recordings (Figure 5). When the tibial nerve is stimulated at the knee at least three distinct potentials can be recorded in the epidural space, and traced as they ascend the spinal cord. The first component, probably generated in the dorsal spinocerebellar tract (Figure 7), travels at up to 85 m/s. Later components, probably conducted in the dorsal columns, travel more slowly at 35–60 m/s. The first component seems to be the most vulnerable to problems associated with surgery, possibly because, unlike the dorsal columns, the dorsal spinocerebellar tracts are composed of second order sensory fibres and the nerve volley must first cross a synapse at the level of the lumbar enlargement. Equivalent 'travelling' spinal SEP components can be recorded at cervical levels following upper limb stimulation, but tend to be obscured by the segmentally-generated activity (N13) which is of much larger amplitude.

Subcortical pathways

Following N13 or N21, the next potential which can be consistently recorded from the skin is a P14/15 to median nerve stimulation at the wrist (Figure 6) or a P32 to tibial nerve stimulation at the ankle. These potentials are generated close to the foramen magnum and are widely distributed all over the scalp. It is not entirely certain whether the peaks are due to dorsal column fibres terminating in the cuneate and gracile nuclei, or to the change in the characteristics of the volume conductor as the volley enters the less constricted space of the cranium through the foramen magnum. Post-synaptic potentials of the dorsal column nuclei and the next sensory relay in the thalamus can be recorded with implanted electrodes. They are predominantly of negative polarity and prolonged (10 ms or more) duration, and can sometimes be seen on the scalp as a broad negative wave overlapping the initial cortically generated potentials.

Cerebral cortex

The cortically generated potentials following median nerve stimulation at the wrist commence with an N20 which is recorded over the parietal cortex contralateral to the stimulated hand (Figure 6). N20 is not recorded anterior to the central sulcus; at frontal electrodes a small positive potential of similar or slightly longer latency may be seen when the reference electrode is on a relatively neutral site such as the earlobe. Both potentials are believed to be generated in the primary sensory cortex, which in man is located on the posterior wall of the central sulcus and the postcentral gyrus. The fact that the cortex consists of columns of cells and that on the posterior wall of the central sulcus these are oriented *tangentially* to the surface of the scalp means that the initial electrical response to sensory input, which is positive on the surface of the cortex and negative at depth, is recorded as a positive potential at pre-central and a negativity at post-central regions of the scalp. As with the cervical N13, the amplitude of N20 and P20 can be combined by having the active electrode over the parietal region and the reference over the frontal cortex. The waves which follow N20 are rather complex and come from many regions of the sensory-motor cortex. For monitoring purposes, however, it is only necessary to record the first positivity following N20 at the same electrode. This may show quite marked spontaneous fluctuations of amplitude and latency related to blood pressure and anaesthetic levels, so great care must be taken to exclude such factors when using cortical SEPs as a monitor of sensory conduction in the spinal cord.

The cortical response to tibial nerve stimulation commences with a P40 which is maximal just posterior to the vertex. As with the median N20, a potential of similar latency but opposite polarity is recorded in the frontal region, so the P40 can be enhanced by locating the reference electrode there. On account of the active region of cortex being located partly on the mesial surface of the hemisphere, where the neuronal columns have a transverse orientation, P40 tends to spread to parietal scalp locations *ipsilateral* to the stimulated leg, while the frontal negative potential spreads more to the contralateral side. Consequently the P40 can also be recorded between the left- and right-sided 'hand area' electrodes, and this montage may be found useful in confirming the identity of the P40 when additional recording channels are available. P40 is even more vulnerable to anaesthetic factors than N20, and is also more likely to be absent in patients with myelopathy.

SEP CHANGES DURING SPINAL CORD MONITORING

The criteria which have been adopted to define a significant and possibly pathological change in the SEP whilst monitoring have mostly been in terms of the amplitude of the response. Latency criteria are also sometimes quoted, but it appears that physiologically significant latency changes rarely if ever occur without also a significant decrease in amplitude. The amplitude of cortical SEPs can vary quite markedly from minute to minute, and it is difficult to define a categorical level below which the likelihood of spinal cord impairment is sufficiently great to cause the surgeon to take corrective action. Many groups have adopted a criterion of 50% amplitude reduction, but others have reported a high incidence of changes of this magnitude occurring without any adverse consequences. Obviously it is undesirable for any monitoring technique to yield a high proportion of 'false positives', where the surgeon is alerted to a problem which in fact does not exist. For cortical SEPs an amplitude reduction exceeding 70% is probably a better compromise, although of course the more stringent the criterion the greater the likelihood that a partial but real failure of spinal cord conduction may be missed ('false negative'). In order to be sure that the decrement is not due to effects at cortical level (deepening anaesthesia or a drop in blood pressure) it is recommended that the neck potentials be recorded also. These tend to be of low amplitude, particularly in older patients, but are much less influenced by anaesthetic factors.

For epidurally recorded spinal SEPs it is generally agreed that an amplitude reduction exceeding 50% is a useful criterion, being most unlikely to occur by chance once steps have been taken to exclude possible technical reasons. It should of course be established that the stimulating electrodes are still delivering an adequate impulse (placing a moistened fingertip on each provides a useful check!) and that the recording electrodes are still making good contact. Problems are sometimes encountered with epidural electrodes when they come into contact with fluids in the operating field, and sometimes it may be advisable for the surgeon to remove the electrode to clean and reposition it. Once these checks have been made, if the potentials remain consistently more than 50% below their baseline amplitude then a physiological problem must be considered a real possibility. Under such circumstances it is extremely useful to have an additional channel recording SEP activity from the cord below the segments at risk, to establish that there is still a satisfactory response at that level.

In a series of 1168 monitored operations for scoliosis[2], all the patients who experienced mild or moderate neurological symptoms postoperatively had an SEP whose amplitude fell by more than 50% at some stage and did not improve rapidly following a technical check. The incidence of problems was greatest in those whose SEPs fell below 20% of their initial amplitude, but was less in those for whom the response subsequently recovered to above 50%. Although most patients with changes exceeding 50% survived without clinical deficits, the incidence of such 'false positives' was only slightly greater than that of correctly detected deficits. Therefore it was felt that the increased risk associated with an SEP fall exceeding 50% (which can only be demonstrated in such a large series) should be sufficient to prompt the surgeon at least to pause until the response recovers. If this fails it is clearly desirable to try to correct whatever might be the source of the problem. In this series the insertion and tightening of sublaminar wires was found to be the most frequent precipitating event, although distraction of the spine was also implicated in a substantial number of cases.

Questions are still being asked about the reliability and cost effectiveness of spinal cord monitoring. Clearly it is unrealistic to expect any technique to be infallible, and it does not invalidate the principle of monitoring that there have been some reports of 'false negatives' using cortical SEPs[3]. As to cost, it might well be argued that one patient in whom paraplegia is averted outweighs several hundred who are monitored unnecessarily. Finally there is the issue as to whether controlled trials ought to be performed. When the question is whether a particular treatment may be of some benefit to a sick individual, then a controlled trial is mandatory and ethically justifiable, but when the failure to monitor may have disastrous consequences for a patient who was previously neurologically normal, then there is a strong case for accepting spinal cord monitoring as a 'clearly perceived benefit' in the same way as cardiac monitoring is now perceived.

References

1. Halonen JP, Jones SJ, Edgar MA, Ransford AO. Conduction properties of epidurally recorded spinal cord potentials following lower limb stimulation in man. Electroenceph Clin Neurophysiol. 1989;74:161–174.
2. Forbes HJ, Allen PA, Waller CS, Jones SJ, Edgar MA, Bentley G, Webb PJ, Ransford AO. Spinial cord monitoring: experience with 1168 cases. J Bone Joint Surg (Br). 1991;73B: 487–491.
3. Lesser RP, Raudzens P, Lueders H, Nuwer M, Goldie WD, Morris HH, Dinner DS, Klem G, Hahn JF, Shetter AG, Ginsburg HH, Gurd AR. Postoperative neurological deficits may occur despite unchanged intraoperative somatosensory evoked potentials. Ann Neurol. 1986;19:22–25.

2
Human spinal cord potentials recorded from the epidural space

K. SHIMOJI, M. TOMITA, T. HOKARI, S. DENDA, T. TOBITA, H. BABA and S. FUKUDA

INTRODUCTION

Since Gasser and Graham[1] first recorded the spinal cord potential (SCP) from the dorsal surface of a cat's spinal cord in response to stimulation of the dorsal roots of the same segment, many investigators have studied its origin in various animals. The fundamental pattern of the segmental SCP consists of an initially positive spike followed by slow negative (N) and positive (P) potentials[2–4]. It is generally agreed that the spike, N and P potentials reflect the incoming afferent volleys along the root, the synchronized activities of interneurons in the dorsal horn and primary afferent depolarization (PAD), respectively[5–8]. On the other hand, when the recording electrode was placed on the dorsal root in the vicinity of the dorsal root entry zone, the potential (dorsal root potential; DRP) shows a mirror image against SCP recorded from the dorsal surface of the cord (Figure 1). The DRP IV and DRP V by Lloyd's terminology[9] correspond to the N and P waves, respectively, with a similar time-course but with an opposite polarity. Magladery et al.[10] first recorded human spinal cord potentials from the intrathecal space in volunteers, but they did not analyse the origin of the potential change, seemingly due to concerns of exposing subjects to undue risk with their methods.

METHODOLOGY

Figure 2 schematically demonstrates our original methods of introducing an epidural catheter electrode into the epidural space using the Tuohy needle. We[11] demonstrated that human spinal cord potentials could be easily and safely recorded from the epidural space (ES) using the catheter electrode, based on the method of continuous epidural anaesthesia. After gaining familiarity with the method, one could also introduce the catheter electrode

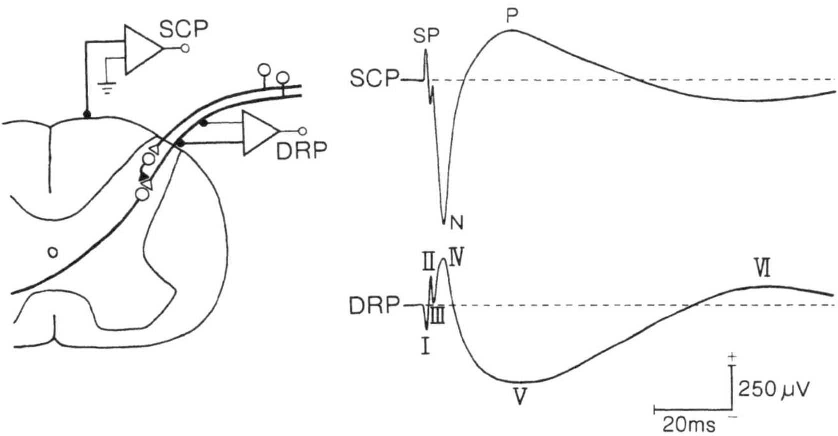

Figure 1 Schematic presentation of a spinal cord potential (SCP) recorded from the cord dorsum (right upper sweep) and a dorsal root potential (DRP) (right lower sweep) in response to stimulation of a nearby dorsal root or a peripheral nerve of the same segment. Placements of the recording electrodes are shown on the left. Upward deflection denotes positivity in this and subsequent figures. DRP I–VI are defined according to Lloyd's terminology[9]

even into the anterior epidural space with a more lateral approach of the needle.

We used an epidural catheter commercially available (Figure 3, A) and slid a stainless steel wire 230 μm in diameter through the inside of the catheter up to 5 mm beyond its tip in our earlier studies. Later, we had a company make epidural catheter electrodes (B). Recently, we have also used the multipurpose epidural catheter which allows us to record electrical activities, epidural pressure, regional blood flow and tissue oxygen pressure. The catheter can also be used for injecting the drugs and stimulating the spinal cord epidurally (C).

FUNDAMENTAL PATTERNS OF THE SCPS

Figure 4, A, B, C demonstrate the simultaneous recording of the segmental SCP and conducting SCP recorded from the ES at the T_{12} and C_6 levels, respectively, in response to cauda equina stimulation during general anaesthesia. A big difference in amplitude between the segmental SCPs and conducting SCPs can be seen. The fundamental pattern of human SCPs recorded from the posterior ES in response to segmental nerve stimulation is found to be the same as that recorded directly from the dorsal surface of spinal animals. It is frequently noted that the slow P (P_2) potential is accompanied by a negative dip which produces the second component of the P_2 in wakeful human subjects (Figure 4, D). The second component of the P_2 wave is highly vulnerable to anaesthetics and also disappears during sleep, suggesting that the component originates from the pathway involving the supraspinal structure[12–15]. A similar second component of the P_2 wave

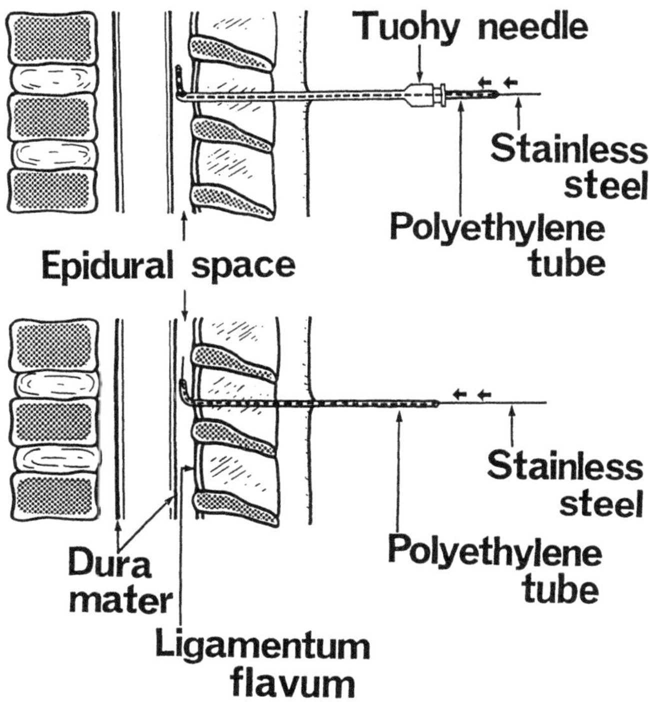

Figure 2 Schematic drawing of the methods of placement of a catheter electrode into the spinal epidural space (ES), based on the continuous epidural anaesthesia. An epidural catheter is introduced into the ES through the Tuohy needle, the tip of which is placed in the ES by the 'loss of resistance' method. Through the epidural catheter a stainless steel wire 230 μm in diameter is advanced approximately 5 mm beyond the top of the catheter

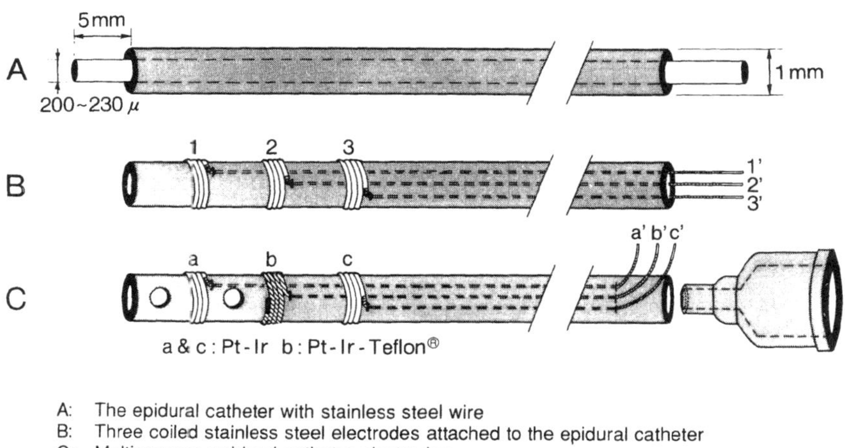

Figure 3 Epidural catheter electrodes we have been using for recording SCPs in man

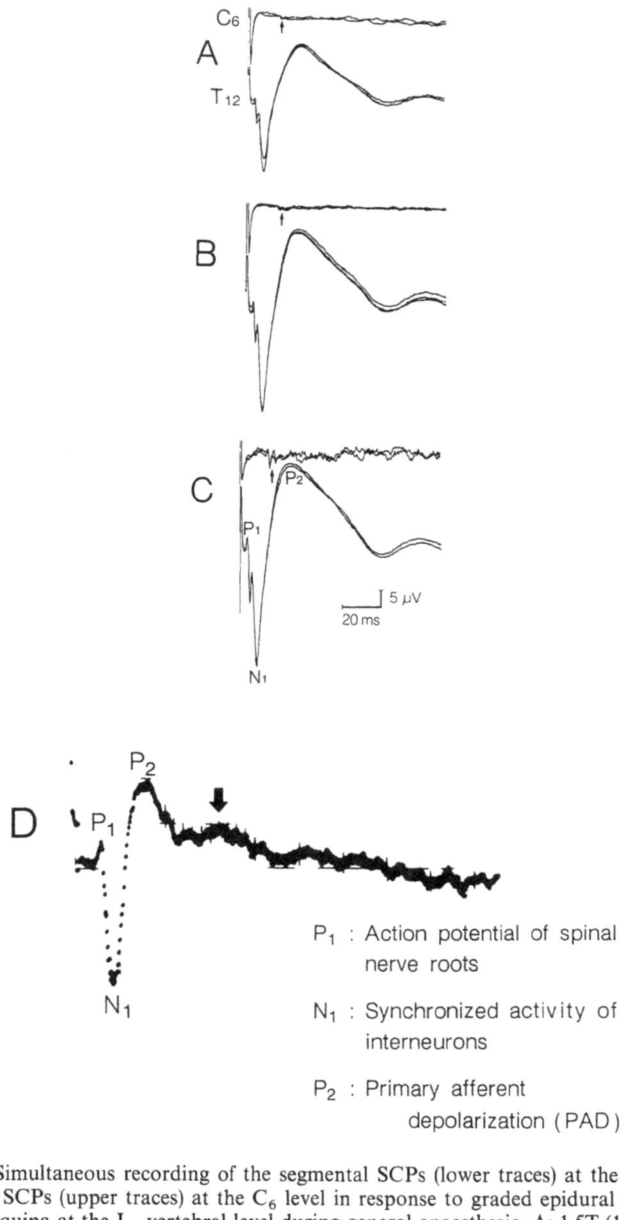

Figure 4 Simultaneous recording of the segmental SCPs (lower traces) at the T_{12} level and conducting SCPs (upper traces) at the C_6 level in response to graded epidural stimulation of the cauda equina at the L_4 vertebral level during general anaesthesia. A: 1.5T (1.5 × threshold strength for evoking the P_1 potential); B: 3.0T; C: 6.0T; D: SCPs produced by stimulation of the segmental nerves (segmental SCPs) during wakefulness

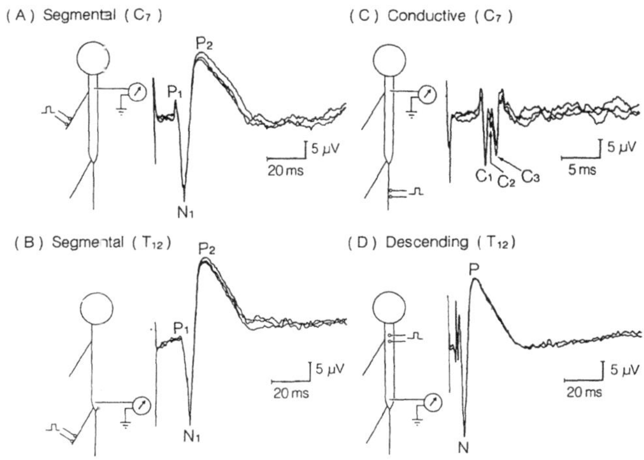

Figure 5 Three types of SCPs recorded from the posterior epidural space in a subject undergoing spine surgery during general anaesthesia

can be produced in the SCP of the rat during ketamine anaesthesia, and is highly vulnerable to inhalational anaesthetics and barbiturates[15–17]. Transection of the rat spinal cord at the $C_{1/2}$ level completely eliminates the second component of the P_2 without significant changes in other components[12–18]. These results in the rat also strongly suggest that the secondary component of the P_2 reflects a potential activated by a feedback loop via supraspinal structures. Further, this secondary component of the P_2 may reflect PAD activated by a feedback loop via supraspinal structures, a feedback inhibition exerted on dorsal horn neurons, which shares several similar characteristics with heterosegmental slow positive potential as discussed later[16–18].

Three types of SCP were recorded from the posterior ES in a subject undergoing spinal cord monitoring during spine surgery under general anaesthesia (nitrous oxide–halothane) (Figure 5). It is noted that the segmental SCP shows fundamentally the same waveform wherever it is recorded. However, there is a lack of the second component of the P_2 during general anaesthesia. This segmental SCP can also be recorded in the upper cervical ES in response to trigeminal nerve stimulation. The second type of SCP is the conducting SCP (upper right), consisting of several spike potentials, usually three components. It can be recorded from the posterior ES of the spinal cord in response to stimulation of a rostral or caudal part of the spinal cord, cauda equina or even a peripheral nerve. The third type of the SCP can be recorded from the posterior ES in response to epidural spinal cord stimulation at the rostral level (bottom right). The spike potentials are followed by a sharp negative and a slow positive wave. Its pattern is very similar to that of the segmental SCP.

Figure 6 shows a comparison of the segmental SCP recorded from the posterior ES at the T_{12} level in response to tibial nerve stimulation with the SCP recorded with the same electrode in response to epidural stimulation

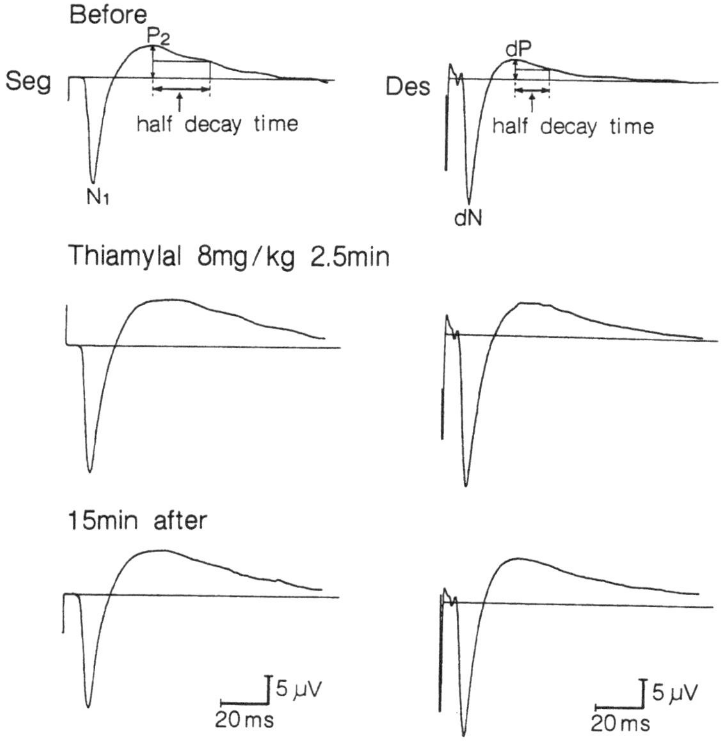

Figure 6 Augmentation by thiamylal of slow negative and positive components of SCPs produced by segmental nerve (Seg) and rostral cord dorsum epidural stimulations (Des)

of the SC at the C_6 level. The similarity in waveform of these two potentials is clearly seen[19]. Also note that there is no secondary component superimposed on the slow P wave in the segmental SCP. As demonstrated in the SCP of the spinal animals, the P_2 potential of human SCP, which is believed to be PAD, is augmented by barbiturates[13]. Figure 6 also shows that the P_2 as well as slow P potentials evoked by descending volleys are augmented by IV injection of thiamylal sodium, an ultrashort-acting barbiturate. Again, the slow negative and positive potentials activated by epidural stimulation of the rostral spinal cord have currently been suggested in our laboratory to originate largely from antidromic volleys of the dorsal column and reflect dorsal horn neuron activities and PAD, respectively. The hypothesis regarding the origins of these negative (dN) and positive (dP) potentials produced by rostral spinal cord stimulation from the posterior ES comes from the similarities of physiological as well as pharmacological characteristics with segmentally activated slow negative (N_1) and positive (P_2) potentials, respectively.

When the conditioning stimulation is applied to the cervical cord to evoke descending SCP and test stimuli to the tibial nerve, there is a substantial interaction between the slow potential changes evoked by descending volleys

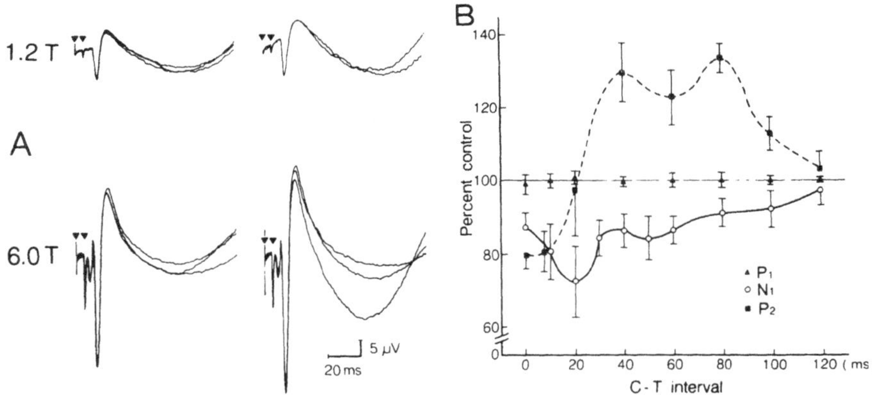

Figure 7 Interactions between the two SCPs produced by segmental nerve stimulation (segmental SCP) and rostral cord dorsum epidural stimulation (descending SCP) during general anaesthesia[19]. Left sweeps in A show interacted potentials between descending and segmental SCPs. Right sweeps in A are the summation of two potentials (segmental and descending SCPs). B: Amplitude changes of the P_1, N_1 and P_2 potentials of segmental SCP by conditioning epidural stimulation of rostral spinal cord as a function of conditioning (C) – testing (segmental stimulation; T) intervals. See the text for explanation

and segmental volleys (Figure 7). The left sweeps in Figure 7 are the summation of the segmental SCP and descending SCP, each evoked separately. The right sweeps in Figure 7 are produced by conditioning stimulation at the cervical cord and test stimulation to the tibial nerve. This interaction indicates that the same components (in other words, the same interneurons) might be involved in the pathways producing these two potentials.

When the forepaw is stimulated in a ketamine-anaesthetized rat, a slow positive potential can be recorded from the dorsal surface of the lumbosacral enlargement. This heterosegmentally activated slow positive potential (HSP) is very vulnerable to anaesthetics and disappears by transection of the spinal cord at the C_1 level. Therefore, the HSP is suggested to be a feedback PAD (Figure 8)[16–18]. On the other hand, we could not record HSP in man in response to radial, median or ulnar nerve stimulation at the wrist. However, when a stimulating electrode is placed at Erb's point, HSP with a long latency could be recorded from the posterior ES in wakeful subjects (Figure 9, A)[20]. When a more intense stimulation is applied to the brachial plexus at Erb's point, two HSPs distinctly different in latency could be demonstrated in some subjects. One can see a relatively early HSP with peak latency of about 18 ms and also another late HSP with peak latency of about 62 ms, recorded from the posterior ES in response to brachial plexus stimulation at Erb's point. Another important finding might be that there are also two distinct slow positive peaks on the decay of the P_2 potential of the segmental SCP (Figure 9, B).

Seventeen patients with chronic pain admitted to the pain unit of our department were subjected to HSP recordings. We could not record HSPs from any of cord-injured or cord-infarcted patients. HSPs could be

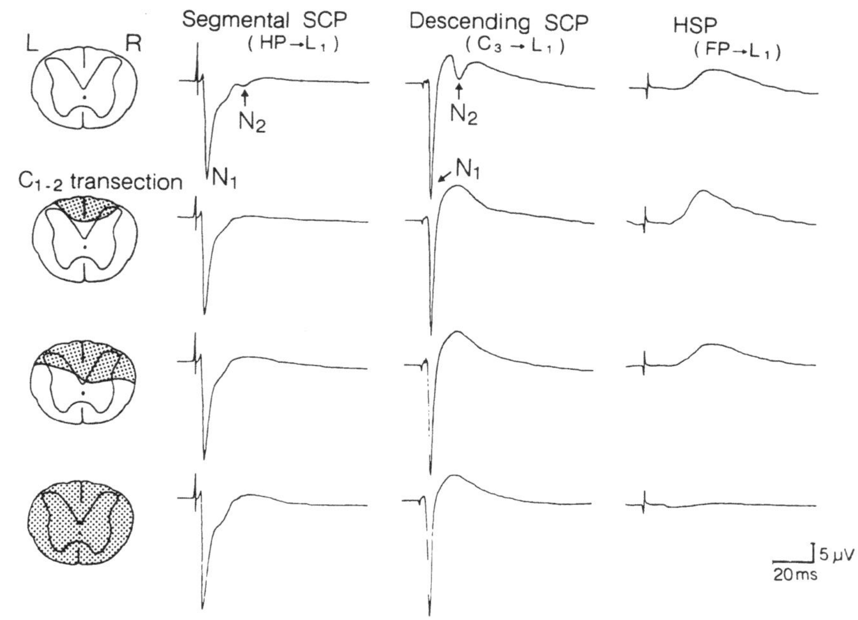

Figure 8 The SCPs produced in the L_5 level in response to stimulations of the hindpaw (segmental SCP), upper dorsal column (descending SCP) and forepaw (hetersegmental SCP; HSP) in the rat. Note the disappearance of the heterosegmental SCP (HSP) without fundamental changes in the segmental and descending SCPs by spinal transection. Note also that the secondary component of the slow positive potential became silent (disappearance of N_2)

demonstrated in 9 of 12 patients who had no evident morphological deficit in the spinal cord.

As mentioned above, two distinct HSPs were recorded in four patients. In four patients, however, only the early or fast HSP could be demonstrated. The late HSP without the early HSP was recorded in one patient. We could not elucidate such individual differences in occurrence of the early or late HSPs. However, the results firstly suggest that there are at least two distinct feedback inhibitory control systems: one, a relatively short-latency feedback and the other, a relatively long-latency feedback. The results secondly suggest that such feedback inhibitory control systems vary with each individual[18].

Transcranial magnetic stimulation of one side of the motor cortex produces electromyograms in the muscles of upper as well as lower extremities of both sides without pain, and also spinal cord potentials in the ES, as several investigators have reported. These motor evoked potentials might provide another tool for monitoring SC function. Motor evoked spinal potentials, however, were variable in waveform. We have so far not been able to characterize the fundamental waveforms of motor evoked spinal cord potentials. Figure 10 shows an example of motor evoked spinal cord potentials recorded from the ES at the level of lumbosacral enlargement with

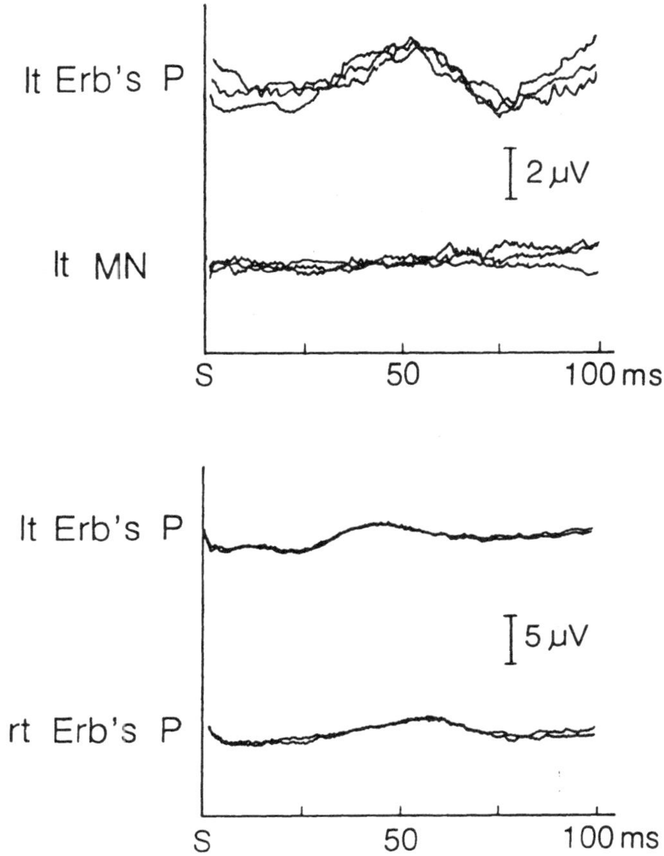

A

Figure 9 Heterosegmental SCPs produced in the posterior epidural space at the T_{12} vertebral level in response to stimulation of the brachial plexus at Erb's point in wakeful subjects at a moderate strength (3 times threshold strength for evoking the segmental SCP; 3T) (A) and at 6T (B). Stimulation to the left of Erb's point (lt Erb's P) produces a slow positive potential, while left median nerve stimulation (lt MN) produces no significant potential changes. See the text for explanation

electromyograms from both sides of upper and lower extremities in response to magnetic stimulation from the scalp in a patient with ossification of the posterior longitudinal ligament (OPLL).

ORIGINS OF SLOW POTENTIAL CHANGES

Table 1 shows a part of our currently published data in the rat[16]. Individual differences in latency can be seen, as indicated by the range of the onset latency of HSPs which were recorded from the C_5 and L_5 segmental levels. There might also be two HSPs in the rat, but this was not noticed during this study. What we did find in the animal study was that the secondary

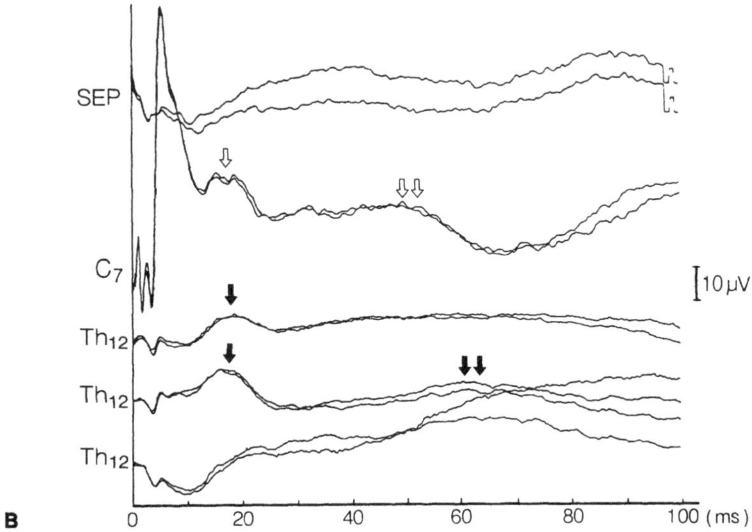

Figure 9 *Continued*

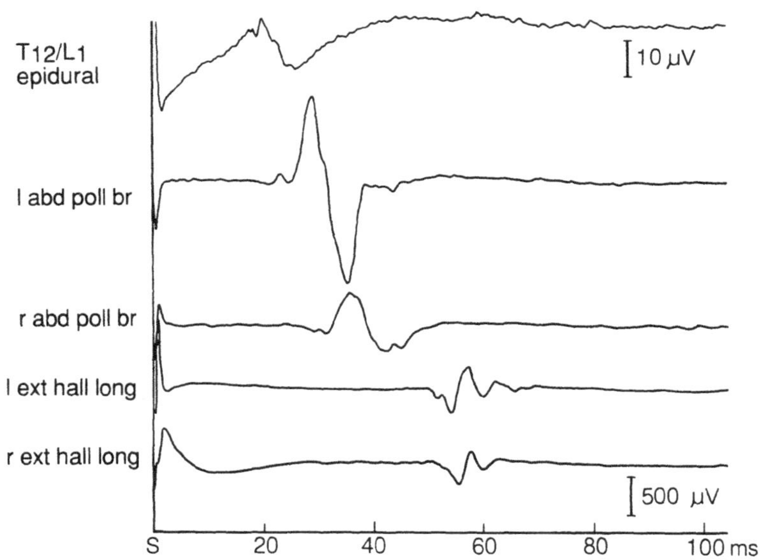

Figure 10 The SCP (the upper sweep) and electromyograms (the second to 5th sweeps) evoked by transcranial magnetic stimulation in a wakeful subject

component of the P_2 in the segmental SCP behaved similarly in terms of vulnerability to anaesthetics, high spinal cord transection and variability in latency[16]. There might be a similar mechanism involved in HSP and in the second component of the P_2 in the segmental SCP.

Table 1 The onset, peak latencies, and duration of the cervical and lumbosacral HSPs, with the peak latency of the second component of the slow positive wave (P_2s) of the segmental spinal cord potentials (SCPs)

	HSPs						*Segmental SCPs*	
	Onset latency (ms)		*Peak latency (ms)*		*Duration (ms)*		*P_2 peak latency (ms)*	
	C5	L5	C5	L5	C5	L5	C5	L5
Mean ± S.E.	11 ± 1	9 ± 1	30 ± 1	31 ± 2	52 ± 4	60 ± 4	32 ± 2	29 ± 1
(Range)	(7–18)	(5–17)	(22–35)	(12–50)	(28–80)	(36–90)	(22–48)	(20–40)

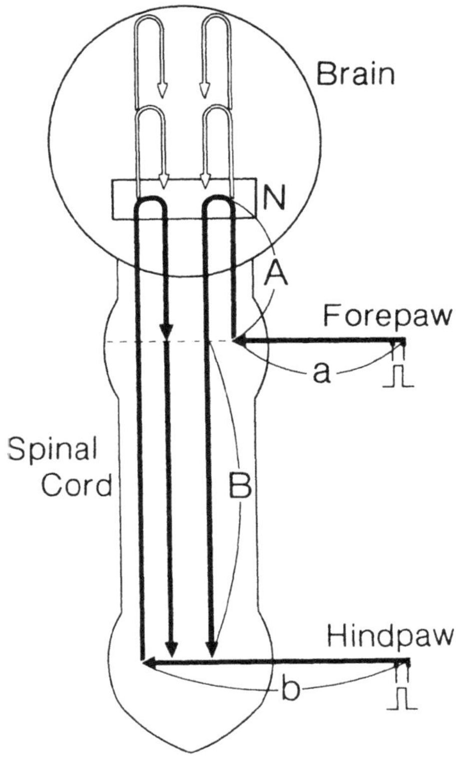

Figure 11 Hypothetical drawing for the mechanisms of individual differences in latency of heterosegmentally evoked spinal cord slow positive potentials (HSP)[18]

Figure 11 shows a schematic drawing of our hypothesis of the mechanism explaining the high variability of the onset of HSPs among recorded individuals. There might be several nuclei rather than one nucleus which send feedback impulses to produce HSPs. Which nucleus plays a dominant role as a feedback relay might be different among subjects.

To test our working hypothesis that HSPs are feedback inhibitory potentials, we recorded the firings of wide dynamic range (WDR) neurons

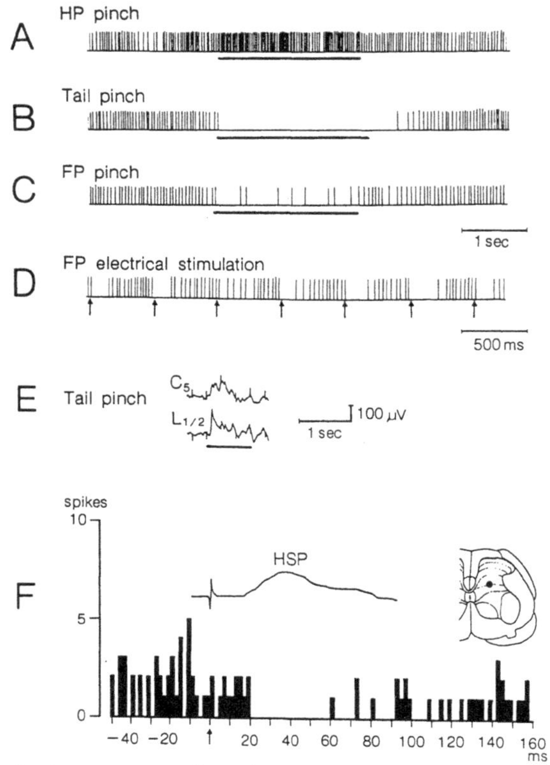

Figure 12 The simultaneous recording of a heterosegmental slow positive cord dorsum potential and firing of a WDR neuron recorded at the level of L_5 in response to stimulation of the forepaw (heterosegmental)[18]

from the lamina V of the dorsal horn in the rat. Firings of the WDR neurons were increased by hindpaw touch as well as pinch. On the other hand, heterosegmental stimuli, such as tail and forepaw touch (7 of 12 neurons) or pinch (all 12 neurons) or electrical shock suppressed the neuronal firing. Both the HSP and peristimulus time histogram of a WDR neuron firing, simultaneously evoked by electrical forepaw stimulation, are displayed on the same time scale on the bottom of Figure 12. The time course of inhibition of the neuronal activity was found to be similar to that of the HSP in all 9 neurons which responded to forepaw electrical stimulation. This suggests that the HSP reflects inhibitory activity, probably primary afferent depolarization (PAD), activated by a feedback loop via supraspinal structures[16].

Further, to prove that the HSP really reflects PAD, we applied the sucrose-gap technique for recording depolarization of the dorsal root *in vivo* in the rat spinal cord. The scheme of recording both dorsal root (DR) depolarization and HSP *in vivo* is shown in Figure 13, A.

Forepaw stimulation clearly produced DR polarization as well as the HSP. Spinal transaction at the $C_{1/2}$ level completely eliminated both potentials in response to forepaw stimulation. By contrast, segmental

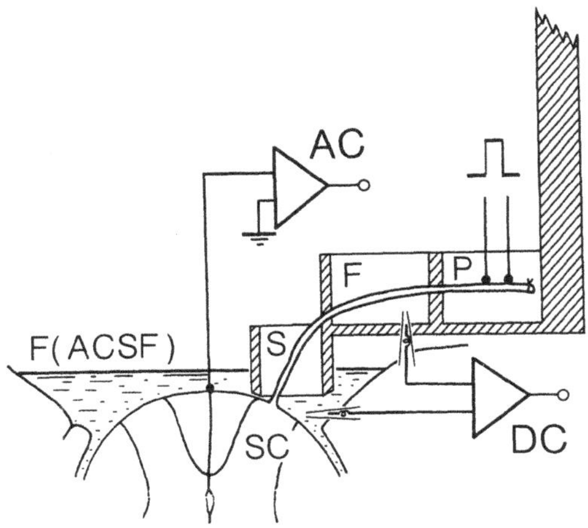

Figure 13 A: Schema showing the methods of simultaneous recordings of both cord dorsum potential and dorsal root depolarization by application of the sucrose-gap technique in vivo. B: Specimen records of cord dorsum potentials (CDP) (upper traces) and dorsal root depolarizations (lower traces) (DRdep) recorded at the L_5 level in response to forepaw (FP) stimulation in the rat during ketamine anaesthesia. Dorsal root (DR) stimulation produces no cord dorsum potential but a monopolar dorsal root spike (an open arrow). See the text for explanation

(hindpaw) stimulation still produced both potentials with a minimal change (disappearance of a negative notch on the slow P_2). Dorsal root stimulation elicited no slow potential changes but a unipolar spike from the peripheral electrode, indicating that the sucrose-gap was complete (Figure 13, B).

Intravenous injection of a GABA antagonist, bicuculline, suppressed both DR depolarization and HSP activated by forepaw stimulation. This was also true in segmentally produced DR depolarization and P (P_2) potential. Intravenous injection of the GABA agonist, pentobarbital, potentiated both segmentally activated DR depolarization and P potential. By contrast, hetero-segmentally activated DR depolarization and P potential were profoundly suppressed. Thus, the results strongly suggest that the HSP in man and rat reflect PAD activated by a feedback loop via supraspinal structures[16–18].

SUMMARY

1. Peripheral nerve stimulation produces initially positive spike(s) (P_1), followed by a sharp negative (N_1) and slow positive (P_2) wave in the posterior epidural space (PES) of man. When the subject is awake, a stronger stimulation evokes the secondary component of the P_2 preceded by a negative dip. Several pieces of physiological and pharmacological evidence suggest that the P_1, N_1 and P_2 reflect the incoming afferent volleys of the roots, the synchronized activity of dorsal horn neurons and primary afferent depolarization (PAD), respectively. It is also suggested that the secondary component of the P_2 (P_2s) reflects PAD

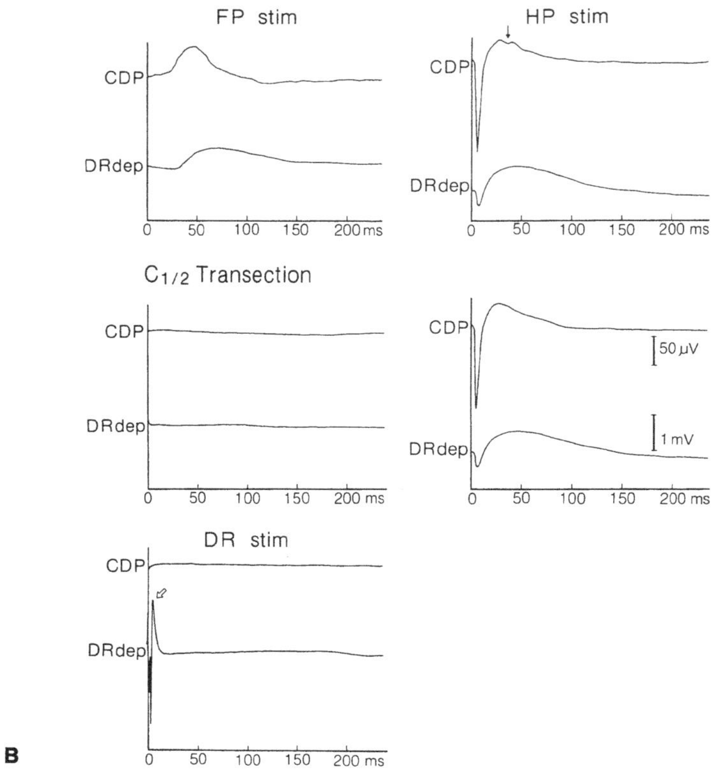

Figure 13 *Continued*

activated by a feedback loop via supraspinal structures. This segmentally activated spinal cord potential (SCP) can be recorded along several spinal segments with a slight latency delay as might be expected.

2. The conducting SCPs along the spinal tracts are small and usually buried in large segmental SCPs. When the recording electrodes are placed at a more rostral site, the conductive SCPs can be clearly recorded, usually consisting of three spikes (C_1,C_2,C_3). These conductive SCPs are hardly recorded from the skin surface.
3. Peripheral nerve stimulation does not evoke the SCP at the heterosegmental area in man. However, when the stimulating electrode was placed at a proximal site such as Erb's point, a heterosegmentally activated slow positive cord dorsum potential (HSP) can be recorded from the posterior ES at the lumbosacral level. This HSP is suggested to be a PAD activated by a feedback loop through the supraspinal structures which might be similar but not the same as the secondary component of the P_2.
4. Another type of SCP can be produced in the lumbosacral enlargement by epidural stimulation of the rostral spinal cord. Its waveform is very

similar to that of the segmental SCP, consisting of the initially positive spikes followed by a sharp negative and a slow positive wave. Physiological and pharmacological evidence suggests that these components reflect mostly antidromic spikes along the dorsal column, the activity of dorsal horn neurons and PAD, respectively. The secondary component of the slow positive wave can also be produced when the subjects are awake and the stimulus strength is high enough.

5. These segmentally or heterosegmentally activated PADs might explain at least in part the mechanisms of pain inhibition produced by transcutaneous nerve stimulation, spinal cord stimulation, and acupuncture, which are also effective by heterosegmental stimulation. These slow components activated by a feedback loop, the second component of the P_2 or an HSP, might provide a tool for studying the feedback system exerted on the dorsal horn neurons in man.
6. Transcranial magnetic stimulation also produces slow potential changes in the human spinal posterior epidural space. The origins of the slow potentials remain to be analysed.

References

1. Gasser HS, Graham HT. Potentials produced in the spinal cord by stimulation of dorsal roots. Am J Physiol. 1933;103:303–320.
2. Barron DH, Matthews BHC. The interpretation of potential changes in the spinal cord. J Physiol (Lond). 1983;92:276–321.
3. Bernhard CG. The spinal cord potentials in leads from the cord dorsum in relation to peripheral source of afferent stimulation. Acta Physiol Scand. 1953;29 (Suppl 106):1–29.
4. Bernhard CG, Widen L. On the origin of the negative and positive cord potentials evoked by stimulation of low threshold cutaneous fibers. Acta Physiol Scand. 1953;29 (Suppl 106):42–54.
5. Koketsu K. Intracellular potential changes of primary afferent nerve fibers in spinal cords of cats. J. Neurophysiol. 1956;19:375–392.
6. Eccles JC, Krnjevic K. Potential changes recorded inside primary afferent fibers within the spinal cord. J Physiol (Lond). 1959;149:250–273.
7. Carpenter DO, Rudomin P. The organization of primary afferent depolarization in the isolated spinal cord of the frog. J. Physiol. (Lond). 1973;229:471–493.
8. Beall JE, Applebaum AE, Foreman RD, Willis WD. Spinal cord potentials evoked by cutaneous afferents in the monkey. J. Neurophysiol. 1977;40:199–211.
9. Lloyd DPC. Electrotonus in dorsal nerve roots. Cold Spr Harb Symp quant Biol. 1952;17:203–219.
10. Magladery JW, Porter WE, Park AM, Teasdall RD. Electrophysiological studies of nerve reflex activity in normal man. IV. The two-neurone reflex and identification of certain action potentials from spinal roots and cord. Bull Johns Hopkins Hosp. 1951;88:499–519.
11. Shimoji K, Higashi H, Kano T. Epidural recording of spinal electrogram in man. Electroenceph Clin Neurophysiol. 1971;30:236–239.
12. Shimoji K, Matsuki M, Shimizu H. Wave-form characteristics and spatial distribution of evoked spinal electrogram in man. J. Neurosurg. 1977;46:304–313.
13. Shimoji K, Higashi H, Kano T. Epidural recording of spinal electrogram in man. Electroenceph Clin Neurophysiol. 1971;30:236–239.
14. Shimoji K, Matsuki M, Ito Y, Masuko K, Maruyama M, Iwane T, Aida S. Interactions of human cord dorsum potential. J Appl Physiol. 1976;40:79–84.
15. Shimoji K, Fujiwara N, Fukuda S, Denda S, Takada T, Maruyama Y. Effects of isoflurane on spinal inhibitory potentials. Anesthesiology. 1990;72:851–857.

16. Shimoji K, Fujiwara N. Slow positive potential recorded from hetrosegemental dorsal cord in man and rat (abstract). Jap J Physiol 1992;Suppl.42:5127.
17. Shimoji K, Fujiwara N, Denda S, Tomita M, Toyama M, Fukuda S. Effects of pentobarbital on heterosegmentally activated dorsal root depolarization in the rat–investigation by sucrose-gap technique *in vivo*. Anesthesiology. 1992;76:958–966.
18. Shimiji K, Sato Y, Denda S, Takada T, Fukuda S, Hokari T. Slow positive dorsal cord potentials activated by heterosegmental stimuli. Electroenceph Clin Neurophysiol. 1992;85:72–80.
19. Shimizu H, Shimoji K, Maruyama Y, Matsuki M, Kuribayashi H, Fujioka H. Human spinal cord potentials produced in lumbo-sacral enlargement by descending volleys. J Neurophysiol. 1982;48:1108–1120.
20. Shimoji K, Fujioka H, Maruyama Y, Shimizu H, Hokari, T, Takada T. Spinal cord potentials (SCPs) produced by descending volleys in man. In Ducker TB, Brown RH, eds. Neurophysiology and Standards of Spinal Cord Monitoring. New York: Springer-Verlag; 1986, 114–121.

3
Indications for the use of evoked potentials in spinal surgery – review of the Nottingham experience and current use

S. L. PAPASTEFANOU and J. K. WEBB

INTRODUCTION

Maintenance of the functional integrity of the spinal cord is a major concern for every spine surgeon. Severe neurologic complications may occur following surgical correction of spinal deformity and recovery is proportional to the speed of removal of the instrumentation[1,2]. In the past, a careful physical examination in the recovery room was the earliest warning of cord injury. During the last two decades, gross physical examination of the partially awakened patient during the operation – the wake-up test – has been widely used. The desire of many researchers and clinicians to find a better method of evaluating cord function intraoperatively, has resulted in renewed interest in the techniques of evoked response monitoring.

In 1947, Dawson[3] first described a technique for the recording of cerebral responses to electrical stimulation of peripheral nerves in man, using scalp electrodes. However it was only in the last 15 years that technology has advanced the point that evoked response monitoring can be implemented reliably to investigate human neural pathways.

Traditionally there have been five methods for monitoring spinal cord function intraoperatively: somatosensory cortical evoked potentials, somatosensory epidural evoked potentials, spinal spinal evoked potentials, motor evoked potentials and the wake-up test of Stagnara. Each method of monitoring has its own advantages and disadvantages[4,5].

Controversy continues to surround the routine use of spinal cord monitoring for spinal surgery[6–8]. Major issues in question include whether its use affects the operating practice and whether the wake-up test can be replaced. Furthermore a standardized approach to spinal cord monitoring has not yet been established. What has changed over the years is the sophistication and reliability of the equipment and the variety of modalities available.

STANDARDIZATION

Several important general concepts about the standards of performance and reporting of the results of spinal cord monitoring have been reported so far[8]. It is essential for each centre to develop a well defined and consistent protocol. There are many variables concerning the equipment, electronic parameters and physiology, which can affect the recordings[9–11]. It is imperative to establish a system of intraoperative baseline data, using a particular protocol, and common terminology has to develop in order to permit comparative studies which will allow a proper assessment of the method(s) used.

CURRENT APPLICATIONS OF SPINAL CORD MONITORING

Since the early 1980s the number of centres using spinal cord monitoring has markedly increased. It has been a common practice to conduct some form of spinal cord monitoring, when performing any spinal operation, especially if there is a high risk of neurologic injury. Such operations include procedures where corrective forces are applied to the spine, the patient has a pre-existing neurological deficit, the cord has been invaded, or an osteotomy is carried out in the immediate juxtaposition of the spinal cord[2,8]. Although the necessary equipment has been widely available and used, the approach has not progressed to the point of replacing the wake-up test as the standard of care[2]. The latter is still claimed by some to be the gold standard[2], because, if performed properly, it is associated with little morbidity, reveals the status of the patient's motor function at the moment at which it is performed and is the cheapest monitoring method available. The logic of having the capability to evaluate both the sensory and motor component of cord function during risky procedures has been evident, therefore the need for the development and further evolution of the motor evoked potentials.

CURRENT PROBLEMS

Electronic monitoring systems are now developed to monitor complex, higher risk cases, because of their sensitivity, specificity, versatility and reproducible results. Of course these systems require a more demanding and complex technique, and are more expensive, but the obtainable information can be significantly more useful. There are still many questions regarding the effectiveness of such systems in detecting neurologic complications. Even today, the sensitivity of the monitoring response to intraoperative events has not been clearly established. A continuum of insults exists from mild ischaemia to paraplegia, following different operative insults of the cord. It has not been shown effectively whether this correlates directly to variations of the response waveform[12]. One of the reasons for such inconsistency is that the terminology we use to assess the results of the method only takes into account the postoperative neurological status of the patient and neither the degree, nor the pattern of change of the response intraoperatively are considered.

In different studies false negative results are described as cases in which the recordings were well within normal ranges, but the patient awoke with neural deficit postoperatively[13]. Their numbers are usually very small. False positives are cases in which significant changes occured intraoperatively, without any new postoperative neurological deficit. The reported percentages vary between 5–25% in different studies. Conversely true positives and true negatives are reported in 1% and 60% respectively[14]. From the above the main dilemma in the use of SEPs occurs: the sensitivity is high, because of the low false negative rate, but specificity levels are modest, due to the higher incidence of false positives. Up to now, those levels are considered acceptable, because the consequences of false positives are less grave than those of false negatives.

THE NOTTINGHAM EXPERIENCE

In Nottingham for the last 10 years we have used a rather unusual combined method of somatosensory evoked potentials, recorded at surface cervical and cortical electrodes[15] in 308 spinal procedures (deformity, trauma and tumours). The responses were designated intraoperatively, with the consultation of a neurophysiologist, according to the following definitions, and were directly correlated to intraoperative events.

True positives had a 50% or greater drop in amplitude with an identifiable cause, irrespective of neurological outcome.

False positives had a 50% or greater drop in amplitude, without a recognizable cause and no postoperative neurological deterioration.

True negatives had no drop in amplitude greater or equal to 50% and had no neurological deterioration.

False negatives had no drop in amplitude to the 50% level or greater but presented neurological deficit afterwards.

Technical failures were defined as responses which, despite adequate preoperative recordings, had no measurable signal throughout the procedure (possibly due to technical cause).

For the purpose of defining true positives, a recognizable cause was one which, on correcting, resulted in an improvement in the response. True positives were further subdivided into those *with* and those *without* postoperative neurology. We ended up with a bigger number of true positive events, which however could be related to a physiological and surgical identifiable cause, thus likely to be correctable intraoperatively. Our false positives were a very small percentage (about 2%), and remain an enigma. The size of the drop was severe in all cases and no apparent cause was identified. In 50% of them this amplitude drop persisted for prolonged periods, often to the conclusion of surgery. There was no predilection for approach or preoperative diagnosis. The series presented an acceptable rate of technical failures (6%); preoperative neurological deficits and electrical interference contributed the most to the failure to obtain a response.

As a general rule, if the fall of amplitude occured for less than 30 minutes and was seen to be recovering spontaneously at an early stage, or after correction of an obvious physiological or pharmacological cause, a wake-up test was not performed. If it was prolonged more than 30 minutes, or a complete loss of response occured, with no recovery after correction of any other obvious physiological or pharmacological cause, a wake-up test was done. In fact, only the patients who were considered true positives with neurology had abnormal SEP response. In this respect, SEPs proved a more sensitive indicator than the traditional wake-up test.

IMPLICATIONS–CONCLUSIONS

Unmonitored spinal procedures are rapidly becoming a thing of the past. Spine surgery with a known risk of potential neurological damage requires spinal cord monitoring. In the survey conducted by the Scoliosis Research Society[2] surgical correction of scoliosis using distraction techniques, such as Harrington rod instrumentation, carried a small but significant risk of about 1%. Practically spinal cord monitoring can be applied to every spinal procedure. However in every method, there exists a percentage of technical errors and cases difficult to interpret. For those cases, more conventional methods of checking should be employed. For the rest, spinal cord monitoring might give intraoperative information that could possibly affect the surgeon's decision-making.

Certain conditions are associated with increased risk of neurologic injury, including kyphosis, congenital scoliosis, scoliosis with a severe degree of deformity preoperatively, pre-existing neurological deficit and deficit acquired in skeletal traction. Furthermore certain procedures appear to increase the risk of neurologic insult, such as skeletal traction, spinal osteotomy or vertebrectomy[16], Harrington rod instrumentation without postoperative traction in congenital scoliosis and Harrington rod instrumentation after preoperative skeletal traction[14].

Surgical procedures in relation to the spinal cord itself are an important indication of spinal cord monitoring. Surgery for cervical spondylosis, herniated disk in the cervical spine, syringomyelia, spinal cord tumour, and arteriovenous malformation are the major indications for its use. Another potential risk has been described also in surgical procedures requiring temporary occlusion of the thoracic aorta or interruption of important spinal radicular vessels, such as in aortic aneurysm surgery[14].

The wake-up and the clonus test are appropriate in lower risk cases, such as correction of idiopathic adolescent scoliosis in a teenager. Although procedures using the anterior approach to the spine are generally low risk, monitoring should be considered desirable if instrumentation or peridural procedures are utilized[14]. Furthermore spinal cord monitoring has been used for the diagnosis and assessment of other neurological conditions, especially low back pain syndromes[17–19]. It has also proved useful in the surgical treatment of cervical and lumbar spine diseases, such as spondylosis and spinal stenosis, where extensive dissection and decompression are needed. In such cases monitoring is both protective and diagnostic, in the form of improving or deteriorating recordings during decompression. In this respect the preoperative recordings provide both control data and information about the specific location or level of abnormality.

In the case of spinal trauma, especially when instability and/or neurologic deficit is present, the wake-up test can provide limited information. Preoperative baseline recordings are essential and monitoring right and left differences, frequent repetitions and comparing the findings with those for the unaffected extremities are valuable. It should be stressed however that, whereas spinal cord monitoring has correlated well with recovery in incomplete injuries, it has not been predictive in cases of complete cord injury[20].

Other orthopaedic uses for evoked potential monitoring are now identified. An example is monitoring of the sciatic nerve in hip surgery, particularly in revision cases, where such a palsy has high incidence. Recent reports show altered response in more than 50% of cases[20]. Evoked potentials have also proved helpful in brachial plexus injury and can be used to determine the location and the extent of lesions.

The standards of intraoperative spinal cord monitoring are, so far, poorly defined[2]. 'False positive' in monitoring, as currently defined[6], is an unsatisfactory term designating at least three separate sub-groups, which are vitally important to distinguish. These are monitoring changes due to identifiable and correctable non-surgical factors (anaesthesia, arterial blood pressure changes, etc.); changes associated with instrumentation or other critical surgical manoeuvre, which should be reported to the surgeon who may or may not modify his strategy; and finally those potentially misleading changes in monitoring which cannot be related to either surgical or physiological factors. It is possible that changes in monitoring related to surgical parameters, but not followed by a postoperative neurological deficit, do reflect a real but reversible transient cord dysfunction, and the term 'false positive', as applied to this group, is highly misleading, as is confirmed by the survey of Dawson et al.[2].

In our experience, following modified 'different from the conventional' definitions proved useful and safer for intraoperative routine. Changes of the waveform, either in amplitude or in response delay, were considered as 'events', which needed an immediate intraoperative explanation, and therefore were considered true positives until proved otherwise. The term 'false positive' was restricted only to the third of the above mentioned groups (changes due to unidentified factors, irrelevant to surgical manoeuvre). Thus the specificity for our series has risen to almost 98%, a percentage definitely higher than previous reported series, using more invasive techniques[21].

A small number of wake-up tests is however still required for the following groups: *For patients who sustained suppression of spinal cord monitoring despite correction of all physiological causes*, to decide whether any remaining metal work should be removed; *for unresolved false positive cases*, which will require it if the indications for monitoring are correct in the first place; *for the technical failures*, where there is no other means of monitoring and the cord is in peril. The misadventure of reporting false positive or false negative results, discounted as mainly technical or intrinsic methodological problems[8], can be avoided by addressing the definitions under current use and subsequently defining the indications for spinal cord monitoring. Only patients in which the result of monitoring is going to be acted on should be considered as candidates.

Exclusions should include:

I. Spinal tumours, where neurological deterioration will not alter the surgical decision for complete tumour removal.

II. Any patients with poor neurological function preoperatively (not related to the condition operated as a cause), such as severe motor or sensorimotor disability patients, with inability to walk, for whom the purpose of the operation would be to provide them with a better sitting posture, since these are more likely to be intraoperative technical failures for monitoring.

III. Anterior discectomy cases for spinal release, where the risk is low, and any loss of monitoring is likely to be due to a non-recoverable surgical event; these patients need further investigation, as far as their likelihood to be monitored in the future is concerned.

FUTURE TRENDS

The use of spinal cord monitoring to aid diagnosis is possibly one of the most promising areas of future evolution. There are encouraging reports on the use of a dermatomal evoked potential system to define the specific nerve roots involved in lumbar radicular pain[17–19,22]. Specific dermatomal testing montages are selected, and information can be gained about the status of transmission along a specific nerve root. This approach can be used to obtain data from as low as S4 or S5, proximally to the upper cervical area. Recordings can be made from the spinal areas intraoperatively, but cortical electrodes placed over the appropriate contralateral cortex are diagnostically useful because individual roots can be assessed to determine specific deficits and correlate them with other clinical and diagnostic findings. This is especially important in cases with conflicting or vague preoperative information. Those recordings can provide also a baseline for monitoring the effect of decompressive surgery. Similar root level diagnostic testing is currently used in Japan in the cervical area via the dermatomal approach or the more invasive method of inserting an extradural stimulating electrode in the lower cervical spine. As a general conclusion, the capacity to assess cord function, both diagnostically and protectively, in a variety of ways, upon the need of the individual patient will most likely be the configuration of the future in the field.

References

1. MacEwen GD, Bunnel WP, Sriram K. Acute neurological complications in the treatment of scoliosis. A report of the SRS. J. Bone Joint Surg. 1975;57A:404–408.
2. Dawson EG, Sherman JE, Kanim LE, Nuwer MR. Spinal cord monitoring: results of the

scoliosis research society and the European spinal deformity society survey. Spine. 1991;16 (Suppl 8):361–364.

3. Dawson GD. Cerebral responses to electrical stimulation of peripheral nerve in man. J Neurol Neurosurg Psychiatr. 1947;10:134–140.
4. Bunch WH, Scarff TB, Trimble J. Spinal cord monitoring – current concepts review. J Bone Joint Surg. 1983;65A:707–710.
5. Ben-David B. Spinal cord monitoring. Orthop Clin North Am. 1988;19:427–448.
6. Evoked Potentials Committee of the American Electroencephalographic Society. Guidelines for intraoperative monitoring of sensory evoked potentials. J Clin Neurophysiol. 1987;4: 397–416.
7. Lubicky JP, Spadarao JA, Yuan HA, Frederickson BF, Henderson N. Variability of somatosensory cortical evoked potential monitoring during spinal surgery. Spine. 1989;14:790–798.
8. Nash CL, Brown RH. Current concepts review: Spinal Cord Monitoring. J Bone Joint Surg 1989;71A:627–630.
9. Engler GL, Spielholz NI, Bernhard WN, Danzinger F, Merkin BA, Wolff EE. Somatosensory evoked potentials during Harrington instrumentation for scoliosis. J Bone Joint Surg. 1978;60A:528–532.
10. Gollehon D, Kahanovitz N, Happel LT. Temperature effects on feline cortical and spinal evoked potentials. Spine. 1983;8:443–446.
11. Grundy BL. Intraoperative monitoring of sensory evoked potentials. Anaesthesiology. 1983;58:72–87.
12. Szalay EA, Carollo JJ, Roach JW. Sensitivity of Spinal Cord Monitoring to intraoperative Events. J Ped Orthop. 1986;6:437–441.
13. Lesser RP, Raudzens P, Lüders H et al. Postoperative neurological deficits may occur despite unchanged intraoperative somatosensory evoked potentials. Ann Neurol. 1986;19:22–25.
14. Dinner DS, Shields RW, Lüders H. Intraoperative spinal cord monitoring. In: Rothman RH and Simeone FA, (eds). The Spine, 3rd Edition, Philadelphia–London: W.B. Saunders;1992:1801–1814.
15. Dunn C, Jardine AD, Smith NJ, Webb JK, Yeomans PM. Electrophysiological monitoring of spinal cord function during scoliosis surgery. Electroencephalogr Clin Neurophysiol. 1984;58:15P.
16. Hamilton A, Webb JK. The role of vertebrectomy in correction of spinal deformity. ESDS Meeting, Lyon, France. 1992.
17. Aminoff MJ, Goodin DS, Barbaro NM et al. Dermatomal somatosensory evoked potentials in unilateral lumbosacral radiculopathy. Ann Neurol. 1985;17:171–176.
18. Machida M, Asai T, Sato K et al. New approach for diagnosis in herniated lumbosacral disc: Dermatomal somatosensory evoked potentials (DSSEPs). Spine. 1986;11:380–384.
19. Herron LD, Trippi AC, Gonyeau M. Intraoperative use of dermatomal somatosensory evoked potentials in lumbar stenosis surgery. Spine. 1987;12:379–383.
20. Nash CL. Spinal cord monitoring. In Orthopaedic Knowledge Update No 3, AAOS;1990:227–234.
21. Forbes HJ, Allen PW, Waller CS, Jones SJ, Edgar MA, Webb PJ, Ransford AO. Spinal Cord Monitoring in scoliosis surgery. J Bone Joint Surgery. 1991;73B:487–491.
22. Perlik S, Fisher MA, Patel DV et al. On the usefulness of somatosensory evoked potentials for the evaluation of lower back pain. Arch Neurol. 1986;43:907–913.

4
Anaesthetic requirements for spinal surgery and effective cord monitoring

W. T. FRAZIER

INTRODUCTION

The main issue to be addressed is the effect of general anaesthetics on the quality of the somatosensory evoked potentials recorded from the scalp during major spine surgery (i.e. cortical somatosensory evoked potentials [C-SSEPs]). Although many clinical neurophysiological teams utilize other techniques (e.g. recording at 'lower' levels such as spinal cord or brain stem), many others continue to find it useful to stimulate peripheral nerves and record from scalp (i.e. 'cortical') sites. The comparative success rate of various sensory evoked potential techniques (e.g. cortical versus brainstem) and the relative roles of sensory and motor evoked potentials are beyond the scope of this presentation.

The major concentration will be on a brief review of the status of an anaesthetic technique at the time of the IVth International Symposium on Spinal Cord Monitoring SCMIV, (Niigata, Japan – 1989[1]) and particularly a summary of the new information reported since that time.

In comparing approaches to spinal cord monitoring (SCM) around the world in 1992, it is still evident that there is a good deal of variability in the management of the amount and types of anaesthetic drugs used. Perhaps even more varied and complex are the working relationships between the neurophysiologists, anaesthesiologists, and surgeons. Even in those situations where the anaesthesia team is managing both the anaesthetic plan and the neurophysiological monitoring, it is often a difficult task to balance the needs for an adequate level and type of anaesthesia with the need to keep the level 'light' in order to facilitate SCM. It is sometimes particularly difficult to assure a level adequate to suppress awareness and/or postoperative recall, since in the presence of muscle relaxants, we still do not have universally accepted criteria for easily measuring anaesthetic level. Although computerized EEG analysis still holds promise, there is no easily identifiable 'set-

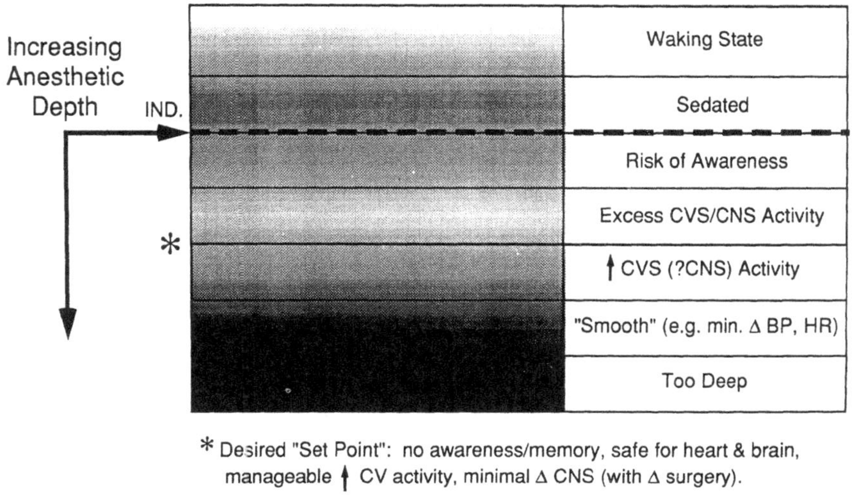

Figure 1 The horizontal dotted line represents the 'sleep' threshold (i.e. loss of responsiveness at the time of induction of anaesthesia; in essence, MAC-Awake). At this level (and slightly deeper), there is some risk of awareness and/or recall. 'CNS Activity' indicates that surgical stimuli may increase signs of not only cardiovascular (CV) activity, but CNS activity also (e.g. changes in EEG amplitude or spectrum). In many instances not involving SCM, anaesthesiologists may deepen the anaesthetic to the 'smooth' area (mainly for control of the CV system); C-SSEPs are likely to be very depressed at this depth. 'Too deep' would usually be excessive cardiovascular depression (e.g. with anaesthetic gases) and extremely depressed C-SSEPs

point', particularly with varying levels and types of drugs and with variable surgical stimuli. For example, it is left to the experience and judgment of the anaesthesiologist to determine if a significant rise in heartrate and/or blood pressure should be managed by 'deepening' the anaesthetic or by the use of peripherally acting drugs (e.g. beta blockers and/or vasodilators). All of these problems are more difficult to handle in older patients and/or those with pre-existing neurological deficits. It is particularly difficult to decide which instances may require a change from the routine/standard anaesthetic to a 'special' technique (e.g. the use of etomidate or ketamine). When the anaesthetic and the monitoring are being handled by different individuals, it is an even more difficult 'balancing act' due to the complexities of communication between team members who may not fully appreciate the technical difficulties faced by each group. Of course, this situation is further complicated by the necessity to communicate with the surgeons regarding the status of the neurophysiological monitoring and how it can impact on the surgical strategy in order to optimize the surgical outcome. To say the least, this total effort is one of the more complex medical team interactions and requires a good deal of mutual understanding of terminology, problems, expectations, and ground rules.

In Figure 1, there is a pictorial summary of the progression from the 'awake state' through a 'sedated state' (e.g. a typical presurgical premedication), and then on to varying degrees of anaesthetic depression of the central nervous system. The various actions of anaesthetic drugs do not cause such a simple, one-dimensional continuum, but such a simplified graphical progression is useful. It is commonly recognized that some patients can have some recall when apparently anaesthetized, even in the absence of muscle relaxants. It is also clear that at a deeper level (with no possibility of recall), there are cardiovascular responses to surgical stimuli (e.g. hypertension) which must be managed either by deepening the anaesthetic (e.g. increased percent Forane) and/or by treating the cardiovascular (CV) system directly (e.g. add a vasodilator). As a slightly deeper level, such CV reactions are less likely and are more easily manageable. At even deeper levels (of most agents), the price paid for a 'smooth' course is likely to be very depressed C-SSEPs. There are also clinical instances (e.g. hypotension secondary to blood loss) where in the absence of SCM, the anaesthesiologist would usually 'lighten' the anaesthetic to take away some depression of the CV system. This action can, of course, change the baseline CNS drug induced changes and, therefore, the baseline C-SSEPs, also.

In Figure 2A, there is illustrated an idealized typical anaesthetic over time with a level just sufficient to give a margin of safety and hopefully adequate SCM conditions. Figure 2B is a slightly more realistic diagram, illustrating that it takes actually more time to progress through the induction period to the maintenance period (e.g. transition from i.v. induction agents to anaesthetic gases, etc.), and to adjust the level appropriate for this particular patient. Figure 2C begins to give a picture more like an actual surgical/anaesthetic case, where the need for anaesthetic drugs varies with the degree and type of surgical stimulation. It is these instances that require the anaesthesiologist to make a judgement call on whether to increase the level of one or more of the anaesthetic drugs or to use direct acting CV drugs. It is at these times that there needs to be a maximum of communication between the neurophysiologist and the anaesthesiologist since, on occasion, there may be no ideal management and compromises may have to be made.

I think it is fair to say that we have come a long way in making better anaesthetic/SCM compromises since the first International Symposium (Tokyo, Japan, 1981), but we still have a long way to go to assure that we can safely monitor all patients who could benefit from SCM (via C-SSEPs) under general anaesthesia.

BASIC QUESTIONS TO BE REVIEWED

A. Do we have a consensus in regard to an anaesthetic plan and/or the residual major problems caused by anaesthetic drugs? I think it is fair to say that by reading the literature and by personal communications (e.g. at national and international conferences), we do not yet have a single plan on which most groups agree. However, there is an evolving consensus in that most groups use a 'balanced technique' involving some anaesthetic gases, some intravenous anaesthetic/analgesic drugs and muscle relaxants. However, there is a great deal of variability on which anaesthetic agents to use and what ratio, to say nothing of how to

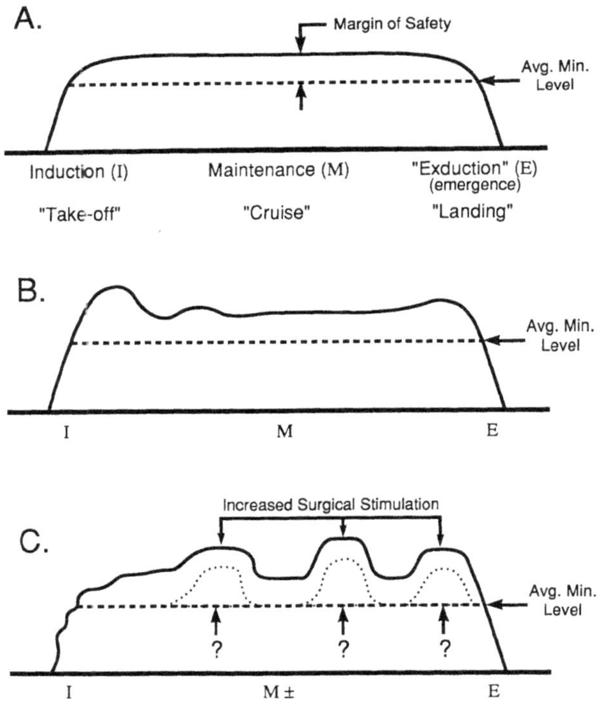

Figure 2 The solid line in A, B, and C represents the amount of total anaesthetic drugs and the dotted line represents the needed anaesthetic coverage (and the difference equals 'margin of safety').The '?' indicates the somewhat contradictory goals of a steady level of anaesthesia (and hopefully a steady level of minimal C-SSEP depression) and a varying need, depending on variable surgical stimulation. A. A simplified representation, where there is steady level of anaesthesia, except for brief periods at induction and emergence; the level achieved allows a margin of safety deeper than that which would allow for awareness and/or recall or excessive CV activity. B. A more accurate representation, indicating longer periods at induction and emergence when anaesthetic levels may vary (and thus influence C-SSEPs). C. An even more accurate picture, showing periodic need to deal with varying levels of surgical stimulation – either by increasing the anaesthetic (to keep some 'margin of safety') or by using other means to manage increased CV activity

change these variables at the time of important events during the case (e.g. hypertension, decreased C-SSEP signals, etc.).

B. Do we have more choices of anaesthetics than at the time of the 1989 conference? Yes and no. There are no new clinically approved drugs, but we do have more information available on the utility of drugs which were relatively new in 1989 (e.g. propofol, etomidate, ketamine, and midazolam). A new anaesthetic gas (desflurane) is just becoming available and several new intravenous drugs are 'just around the corner'. By 1995 (SCM-VI), new laboratory and clinical data should be available on these new agents.

REVIEW OF THE STATUS OF ANAESTHETIC PLANS AS OF 1989 (SCM-IV)

I think it is fair to say that there was wide agreement that all anaesthetic gases produce a dose-dependent depression of C-SSEPs (e.g. N_2O, Forane [F], Ethrane [E], and halothane [H]) whereas other agents, particularly etomidate, produce augmentation. It is likely that choices between the various halogenated agents (i.e. F, E, and H) have been on the grounds of economics and personal preference. In the US it is probable that Forane has been and will be the dominant volatile agent for the next few years. There was still considerable debate on the use of N_2O, its main advantages being fast kinetics (i.e. quick onset and offset) and minimal cardiovascular effects. In my opinion, differences in the C-SSEP effects of N_2O, H, E, and F on C-SSEPs are small ($\sim \pm 10\%$), whereas the differences in kinetics and cardiovascular effects are relatively large. Therefore, it was and still is a common anaesthetic practice to use N_2O (e.g. $\sim 50\%$), in combination with lower doses of a volatile agent (e.g. $\sim$0.25–0.3% Forane), having reduced the overall need for anaesthetic gases by using a continuous level of an intravenous analgesic (e.g. fentanyl infusion[3,4]). Other 'major variants' were plans which omitted N_2O (and used a higher concentration of an agent such as Forane or an i.v. infusion of a drug such as midazolam) and those which used other i.v. analgesics (e.g. sufentanyl or alfentanil) to achieve varying degrees of reduction of the need for anaesthetic gases.

At the time of the 1989 meeting (SCM-IV), the major new data presented was on etomidate[5] and ketamine[6], either as major components of alternate anaesthetic plans or as 'bail-out' plans to be used at certain times of diminished C-SSEP signals. A more detailed summary of these two presentations will follow.

ETOMIDATE USE IN SPINAL CORD MONITORING

McPherson and Levitt[5] presented additional experience building on the presentation at the 1986 meeting[7,8] and other reported data[9,10]. In their plan, etomidate served the dual function of being a major part of the general anaesthetic as well as a means to augment the size of the C-SSEPs under certain circumstances. Their use of etomidate was complimented by Forane (0.4–0.8% in O_2) and fentanyl (a large loading bolus of 15–20 μg/kg, plus 50–100 μg 'prn'). They noted that C-SSEPs partially depressed by Forane could be augmented by an intermittent bolus of etomidate at a time of further significant decrease of C-SSEPs (e.g. perhaps secondary to a surgical event) and that the augmentation was dose dependent. The rationale of this strategy seemed to be to prove that the appropriate spinal cord/cortex connections were still present over the time of action of the new drug bolus. Their patient population was presumably neurosurgical and involved patients probably already on steroids (recommended after more extended use of etomidate). They preferred an intermittent bolus technique (rather than a continuous infusion) in order to minimize a predicted longer waking time.

In my opinion, residual difficulties with etomidate continue to be the required steroid support in the postoperative period and some reports of postoperative seizure-like events[11-16].

KETAMINE USE IN SPINAL CORD MONITORING

Frazier[17] presented more extensive data (since the initial report of Frazier *et al.*[6] on the use of an 'intervention plan' to be used when EP signals had gradually grown smaller over a relatively long period of time (particularly in elderly patients with pre-existing neurological deficits). The logic of this intervention was to (a) prove that the spinal cord–cortex connection was still there, and (b) to re-establish a new control level of C-SSEP monitoring for another 1–2 hours.

The base plan was: induction with (e.g.) thiopental 2.5–3.5 mg/kg and fentanyl 5 μg/kg (up to 10 μg/kg in patients with pre-existing tolerance); maintenance with 50% N_2O, ~0.25% Forane, and fentanyl infusion at 2 μg/kg/min (up to 4 in patients with tolerance). In the presence of a continuing diminution of C-SSEPs, the N_2O was discontinued and the Forane and fentanyl were continued. As the N_2O was washed out, ketamine was concurrently administered, usually 3 divided boluses over ~5 minutes (total of ~1 mg/kg) followed by an infusion of ~3 mg/kg/h. Augmentation of C-SSEPs could be achieved immediately (~3–5 minutes) and maintained with this technique over 1–2 hours, the results being correlated with the quality of the original control signals and usually with the age and pre-existing neurological status of the patient.

FUNDAMENTAL ANAESTHETIC CONSIDERATIONS

Inherent in all discussions of anaesthetic plans is the fundamental notion of a level of anaesthetic which is enough to prevent awareness and recall (especially of pain and extreme anxiety) and to minimize significant cardiovascular response (i.e. increases in blood pressure and heart rate). Although the response to surgical stimuli may be in some respects a continuum, for gaseous anaesthetics there is a reference level called MAC (i.e. minimal alveolar concentration) which is related to the blood concentration of an anaesthetic gas. MAC is an empirically determined value which is defined as the concentration at which 50% of experimental subjects move in response to a 'standard' stimulus (i.e. MAC_{50})[18]. MAC is in turn dependent on several variables such as species, temperature, age, etc. Although not routinely acknowledged, MAC is probably also dependent on the type and strength of the stimulus, site of application, sensitivity of the 'motion detector', etc. Nevertheless, MAC is an extremely useful concept and set of numbers. Not only is there a MAC for each anaesthetic gas in humans, but there is also a MAC for various gas mixtures (e.g. Forane in 50% N_2O), and 'MACs' for various mixtures of gases and i.v. agents. There are also other types of 'MACs' – one of the most useful being the level at which 5% (rather than

Table 1 MAC definitions

1. MAC Awake	The level at which the eyes just spontaneously close or open with minimal stimulation (e.g. 'open your eyes')	~0.5 MAC
2. MAC – Aware/Memory	Estimated, not known	< ~0.8 MAC
3. MAC_{50}	As 'classically defined'	1.0 MAC
4. MAC_{95}	~95% of subjects do not respond	~1.3 MAC
5. MAC-BAR	BAR = block of autonomic response	~1.5 MAC

50%) of the subjects move (more representative of a surgically relevant level) – the MAC_{95}. (See Table 1 for a summary of the MACs as there is a progressive and increasing level of anaesthetic.)

It is commonly recognized (and now quantitated[3,4]) that concurrent use of intravenous analgesics such as fentanyl results in a reduction of MAC for an anaesthetic gas. Furthermore (and perhaps even more interestingly), this MAC reduction by i.v. analgesics has a ceiling effect. That is, high dose continuous blood levels of fentanyl reduce the MAC of Forane by 65–70% and higher levels cause little if any further reduction in MAC[3,4]. Many currently used anaesthetic plans take advantage of this analgesic MAC reduction since using less volatile agent can result in faster awakening, particularly in long cases. Furthermore, for SCM, i.v. analgesics seem to interfere with C-SSEPs minimally and all anaesthetic gases have a dose dependent depressive effect. A level of 1.3–1.5 MAC (approximately MAC_{95} and MAC-BAR) of any anaesthetic gas (alone) results in very depressed C-SSEP signals.

An example of an estimated MAC reduction anaesthetic plan which we commonly use is a reduction of the Forane MAC from ~1.15% to 0.25–0.3% by the concurrent use of 50% N_2O and a 'steady-state' blood level of fentanyl supported by a continuous infusion of fentanyl at 2 μg/kg/h. In most adults without marked pre-existing neurological deficits, this combined anaesthetic effect (level) allows acceptable C-SSEP SCM.

The practical upper level of MAC reduction by i.v. analgesics (or for that matter i.v. sedatives such as thiopental, midazolam, propofol, etomidate, or ketamine) relates to the residual (and slowly subsiding) effects which can interfere with the possibility of doing a prompt 'wake-up' test during the case or accomplishing a relatively rapid anaesthetic emergence at the end (so as to do a prompt neurological exam). Very high dose i.v. analgesics also commit one to prolonged postoperative ventilation unless the dose is tapered very early, the need for which is frequently hard to predict and track. Also, tapering anaesthetics early results in a non-steady-state drug level which interferes with the fundamental notion of keeping drug levels constant so as to provide a stable plateau from which to evaluate changes in C-SSEPs (possibly caused by surgical misadventures). In the final analysis, a series of compromises are made, using combinations of drugs and slowly varying blood levels – which hopefully do not change significantly at times when critical events are likely.

Table 2 Summary of pros and cons of anaesthetics for SCM

Drug	*Pros*	*Cons*
1. Nitrous oxide	Fast kinetics, cardiovascular (CV) stability, no accumulation, acceptable at $\sim 50\%$	Dose related C-SSEP depression, MAC > 100%, displaces O_2, supplement only, residual analgesia
2. Forane	Moderate CV control, bronchodilation, acceptable C-SSEP depression at $\sim 1/3$ to 1/2 MAC. Can be a complete anaesthetic	Slow kinetics, accumulates in fat in long cases, dose related C-SSEP depression, no residual analgesia
3. Fentanyl (i.v. analgesic)	MAC reduction of other agents such as N_2O, Forane, etc. CV stability, minimal C-SSEP depression. Allows 'painless' wake-up (intra-op and at emergence), minimal negative cumulative effects at 2 $\mu g/kg/h$	Supplement only, postop ventilation required if dose high (e.g. 3–4 $\mu g/kg/h$)
4. Propofol	Fast kinetics, $\pm$ less C-SSEP decrease than other sedatives, allows 100% O_2, complete agent in some instances	Complex kinetics in long cases (accumulates), much CV depression, usually not a complete agent, no residual analgesia
5. Etomidate	Increases C-SSEPs, can be a complete anaesthetic, CV stability, allows 100% O_2	Myoclonus and seizures (postop), adrenal depression, no residual analgesia, may accumulate
6. Ketamine	Minimal depression of C-SSEPs (or augment), can be a complete agent, good analgesic and sedative, minimal CV depression, may help asthma (intra-op), allows 100% O_2	Complex kinetics (accummulates; probably use 1–2 hours only), may require additional CV control, needs supplemental drugs to minimize postop reactions, good residual analgesia
7. Midazolam	Intermediate C-SSEP depression, minimal CV depression, amnesia, blunts side-effects of ketamine	Incomplete agent, complex kinetics (accummulates), no residual analgesia
8. Thiopental	Intermediate C-SSEP depression, allows 100% O_2, fast onset, moderate CV effects	Incomplete agent, cumulative, no residual analgesia
9. Ideal	Increases C-SSEPs (like etomidate-E), stable EPs (like ketamine-K), neutral CV effects like E and K, kinetics of atracurium (no accumulation), kinetics of propofol (fast on and off for intermediate periods), can give either i.v. or as inhaled gas	None

SUMMARY OF NEW DATA SINCE SCM IV ('89)

A. 1988

Sloan et al.[19] extended the etomidate story another significant step by presenting a case where in the presence of one of their usual anaesthetic plans (i.e. infusions of midazolam and alfentanil), C-SSEPs were too small to be of value. Replacing midazolam with etomidate (loading dose, plus continuous infusion) allowed them to re-establish a minimal acceptable level of SCM for approximately 10 hours (the patient being supported by postoperative steroids). The patient was age 60 and the surgery was in the lumbar region with lower limb nerve stimulation.

Koht et al.[20] presented a study investigating the efficacy of a plan combining either etomidate, midazolam, or thiopental with fentanyl and N_2O (in additive steps). They used median nerve stimulation and the study was carried out in the presurgical time. They also confirmed the augmenting effects of etomidate and demonstrated that the addition of fentanyl and then N_2O significantly modified the etomidate augmentation. Midazolam and thiopental (alone) caused remarkably few changes in the C-SSEPs and the addition of fentanyl (to either) produced few effects. The addition of N_2O depressed the signals the most in the etomidate cases, intermediate in the thiopental cases, and minimally with midazolam.

Wolfe et al.[21] studied C-SSEPs produced by posterior tibial nerve (PTN) stimulation and compared various effects of combinations of N_2O and Forane on both cortical and brainstem SSEPs. For C-SSEPs, the depressive effects of N_2O and Forane were additive (at MAC 1.0 and greater), with minimal effects in brainstem signals.

Kaulkman et al.[22] studied posterior tibial nerve C-SSEPs (in cardiothoracic surgery patients) in the presence of high dose opioids (fentanyl or alfentanil or sufentanyl) given as a continuous infusion after a loading bolus and with lorazepam premedication (4–6 mg p.o.). Moderate depression of signals was less than with plans using anaesthetic gases and there were no differences between the various opioids. They noted prolonged recovery.

Maurette et al.[4] studied a plan combining propofol (bolus and infusion) with 50% N_2O during spine surgery and using ankle stimulation. C-SSEPs were reduced by more than 50% (more than when methohexitone was used rather than propofol in a previous study).

B. 1989

Freye et al.[23] studied median nerve C-SSEPs with an anaesthetic plan including propofol and N_2O (66%).

Scheepstra et al.[24] studied median nerve C-SSEPs on non-spine cases in the presence of propofol infusion and fentanyl over 1–2 hours. Minimal changes in N14 and N20 were noted, but all later signals were almost absent.

Pathak et al.[25] studied 28 scoliosis patients with PTN stimulation and with spinal cord and cortical SSEPs. They used various combinations of H, E, F in 60% N_2O, the worst results being with H. They concluded that subcortical SSEPs were promising.

Loughnan et al.[26] studied epidural SSEPs (C_7–T_1) with variable combinations of N_2O and halothane (H) and found that H had minimal effects (up to 1.5%).

Sloan et al.[27] studied median nerve C-SSEPs for 30 minutes prior to surgery in the presence of thiopental and noted only moderate depression of signals.

Russ et al.[28] studied median nerve EPs in the presence of various combinations of N_2O, thiopental, etomidate, and fentanyl. They also noted the augmenting effects of etomidate and stated a preference for i.v. agents over N_2O and volatile agents.

C. 1990

Abel et al.[29] recorded C-spine SSEPs in scoliosis patients anaesthetized with halogenated agents, N_2O, and a narcotic.

Boston et al.[30] performed an animal study (swine) investigating the time related effects of Forane on C-SSEPs over ~2 hours.

Black et al.[31] studied the effects of a new volatile anaesthetic (I-653, desflurane) on C-SSEPs. This drug will be available for clinical use in late 1992.

Ragazzoni et al.[32] studied median nerve C-SSEPs as influenced by thiopental and i.v. analgesics.

Schubert et al.[33] studied median nerve SEPs in patients undergoing abdominal surgery as influenced by ketamine (bolus and infusion) and added N_2O. Ketamine alone increased C-SSEPs ~15–50% over control levels and the N_2O approximately cancelled out this augmentation.

Shimoji et al.[34] studied the effects of Forane on spinal cord signals in rats (with basal anaesthetic of intraperitoneal ketamine).

Sloan et al.[35] used median nerve C-SSEPs and C-spine SSEPs to study the effects of midazolam on patients during the presurgical period. They noted minimal decreases in C-spine signals (~13%) and moderate decreases in C-SSEPs (~51–63%).

D. 1991

Samra[36] studied the effects of etomidate in an animal model (cats) in the presence of a basal anaesthetic (50% N_2O and 0.8–1.8% halothane), recorded both in thalamus and on the cortex, and concluded that the etomidate-induced enhancement of C-SSEPs is due to an effect at the cortical level.

Kalkman et al.[22] used PTN C-SSEPs to study the effects of N_2O and propofol during a continuous infusion of alfentanil. The plan called for replacement of 66% N_2O by propofol, the result being an increase in the early components and a decrease in the later components.

Vandesteen et al.[37] studied the cortical distribution of median nerve C-SSEPs (short latency) in the presence of halogenated anaesthetics.

E. OTHER SIGNIFICANT SCM/EP LITERATURE

Gugino and Chabot[38] published an extensive review of the SSEP literature and anaesthetic effects for the period of 1974–1988.

Thakor et al.[39] and John et al.[40] presented new and promising signal processing techniques.

Van Hemelrijk et al.[41] presented an EEG study demonstrating the difficulties in achieving target level effects of i.v. anaesthetics (using computer driven i.v. pumps) while administering combinations of propofol, N_2O and Forane.

Browning et al.[42] studied the effects of temperature on C-SSEPs in cats under halothane anaesthesia and showed an interesting maximum between 34–36°C.

Guit et al.[43] studied a combination of ketamine and propofol as a total i.v. anaesthetic (in comparison to propofol and fentanyl). They preferred the ketamine/propofol plan due to more stable haemodynamics and noted no postoperative psychological problems – concluding that propofol prevented any such problems potentially due to ketamine.

Fruergaard et al.[44] studied a total i.v. anaesthetic plan consisting of fentanyl with either propofol or etomidate and noted more nausea and vomiting after propofol, but 'mental side-effects' only after etomidate. They also noted complaints of 'reduced power of concentration' in the etomidate group ($\sim 1/2$ vs. 10% in the propofol group) at a three month follow-up interview.

White[45] reviewed the current clinical status of ketamine and summarized the use of other i.v. agents to minimize potential side-effects.

OVERVIEW OF ANAESTHETIC PLAN AND MANAGEMENT CHOICES

A. Drug choices

There seems to be a definite drift toward Forane (at least in lower concentrations) as the drug of choice as a halogenated agent (more helpful in managing blood pressure, bronchospasm, etc. than N_2O only and/or total i.v. anaesthesia). MAC reduction by the utilization of i.v. agents is generally being accomplished with i.v. sedatives (e.g. thiopental, midazolam, or propofol) and/or i.v. analgesics (e.g. fentanyl, alfentanil, or sufentanyl); the goal is to achieve less depression of C-SSEPs while achieving an acceptable anaesthetic level. Etomidate and ketamine are still most likely to be 'bail out' choices to be used in especially difficult situations.

B. Technique choices

With regard to the choice between repeated bolus doses (either intermittent or one large dose at the beginning, with a slow changing decay phase) or continuously infused i.v. agents (after loading bolus), the latter seems to be predominating. Mixed or 'balanced' plans are replacing single agent plans (usually as gas and i.v. and/or multiple i.v. agents).

C. Strategy choices

When there is a significant diminution of the C-SSEPs, there are choices which require joint decisions on the part of the total team (surgery, anaesthesiology, and neurophysiology). Has there been an important surgical event? Has the anaesthesia caused the decrease or is something wrong with the neurophysiology techniques (e.g. stimulators, amplifiers, computer, etc.)? The most difficult situation is when C-SSEPs have diminished to a level where they are of borderline significance. These times require candour on the part of all participating team members. If other technical factors can be ruled out, then I think it is appropriate to change the anaesthetic plan by using either etomidate (see McPherson and Levitt[5] or Sloan et al.[19]) or ketamine (see Frazier et al.[6] or Schubert et al.[33]) as a major component in order to establish the degree of remaining function and to attempt to achieve a new control level for at least a few hours.

SUMMARY AND FUTURE GOALS

The last 3–4 years have probably been a period of consolidation for most SCM teams – as they have refined their neurophysiological techniques and have evolved 'convergently' to 'balanced' anaesthetic techniques.

However, a great deal of work remains to be done even with the drugs available to us at this time (assuming no new agents become available). Especially studies need to be done to explore which combinations and levels of drugs produce the best overall results and particularly work needs to be done on how to judge the level of the total anaesthetic (to assure prevention of awareness and recall) and how to deal most effectively with CV responses (e.g. control hypertension with 'deeper' anaesthetic level or with drugs acting only in the CV system).

In designing future studies on the anaesthetic management of patients undergoing SCM, particular attention should be paid to the following:

1. Older age groups: It is commonly acknowledged that older patients are more difficult.
2. Patients with pre-existing neurological deficits: Again, well known to present a special challenge.
3. Study patients undergoing spine surgery: Lower extremity signals are more difficult and require less drug depression than studies using ulnar/median nerve C-SSEPs.
4. Study techniques which are likely to work in patients with co-existing disease and during periods of major surgical stress (e.g. large blood loss).
5. Deal with long cases (i.e. greater than 4 hours).

6. Specify whether or not technique allows for a 'wake-up' test (especially on short notice).
7. Develop methods to grade/evaluate control and monitored signals over long periods (e.g. 4 hours or more).
8. Develop and document strategies for dealing with negative surgical events (e.g. versus failed equipment, etc.; change anaesthetic plan?).
9. Document postop course, waking time, etc. (e.g. time to adequate neurological exam).
10. Document postop follow-up several days later – especially new neurological findings or changes and correlate with sensitivity of SCM to detect intra-operative surgical events.

References

1. Shimoji K, Kurokawa T, Tamaki T, Willis WD, eds, Spinal cord monitoring and electrodiagnosis. New York: Springer Verlag;1991.
2. Homma S, Tamaki T, Shimoji K, Kurakawa T. Fundamentals and clinical application of spinal cord monitoring. Tokyo: Saikan Publishing Co Ltd; 1984.
3. Murphy MR, Hug CC. The anesthetic potency of fentanyl in terms of its reduction of enflurane MAC. Anesthesiology. 1982;57:485.
4. Murphy MR, Hug CC. The anesthetic potency of fentanyl in reducing isoflurane MAC. Anesthesiology. 1983;59:A338.
5. McPherson RW, Levitt RC. Etomidate augmentation and scalp recorded somatosensory waves: Time course, reproducibility, and dose effect. In: Shimoji K, Kurokawa T, Tamaki T, Willis WD, eds, Spinal cord monitoring and electrodiagnosis. New York: Springer-Verlag;1991:163–170.
6. Frazier WT, Biggs DB, Odom SH, Jenkins SE. Ketamine as a major component of an anesthetic plan for spinal cord monitoring. In: Shimoji K, Kurokawa T, Tamaki T, Willis WD, eds, Spinal cord monitoring and electrodiagnosis. New York: Springer-Verlag; 1991:325–334.
7. McPherson RW, Sell B, Traystman RJ. Effects of intravenous induction agents on somatosensory evoked potentials: thiopental, fentanyl, and etomidate. In: Ducker TB, Brown RH, eds, Neurophysiology and standards of spinal cord monitoring. New York: Springer-Verlag;1988:163–167.
8. McPherson RW, Sell B, Traystman RJ. Effects of thiopental, fentanyl, and etomidate on upper extremity somatosensory evoked potentials in humans. Anesthesiology. 1986;65: 584–589.
9. Kalkman CJ, van Rheineck AT, Hesselink EM, Bovill JG. Effects of etomidate or midazolam on median nerve somatosensory evoked potentials. Anesthesiology. 1986;65(3A):A356.
10. Kochs E, Treede RD, Schulte-am-Esch. Increase of somatosensory evoked potentials during anesthesia induction with etomidate. Anaesthetist. 1986;35:359–364.
11. Gancher S, Laxer KD, Krieger W. Activation of epileptogenic activity by etomidate. Anesthesiology. 1984;61:616–618.
12. Goroszeniuk T, Albin M, Jones RM. Generalized grand mal seizure after recovery from uncomplicated fentanyl-etomidate anesthesia. 1986;65:979–981.
13. Laughlin TP, Newberg LA. Prolonged myoclonus after etomidate anesthesia. Anesth Analg 1985;64:80–82.
14. Parker CJR. Respiratory disturbance during recovery from etomidate anesthesia. Anaesthesia. 1988;43:16–17.
15. Wanscher M, Tonneson E, Huttel M, Larson K. Etomidate infusion and adrenocortical function – A study of elective surgery. Acta Anes Scan, 1985;29:483–485.
16. Preziosi R, Vacca M. Adrenocortical suppression and other endocrine effects of etomidate. Life Science. 1988;42(5):477–489.

17. Frazier WT. Effects of anesthetic drugs on spinal cord monitoring: An update. In: Ducker TB, Brown RH, eds, Neurophysiology and standards of spinal cord monitoring. New York: Springer-Verlag;1988:125–139.
18. Eger EI, Saidman LJ, Branstater B. Minimum alveolar concentration: A standard of anesthetic potency. Anesthesiology. 1965;26:756.
19. Sloan TB, Ronai AK, Toleikis JR, Koht A. Improvement of intraoperative somatosensory evoked potentials by etomidate. Anesth Analg. 1988;67:582–585.
20. Koht A, Shutz W, Schmidt G, Schramm J, Watanabe E. Effects of etomidate, midazolam, and thiopental on median nerve somatosensory evoked potentials and the additive effects of fentanyl and nitrous oxide. Anesth Analg. 1988;67:435–431.
21. Wolfe, D, Drummond JC. Differential effects of Isoflurane/nitrous oxide on posterior tibial somatosensory evoked responses of cortical subcortical origin. Anesth Analg. 1988;67: 852–859.
22. Kalkman CJ, Traast H, Zuurmond WWA, Bovill JG. Differential effects of propofol and nitrous oxide on posterior tibial nerve somatosensory cortical evoked potentials during alfentanyl anaesthesia. Br J Anaesth. 1991;66:483–489.
23. Freye E, Hartung E, Schenk GK. Somatosensory-evoked potentials during block of surgical stimulation with propofol. Br J Anaesth. 1989;63:357–359.
24. Scheepstra GL, DeLange JJ, Booij LHDJ, Ros HH. Median nerve evoked potentials during propofol anaesthesia. Br J Anaeth. 1989;62:92–94.
25. Pathak KS, Amaddic MD, Scoles PV, Shaffer JW, Mackay W. Effects of halothane, enflurane, and isoflurane in nitrous oxide on multilevel somatosensory evoked potentials. Anaesthesiology. 1989 70:207–212.
26. Loughnan BA, King MJ, Grundy EM, Young DL, Hall GM. Effects of halothane on somatosensory evoked potentials recorded in the extradural space. Br J Anaesth. 1989;62:297–300.
27. Sloan TB, Kimovec MA, Serpico LC. Effects of thiopentone on median nerve somatosensory evoked potentials. Br J Anaesth. 1989;62:51–55.
28. Russ W, Kling D, Schwandt HKJ, Krumholt W, Hempelmann G. Effects of thiopentone and etomidate on median nerve somatosensory evoked responses. Neurological Research. 1989;11:51–56.
29. Abel MF, Mubarak SJ, Wenger DR, Costello J, Hicks GE. Brainstem evoked potentials for scoliosis surgery: A reliable method allowing use of halogenated anesthetic agents. Journal of Pediatric Orthopaedics. 1990;10:208–213.
30. Boston JR, Davis PJ, Brandom BW, Roeber CM. Rate of change of somatosensory evoked potentials during isoflurane anesthesia in newborn piglets. Anesth Analg. 1990;70:275–283.
31. Black S, Konstadt SN, Sami H, Rao TK. Effect of I-653 (desflurane) on somatosensory evoked potentials. Anesthesiology. 1990;73(3A):A181.
32. Ragazzoni PA, Amantini A, deScisciolo G, Bartelli M, Rossi R, Pieraccioli E. Anesthetic induction with thiopental: Its effect on scalp topography of median nerve somatosensory evoked potentials. Anaesthesiol Scand. 1990;34:389–392.
33. Schubert A, Licina MG, Lineberry PJ. The effect of ketamine on human somatosensory evoked potentials and its modification by nitrous oxide. Anesthesiology. 1990;72:33–39.
34. Shimoji K, Fujiwara N, Fukuda S, Denda S, Takada T, Maruyama Y. Effects of isoflurane on spinal inhibitory potentials. Anesthesiology. 1990;72:851–857.
35. Sloan TB, Fugina ML, Toleikis JR. Effects of midazolam on median nerve somatosensory evoked potentials. Br J Anaesth. 1990;64:590–593.
36. Samra SK, Sorkin LS. Enhancement of somatosensory evoked potentials by etomidate in cats: An investigation of its site of action. Anesthesiology. 1991;74:499–503.
37. Vandesteen A, Nogueira MC, Mavroudakis N, Defevrimont M, Brunko E, Zegers de Beyl D. Synaptic effects of halogenated anesthetics on short-latency SEP. Neurology. 1991;41: 913–918.
38. Gugino V, Chabot RJ. Somatosensory evoked potentials. International Anesthesiology Clinics. 1990;28(3):154–164.
39. Thakor NV, Vas CA, McPherson RW, Hanley DF. Adaptive Fourier series modeling of time-varying evoked potentials: Study of human somatosensory evoked response to etomidate anesthetic. Electroencephalography and Clinical Neurophysiology. 1991;80: 108–118.

40. John ER, Chabot RJ, Prichep LS, Ransohoff J, Epstein F, Berenstein A. Real-time intraoperative monitoring during neurosurgical and neuroradiological procedures. J Clin Neurophysiology. 1989;6(2):125–158.
41. Van Hemelrijck J, Tempelhoff R, White PF, Jellish WS. EEG-Assisted titration of propofol infusion during neuroanesthesia: Effect of nitrous oxide. J Neurosurg Anesth. 1992;4(1): 11–20.
42. Browning JL, Heizer ML, Baskin DS. Variations in corticomotor and somatosensory evoked potentials: Effects of temperature, halothane anesthesia, and arterial partial pressure of CO_2. Anesth Analg. 1992;74:643–648.
43. Guit JBM, Koning HM, Coster ML, Niemeijer RPE, Mackie DP. Ketamine as analgesic for total intravenous anaesthesia with propofol. Anaesthesia. 1991;46:24–27.
44. Fruergaard K, Jenstrup M, Schierbeck J, Wiberg-Jorgensen F. Total intravenous anaesthesia with propofol or etomidate. European J Anaesth. 1991;8:385–391.
45. White PF. Ketamine update: Its clinical uses in anesthesia. In: Domino EF, ed, Status of ketamine in anesthesiology. Ann Arbor, MI: NPP Books;1990:343–366.

5
Application of intraoperative spinal cord monitoring to neurosurgery

M. KURTHEN and J. SCHRAMM

INTRODUCTION: THE CONCEPT OF 'REVERSIBLE DAMAGE' DETECTION

Electrophysiological monitoring of spinal cord functions is relevant to the surgical management of spinal lesions in three respects:

- Preoperatively, it contributes to the determination of the localization and degree of dysfunction and, by that, to the clarification of candidacy for surgery.
- Postoperatively, it becomes part of a follow-up programme for the control of effects of surgical treatment.
- Intraoperatively, it is employed to detect acute additional spinal cord lesions before they become irreversible. If acute intraoperative deterioration of spinal cord function is suspected to be due to risky (though sometimes unavoidable) surgical manoeuvres, the occurrence of monitoring 'events'[1] may motivate *ad hoc* changes in operative strategy. Intraoperative monitoring is thus intended to prevent additional postoperative neurological deficits.

The present overview is concerned only with this third – intraoperative – aspect of spinal cord monitoring (SCM) with a special emphasis on the conception of 'reversible damage' detection mentioned above. In other fields of clinical neurosurgery like aneurysm surgery, there is evidence that electrophysiological monitoring yields an effective reversible damage detection that may help to improve outcome from surgery[2]. The crucial question is whether spinal cord monitoring can achieve comparable effectiveness in order to legitimize routine application in spinal cord surgery. The last years' intensive clinical and experimental research in the field and an increasing number of relevant publications may give rise to an attempt of at least an interim stock-taking.

SPINAL CORD MONITORING: DIFFERENT TECHNIQUES AND APPLICATIONS

For an overview, the widespread work on SCM can be structured by the pairwise differentiation of:

(a) SCM modalities: somatosensory (SEP) vs. motor (MEP) evoked potentials;

(b) categories of research: experimental (animal) vs. clinical (human);

(c) patient characteristics: orthopaedic vs. neurosurgical diseases;

(d) stimulation sites: peripheral or spinal in SEP, cortical or spinal in MEP;

(e) recording sites: cortical or spinal in SEP, spinal or peripheral in MEP.

In the present review, which focuses on clinical neurosurgical SCM, orthopaedic and experimental studies are left out. In the next section, the recent work on genuinely neurosurgical MEP monitoring will be reviewed, followed by an evaluation of the current status of SEP monitoring. Evoked spinal cord potentials (ESCP) – that is, responses from the spinal cord after spinal stimulation – will then be treated separately, since this technique often involves both ascending and descending fibre systems. Finally, the effectiveness of all these procedures concerning reversible damage detection will be considered.

MOTOR TRACT MONITORING

Methods and techniques

The intraoperative application of motor tract stimulation can be expected to improve the sensitivity of SCM, since it yields a direct monitoring of anterior spinal pathways, which is not possible within the traditional SEP monitoring framework[3]. Transcranial motor tract stimulation can be *electrical* or *magnetic*, but in the operating room electrical stimuli are preferred since their more direct impact on the pyramidal tract makes them less susceptible to inhibitory anaesthetic effects (see[4] for review). Responses after transcranial stimulation can be obtained from selected muscles as well as the peripheral nerves or the spinal cord. Furthermore, stimulation can be performed at the spinal rather than the cortical level, with recording either from the muscles or peripheral nerves or from more caudal spinal levels[4] (see below 'ESCP monitoring'). Although even transcranial electrical stimulation techniques are easily compromised by negative effects of general anaesthesia and relaxants, responses under nitrous oxide[5] or propofol[6,7] anaesthesia have been described as sufficient for a regular monitoring procedure. Furthermore, additional peripheral nerve stimulation can facilitate muscle responses, although the quantification of the degree of facilitation remains difficult[8].

Results from clinical studies

Despite the great interest in motor tract monitoring, *clinical* studies on genuinely neurosurgical SCM with MEP are rare. In a study on eight patients with spinal tumours or arteriovenous malformations[6,7], it was found that permanent loss of potentials recorded from hand muscles after transcranial electrical stimulation was associated with new postoperative neurological deficits (one patient), while reversible amplitude and latency changes did not indicate deterioration in neurological function. In a larger series of 50 patients with various spinal cord diseases[9], responses were obtained from both thenar and anterior tibial muscles after transcranial electrical stimulation. Monitoring was successful in 43 cases (86%). By convention, amplitude decrement up to 50% was considered acceptable. On the basis of intraoperative SCM data, postoperative neurological status was correctly predicted in 81.4% (anterior tibial muscle) and 76.2% (thenar muscle) of cases; false negative results were not obtained. A major disadvantage of this technique is the unwelcome occurrence of muscle contractions as a result of high intensity electrical stimulation. Furthermore, the special anaesthesia required for regular recording might lead to undesirable recall phenomena postoperatively. In a further study in forty patients, again with various spinal cord diseases, responses from the spinal epidural space (successful in 65.2%) and from the cauda equina (successful in 82.5%) were obtained[10]. Again, there were no false negative results, and in 87% (spinal cord recording) and 85% (cauda equina recording) of cases, postoperative motor status was predicted correctly. In another study on 45 patients with spinal cord impairment, multilevel recordings from the ligamentum flavum, peripheral nerves, and muscles after transcranial electrical stimulation were obtained[11]. In these patients, motor tract monitoring did not miss any postoperative neurological deficits, and in a few cases intraoperative MEP deterioration led to a modification of operative strategy with subsequent recovery of responses.

These studies indicate that SCM with MEP is capable of development and may even work in some patients with impaired baseline recordings, which are frequent in neurosurgery. But none of these clinical studies renders a systematic account of reversible damage detection (see below 'Reversible damage detection: warning and intervention criteria').

SOMATOSENSORY TRACT MONITORING

Stimulation and recording techniques

Traditionally, neurosurgical SEP monitoring has been carried out with non-invasive recording from the scalp or neck after peripheral stimulation of the tibial, peroneal or median nerve[12]. In the last few years, additional or exclusive recording from spinal levels has become more common[13]. Two methods are to be distinguished: *conductive* spinal cord potentials (SCEP) can be obtained from levels clearly above the lesion site after peripheral or spinal (see below 'ESCP monitoring') stimulation[14]; *segmental* SCEP are recorded near the lumbosacral or cervical enlargement after lower or upper extremity nerve stimulation[14]. A disadvantage of spinal recording is the variability of responses dependent on even slight changes of recording electrode position[15]. For some special purposes, dermatomal SEP monitoring has been reported to be useful. Lumbar dermatomal SEP latencies have been found to be reliable indicators of adequate lumbar nerve root decompression intraoperatively[16], and cervical dermatomal SEP and SCEP after finger stimulation may become a helpful device for intraoperative electrodiagnosis or monitoring of decompression procedures in cervical spinal surgery[17].

Results from recent clinical studies

SCM with SEP is a well established procedure in orthopaedic surgery, but – perhaps surprisingly – genuinely neurosurgical studies are also rare for this modality. McDonnell[18] presented four cases with advanced cervical myelopathy who received surgical decompression and osseous stabilization. SEP were recorded from the scalp after median nerve stimulation. Amplitude decreases of more than 50% and 'significant' latency increases were taken as 'events' meaningful to the surgeon. There was a total of 80 events in the eight operations performed in these four patients. Surgical strategy was adapted to the monitoring results as far as possible, and all patients had good neurological outcome. In another study on 43 patients with various affections of the cervical spine, there was evidence of an effective prevention of postoperative neurological deficits in one patient who had progressive reduction of cortical SEP which was reversible after removal of fixation wires[19]. Schramm[20] presented monitoring results from 25 cases with spinal tumours. There were neither irreversible intraoperative SEP changes nor additional postoperative deficits, but intraoperative improvement of SEP responses was associated with clinical improvement postoperatively. In a series of 28 patients with cervical, thoracic, or conus medullaris lesions[13], it was found that while SCEP after peroneal nerve stimulation in thoraco-lumbar spinal cord surgery were variable and often (48%) gave false-positive results, cervical recordings after median nerve stimulation were more stable (0% false-positives). None of the patients suffered additional postoperative deficits. Since spinal recordings are less susceptible to anaesthesia effects and require less time to average than cortical SEP, this approach is worth pursuing. Comparisons of SEP responses from spinal and cortical levels in 40 patients with various spinal cord diseases showed that while cortical responses were sufficiently reliable (one false positive, one false negative), spinal responses were difficult to obtain and displayed poor correlation with neurological outcome[21]. These studies indicate that it is difficult to demonstrate the clinical value of SCM. Evidence for benefits of SCM is often more conclusive in single case reports than in evaluations of larger patient groups, especially when major postoperative deficits do not occur in the larger series. But in the studies presented above, there were at least some attempts to define warning and intervention criteria in addition to the mere analysis of correlations between monitoring data and postsurgical outcome (see below 'Reversible damage detection: warning and intervention criteria').

Combination of SEP and MEP monitoring

Various experimental studies suggest that MEP and SEP indeed reflect the functional status of the respective spinal pathways and can be used as complementary techniques in clinical SCM. To name just one example, in a cordo-/rhizotomy model in hogs with cortical or spinal motor stimulation and sciatic nerve recording, MEP failed to indicate sensory tract lesioning, while SEP remained unchanged with severe motor deficits[22]. As for clinical studies, Matsuda[23] compared transcranially evoked spinal MEP to cauda-equina-evoked spinal SEP in patients with various spinal cord affections. The course of SCM in five operations of intramedullary tumours suggested that SEP and MEP efficiently monitor the respective pathways and that especially the absence of marked changes in MEP – also when combined with major SEP reduction – warrants regular continuation of the operation. In a recent orthopaedic study[24], 40 patients received combined MEP and SEP (conductive SCEP) monitoring during scoliosis surgery. Although there were neither persistent EP changes nor new neurological deficits, the results were illuminating with respect to the general practicability of combined monitoring: in 7 (18%) of the patients, most of them with preoperative motor deficits, SEP but not MEP could be elicited; in two (5%), only MEP were recordable, and two patients showed no potentials at all. Due to further technical problems, combined monitoring was successful in a total of 28 (70%) patients. These results indicate that successful SCM is difficult to obtain especially in patients with abnormal baseline recordings, a problem that has been stressed before[25]. On the one hand, the combination of two monitoring techniques will increase the proportion of partial failures; on the other hand, monitoring of one modality may be helpful in case of failure of the other technique.

Evoked spinal cord potential monitoring

Responses from the spinal cord can be evoked by ascending[26] as well as descending[27] volleys after spinal – mostly epidural – stimulation. These techniques have primarily been used for preoperative diagnostics. Clinical[26] and experimental[28] studies yield some evidence for the involvement of both motor and somatosensory tracts in responses from the spinal cord after spinal stimulation. For example, electrical spinal stimulation at T7 in the cat results in activity in the L7 ventral *and* dorsal root, and epidural recording from L3 yields a three-component response with the first two peaks reflecting activity in the ventral pathways and the third one representing dorsal column activation[28]. These data may raise some optimism concerning a combined ascending/descending SCM with rostral and caudal electrodes that can be used for both stimulation *and* recording[26]. Turning to the few clinical studies on ESCP in neurosurgery, Yokogushi[29] recorded ascending conductive ESCP intraoperatively in 14 cases of spinal cord tumours. While there were neither false-positive results nor any additional postoperative neurological deficits, recordings were found to be of some value since responses improved after decompression in six out of eight cases with abnormal ESCP, thus anticipating postoperative improvement. A similar technique was applied in 36 patients with various cervical spine diseases[30]. In seven cases (19%), SCM was not successful; only two cases showed more than 50% amplitude attenuation, but did not develop major new neurological deficits. As far as these few pioneering studies are concerned, the ideal of a combined ascending/descending SCM has not yet been achieved with the spino-spinal technique. Perhaps one should rather pursue the project of a combined, but separate monitoring of anterior and posterior fibre systems in order to obtain more specific information about motor and sensory functions, respectively. If the ESCP procedure works, however, it is elegant and may even do without time-consuming averaging of signals.

REVERSIBLE DAMAGE DETECTION: DEFINING INTERVENTION CRITERIA

As stated in the introduction above, the primary aim of intraoperative SCM is the prevention of avoidable additional postoperative deficits. To achieve this aim, the detection of acute intraoperative EP changes must lead to modification of surgical strategy according to predefined 'intervention criteria' which are understood as 'changes in various response parameters believed to represent a significant defect of conduction and therefore calling for corrective action'[31]. Due to this primacy of reversible damage detection, results concerning the mere 'prediction' (often stated retrospectively!) of neurological outcome can only be relevant if they are used for the prospective determination of intervention criteria. Unfortunately, very few authors have described the development of criteria in that way[31].

Due to the methodological heterogeneity of SCM studies and the variability of intraoperative recordings, intervention criteria are not easily determined. Purely 'statistical' criteria (e.g. amplitude decrement of more than two or three standard deviations from the mean baseline amplitude) are not necessarily useful for clinical purposes[31]. Retrospective analyses of monitoring procedures show that clinically oriented criteria mostly exceed purely statistical criteria[31]. In Tables 1 and 2, criteria of 'significant change' in some representative studies are listed together with other parameters relevant to reversible damage detection. These are not necessarily *intervention* criteria, since in some studies it is not clear whether EP changes had modified surgical procedures at all. The most common criterion is an amplitude decrement of more than 50% until the end of the operation; latency changes are not considered equally important, and further parameters like wave configuration[31] or conduction of repetitive stimuli[33] are not taken into account. The results presented in the two tables are somewhat ambiguous: for example, the lack of false-positives and false-negatives together with a 100%-correlation of SCM results with neurological outcome does not necessarily indicate perfect monitoring, since such a constellation will also occur when SCM is simply *superfluous* in the operation concerned (which would be the case if there were neither any EP changes nor any postoperative deficits in the

Table 1 Recent studies on neurosurgical spinal cord monitoring with motor evoked potentials: retrospective evaluation of diagnostic and prognostic relevance (* = patients with lumbar disc surgery; ** = patients with spinal tumours from the same study)

References	*Criteria of significant change*	*Successful monitoring*	*False positives*	*False negatives*	*Positive correlation with outcome*
32	Not clearly stated	79%	5%	5%	84%
6, 7*	Amplitude reduction 50% plus latency increase 3 ms	88.5%	0%	0%	100%
6, 7**	Amplitude reduction 50% plus latency increase 3 ms	87%	0%	0%	100%
9	Amplitude reduction 50%	86%	19%	0%	81%
10	Amplitude reduction 50%	82.5%	15.2%	0%	84.8%

Table 2 Recent studies on neurosurgical spinal cord monitoring with somatosensory evoked potentials: retrospective evaluation of diagnostic and prognostic relevance (* = only cases with stimulation of the median nerve are considered)

References	*Criteria of significant change*	*Successful monitoring*	*False positives*	*False negatives*	*Positive correlation with outcome*
30	Amplitude reduction 50%	81%	7%	0%	Not stated
13*	Amplitude reduction 50% plus latency increase 10%	100%	0%	0%	100%
20	Amplitude reduction 50%	100%	0%	0%	100%
19	Amplitude reduction (not clearly stated)	100%	4.6%	0%	95.4%
21	Amplitude reduction 50% plus latency increase 10%	100%	2.5%	2.5%	92.5%

whole patient group). The determination of intervention criteria is further complicated by the fact that acute interoperative EP changes can only count as monitoring 'events' if they are associated with either a clinical event (development of neurological deficits) or a surgical event (significant, perhaps risky steps in the surgical procedure). Without this constraint, uneventful intraoperative EP-fluctuations would be overestimated[1]. In the prevailing literature, reports on modification of surgical strategy motivated by monitoring events are merely anecdotal. In orthopaedic studies, this may be due to the rare overall occurrence of events[34], but the same cannot hold for an evaluation of neurosurgical SCM procedures, where events are frequent. Since the efficiency of neurosurgical SCM cannot be demonstrated without a systematic work-up of the actual influence of monitoring events on surgical strategy in large series, the question of the qualification of SCM for routine application remains open. Strictly speaking, such a systematic work-up could only be taken as evidence of the clinical usefulness of SCM if there were a series of controls in which monitoring information had not been accessible to the surgeon. For the time being, the only way to come to a decision concerning the efficiency of neurosurgical SCM would be to define intervention criteria (dependent on SCM modality, recording and stimulation sites) on the basis of the data collected thus far, and then to apply these criteria prospectively in a new series of candidates for spinal surgery. Since especially the recent techniques of MEP monitoring still suffer from methodological problems that make the determination of intervention criteria difficult (see also 'Conclusions' below), the realization of such a research programme seems to be a dream of the future.

CONCLUSIONS AND OUTLOOK

Although the work reviewed here is instructive and revealing in several respects, it will not suffice to establish SCM as a routine monitoring procedure in neurosurgery of the spinal cord. While considerable progress has been made concerning some special applications and recording and stimulation techniques, the concept of SCM in *neurosurgery* has still not stood its test of time. This is partly due to the fact that there has been no successful coordinated effort toward standardization and unification of SCM procedures, but mainly because many neurosurgical patients have poor – unmonitorable – potentials to start with. Up to now, clinical studies have been very heterogeneous regarding recording and stimulation sites and techniques and patient selection. Furthermore, there is the problem of the dissociation of relevance and validity of SCM data: in domains like scoliosis surgery, valid results can be obtained in large patient groups without monitoring events and without postoperative neurological deficits; however, in other, genuinely neurosurgical domains like intramedullary tumour surgery, there are smaller series with fluctuating SCM data, sometimes severe pre- and postoperative deficits and deteriorated baseline recordings. But the comparably valid data obtained in the former group are of rather little clinical relevance, while the insecure findings in the latter group *might* play a genuine role in intraoperative decision-making if only they were more reliable and valid. At present, SCM is often completely unsuccessful in problematic patients where intraoperative electrophysiological data are desired. Due to these problems in patients with impaired baseline recordings, the applicability of SCM to neurosurgery remains limited. However, further progress can be expected if stimulation and recording become more standardized and warning and intervention criteria are stated prospectively. More multilevel recording in both SEP and MEP SCM is desirable in order to determine the optimal recording site. Transcranial and peripheral nerve stimulation techniques are more widely applicable and less invasive than spinal techniques and therefore offer themselves as primary methods with the most realistic prospect of standardization. Motor tract monitoring is still constrained by various, specifically intraoperative recording difficulties which considerably limit the application of this technique: intraoperative MEP recording after electrical stimulation needs very special anaesthesia techniques which still have to be developed for transcranial magnetic stimulation, and stimulation methods avoiding strong muscle contractions need improvement.

The above critical remarks should not be viewed as defeatist, since we explicitly appreciate some promising approaches, one of which might well result in a major breakthrough within the next few years. After all, SCM is the only method of continuous intraoperative assessment of spinal functions currently at hand.

References

1. Schramm J. Acute evoked potential changes in operative treatment: a summary. In: Ducker TL, Brown RH, eds. Neurophysiology and standards of spinal cord monitoring. New York, Berlin, Heidelberg: Springer; 1988:268–273.
2. Schramm J, Koht A, Schmidt G, Pechstein U, Taniguchi M, Fahlbusch R. Surgical and electrophysiological observations during clipping of 134 aneurysms with evoked potential monitoring, Neurosurgery. 1990;26:61–70.
3. Rosenberg JN. Somatosensory and magnetic evoked potentials in a postoperative paraparetic patient: case report. Arch Phys Med Rehabil. 1991;72:154–156.
4. Rothwell JC, Day BL, Thompson PD, Boyd SG, Marsden CD. Motor cortical stimulation in intact man: physiological mechanisms and application in intraoperative monitoring. In: Desmedt JE, ed. Neuromonitoring in surgery. Amsterdam, New York, Oxford: Elsevier; 1989:71–98.
5. Schmid UD, Boll J, Liechti S, Schmid J, Hess CW. Influence of some anesthetic agents in muscle responses to transcranial magnetic cortex stimulation: a pilot study in humans. Neurosurgery. 1992;30:85–92.
6. Jellinek D, Jewkes D, Symon L. Noninvasive intraoperative monitoring of motor evoked potentials under propofol anesthesia: effects of spinal surgery on the amplitude and latency of motor evoked potentials. Neurosurgery. 1991a;29:551–557.
7. Jellinek D, Platt M, Jewkes D, Symon L. Effects of nitrous oxide on motor evoked potentials recorded from skeletal muscle in patients under total anesthesia with intravenously administered propofol. Neurosurgery. 1991b;29:558–562.
8. Taniguchi M, Schramm J. Motor evoked potentials facilitated by an additional peripheral nerve stimulation. In: Levy WJ et al., eds. Magnetic motor stimulation: basic principles and clinical experience (EEG Suppl. 43). Amsterdam, New York, Oxford: Elsevier;1991: 202–221.
9. Zentner J. Noninvasive motor evoked potential monitoring during neurosurgical operations on the spinal cord. Neurosurgery. 1989;24:709–712.

10. Zentner J. Motor evoked potential monitoring during neurosurgical operations on the spinal cord. Neurosurg. Rev. 1991;14:29–36.
11. Levy WJ. Clinical experience with motor and cerebellar evoked potential monitoring. Neurosurgery. 1987;20:169–182.
12. Nuwer MR. Spinal cord monitoring. In: Nuwer MR. Evoked potential monitoring in the operating room. New York: Raven Press;1986.
13. Romstöck J, Watanabe E, Schramm J. Variability of epidural SEP from below and above spinal cord lesions – the effect of the lesion on spinal SEP. In: Ducker TL, Brown RH, eds. Neurophysiology and standards of spinal cord monitoring. New York, Berlin, Heidelberg: Springer;1988:261–267.
14. Ertekin C. Intradural spinal recordings (particular reference to invasive methods). In: Ducker TL, Brown RH, eds. Neurophysiology and standards of spinal cord monitoring. New York, Berlin, Heidelberg: Springer;1988:82–99.
15. Goya T, Miyahara S, Mikura T, Kinoshita K. Intraoperative monitoring of segmental evoked spinal cord potentials in cervical lesions. In: Shimoji K, et al. eds. Spinal cord monitoring and electrodiagnosis. New York, Berlin, Heidelberg: Springer;1991:370–380.
16. Herron LD, Trippi AC, Gonyeau M. Intraoperative use of dermatomal somatosensory evoked potentials in lumbar stenosis surgery. Spine. 1987;12:379–383.
17. Takahashi H, Suzuki I, Ishiyima B. Somatotopical evoked responses from the spinal cord and cerebral cortex to finger stimulation. In: Shimoji K et al., eds. Spinal cord monitoring and electrodiagnosis. New York, Berlin, Heidelberg: Springer;1991:177–185.
18. McDonnell DE, Flanagan HF, Sullivan HG. Somatosensory evoked potentials (SEP) intraoperative monitoring during cranial vertebral compression and instability. In: Ducker TL, Brown RH, eds. Neurophysiology and standards of spinal cord monitoring. New York, Berlin, Heldelberg: Springer;1988:251–260.
19. Veilleux M, Daube JR, Cucchiara RF. Monitoring of cortical evoked potentials during surgical procedures on the cervical spine. Mayo Clin Proc. 1987;62:256–264.
20. Schramm J, Romstöck J, Watanabe E. Intraoperatives Rückenmark-monitoring. Eigene Ergebnisse und Bestandsaufnahme. Z Orthop. 1986;124:726–731.
21. Watanabe E, Schramm J. Romstöck J. Cortical and spinal intraoperative recordings in uneventful monitoring and in cases with neurological changes. In: Ducker TL, Brown RH, eds. Neurophysiology and standards and spinal cord monitoring. Springer: New York, Berlin, Heidelberg: 1988:226–233.
22. Owen JH, Jenny AB, Naito M, Weber K, Bridwell KH, McGhee R. Effects of spinal cord lesioning on somatosensory and neurogenic-motor evoked potentials. Spine. 1989;14: 673–682.
23. Matsuda H, Shimazu A. Intraoperative spinal cord monitoring using electric responses to stimulation of caudal spinal cord or motor cortex. In: Desmedt JE, ed. Neuromonitoring in surgery. Amsterdam, New York, Oxford: Elsevier;1989:175–190.
24. Hicks RG, Burke DJ Stephen JPH. Monitoring spinal cord function during scoliosis surgery with Cotrel-Dubousset instrumentation. Med J Austral. 1991;154:82–86.
25. Schramm J. Spinal cord monitoring. Current status and new developments. Cent Nerv System Trauma. 1985;2:207–225.
26. Tamaka T. Spinal cord monitoring with spinal potentials evoked by direct stimulation of the spinal cord. In: Desmedt JE, ed. Neuromonitoring in surgery. Amsterdam, New York, Oxford: Elsevier;1989:139–149.
27. Shimoji K, Fujioka H, Maruyama Y, Shimizu H, Hokari T, Takada T. Spinal cord potentials (SCPs) produced by descending volleys in man. In: Ducker TL, Brown RH, eds Neurophysiology and standards of spinal cord monitoring. New York, Berlin, Heidelberg: Springer;1988:114–121
28. Naznik, G. Transfeldt EE, Shichijo F, Oshima T, Pomeranz B. Spinal sensory and motor tract activation after epidural electrical stimulation in the cat. Spine. 1990;15:623–629.
29. Yokogushi K, Katahira G, Yokozawa H, Kawamura H, Yamashita T, Ogiwara N. Analysis of conductive spinal cord function using spinal cord evoked potentials during surgery of spinal cord tumor. In: Shimoji K. et al., eds. Spinal cord monitoring and electrodiagnosis. New York, Berlin, Heildelberg: Springer;1991:396–402.
30. Baba H, Tomita K, Umeda S, Kawahara N, Nagata S, et al. Clinical study of spinal cord evoked potentials. In: Ducker TL, Brown RH, (eds). Neurophysiology and standards of

spinal cord monitoring. New York, Berlin, Heidelberg: Springer;1988:216–221.

31. Jones SJ, Howard L, Shawkat F. Criteria for detection and pathological significance of response decrement during spinal cord monitoring. In: Ducker TL, Brown RH, eds. Neurophysiology and standards in spinal cord monitoring. New York, Berlin, Heidelberg: Springer; 1988:201–206.
32. Fukui Y, Satomi K, Okuma T, Kenmotsu K, Kamata M, Ikai T, Hirabayashi K. Intraoperative spinal cord monitoring using evoked spinal cord potentials on spinal cord tumors. In: Shimoji K et al. eds. Spinal cord monitoring and electrodiagnosis. New York, Berlin, Heidelberg: Springer;1991:296–304.
33. Sakatani K, Ohta T, Yamagata Y, Shimo-Oku, M. Effects of spinal cord compression on repetitive impulse conduction of ascending fibers in the dorsal column. In: Ducker TL, Brown RH, eds. Neurophyiology and standards of spinal cord monitoring. New York, Berlin, Heildelberg: Springer;1988:53–57.
34. Daube JR. Intraoperative monitoring by evoked potentials for spinal cord surgery: the pros. Electroenceph Clin Neurophysiol. 1989;73:374–377.

6
Lessons from spinal cord monitoring

E. M. SEDGWICK

INTRODUCTION

In this chapter, I will encourage readers to take a broad view of spinal cord monitoring and, with the aid of three case histories, indicate some lessons learned, some yet to be grasped and some directions for future investigation. This kind of 'thought' experiment is unusual in biological science but in physics a whole branch is devoted to it. Einstein was a master of thought experiments both for developing new ideas – the theories of relativity – and for teaching and training the scientific mind. It seems appropriate to look at the wider content of spinal cord monitoring and a scan through the titles in this volume gives eloquent testimony to the variety of activities undertaken in this field.

Monitoring, or the act of being a monitor, implies giving reminders or warnings. This is how we should see our activity. It is unfortunate if some act according to another meaning of the word 'one who admonishes another as to his conduct'; a monitoring act familiar to many of us from schooldays. The monitor lizards incidentally are so called as they are supposed to warn of the vicinity of crocodiles.

I shall argue that monitoring in the first sense above is not sufficient but that the accumulation of experience of when things go wrong should be put to our colleagues in order to plan safer operative procedures. We should go further and use our experience with monitoring to develop hypotheses about the pathophysiology of spinal cord damage and to begin to test them.

Spinal cord potentials were recorded by one of the first applications of the oscilloscope in physiology by Gasser and Graham[1] who worked out most of the characteristics of the cord dorsum potential following peripheral nerve stimulation. They were followed by Bernhard and his co-workers who worked during the dark years of World War II and summarized his Scandinavian publications at the Cold Spring Harbour Symposium in 1952.[2]. His group clearly distinguished the segmental potentials from the travelling waves. This early work has been confirmed as applicable to humans and extended notably by Shimoji[3,4] using epidural recordings and many others using surface electrodes[5].

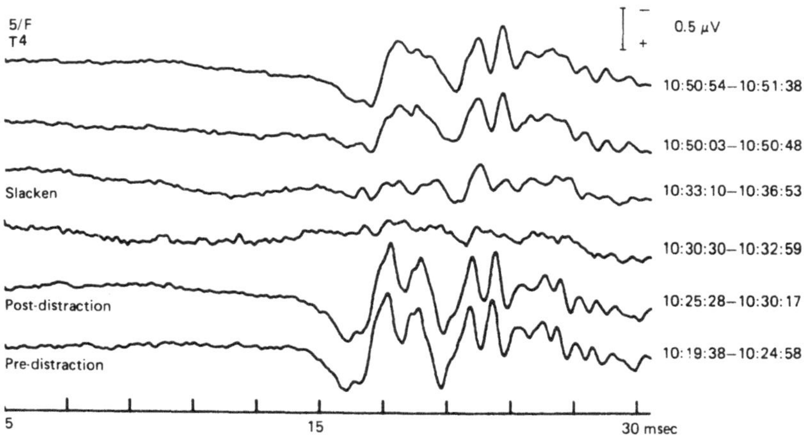

Figure 1 **Case 1** Epidural potentials recorded from T4 level following posterior tibial nerve stimulation at the ankle. The patient (female, aged 13) was undergoing surgery for idiopathic scoliosis. Traces read from bottom to top in time sequence with time indicated on the right (hours: minutes: seconds). The potential disappears during distraction and recovers slowly after slackening of the distraction

Early attempts to record from spinal cord in man were crude in the extreme. Poole[6] inserted a needle into the cord of a paraplegic man using a lumbar puncture technique and recorded mass discharges during a flexor spasm. Sawa[7] used a concentric needle electrode and noted neuronal activity much of which would be recognized as injury discharges today. Neither of these workers were aware of Bernhard's work and neither recorded in association with stimulation. What if they had done so? Where would we be now, one generation later? What questions might have been satisfactorily answered and would the practice of spinal surgery and treatment be any different? This is the thought experiment that you must each do for yourselves. No doubt there is a wide variety of answers and some of them may cause you to alter your approach to spinal cord monitoring over the next year. Could we assume a solution to the problem of what, exactly, are the mechanisms of spinal cord damage during surgery? We know from monitoring that conduction is blocked in some fibres and that this can happen very quickly, in the space of one set of averages. Hypotheses have been advanced suggesting a vascular or hypoxic aetiology but can these be valid in all cases? Let us consider a case history.

Case Report 1

A girl of 13 with idiopathic scoliosis was undergoing surgery for correction of the curvature by distraction, application of Harrington rods and bone graft. Five minutes after distraction the epidurally recorded spinal potential was lost. Immediately some of the distraction was released and a small potential returned. Over the next 20 minutes it returned to almost normal. No wake-up test was done. She made an uneventful recovery with no postoperative neurological manifestations (Figure 1).

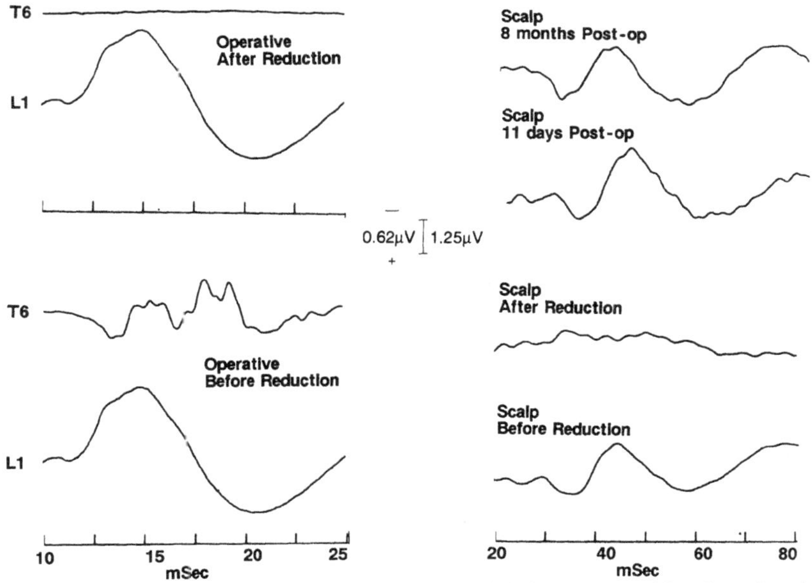

Figure 2 **Case 2** Epidural somatosensory evoked potentials in a 36-year-old male undergoing reduction of a T10/T11 dislocation. The left panels show the dorsal funiculus potentials from T6 level (upper of the two traces) and the cord dorsum potential at L1 level. Lower panel shows the initial recordings and upper panel shows the loss of the T4 dorsal funiculus potential after reduction of the dislocation; the cord dorsum potential was unchanged.

The panel on the right shows that the scalp recorded SEP was lost after reduction but recovered by 11 days with a delayed P40 which had returned to normal at 8 months

This is a common situation and monitoring resulted in corrective action. What is the nature of the cord damage? Does it all occur at one site or is it distributed over the length of the dorsal funiculus? Was slackening of distraction the correct action? Would the potentials have recovered anyway over time or would she have developed paraplegia? There can be little doubt that cases such as this represent true positives; but could some of these patients make a normal recovery without action being taken? This is an experiment one can never undertake. It was one of those instances where corrective action appears to have been effective. That was not so in Case 2.

Case Report 2

A window cleaner aged 36 fell about 6 m from a ladder. He had severe pain and tenderness in his back but there was no head injury and he was neurologically normal. A crush fracture of T12 vertebra was found on radiography. By the next morning he had not passed urine and possible sensory changes in the legs were noted but power and tendon reflexes were normal and perineal sensation was unimpaired. MRI showed a facet dislocation of T10 on T11.

Surgery that day revealed the facet dislocation. The upper part of the facets of T11 were resected and distraction applied until the dislocation slipped suddenly into place. Two minutes later, the epidurally recorded SEP had disappeared (Figure 2). Immediate laminectomy was performed and a thorough search for bony fragments, disk material, etc. revealed nothing to compress the cord. The epidural electrode below the level of surgery confirmed that the lesion was at the surgical T12 level (Figure 2). Having verified that there was nothing protruding into the spinal canal, the vertebral column was stabilized with Harrington rods and bone graft. A wake-up test was not performed as it would not, at that stage, modify treatment.

Postoperatively he had a weak right leg but with normal reflexes and flexor plantar responses. There was no sensory deficit and the left leg was normal. At 6 months he was back to light work and at 9 months was back to normal with no neurological deficit.

The epidural SEP did not recover during surgery but at 11 days post-op the scalp recorded posterior tibial nerve SEP was similar to the one recorded under anaesthesia preoperatively. The P40 latency shortened over the next 8 months.

There was no evidence at all of any pressure applied to the cord in this case. The loss of function was too sudden to be vascular or hypoxic – we know that it takes many minutes to block conduction in a peripheral nerve by means of a sphygmomanometer cuff. The conduction block was localized at T11 as the potential just a few cm below the fracture site at L1 was unaltered. What happened to the nerve membrane? Did the sudden jarring of reduction jostle all the sodium channels out of the membrane and into the axoplasm or perhaps their architecture was deranged causing them to loose their voltage sensitivity. Were they stuck fully closed or fully open? If they had been fully open, one might have expected to record an injury potential. Similar speculation can be made about the nerve membrane in Case 1. One of the methods to discover more about the sequence of events would be the use of multichannel electrodes running from cauda equina up to well above the operative site. This would give information about the site of the lesion, whether any injury current developed and, by using the electrodes for stimulating as well as recording, the length of the non-functioning part of the axons could be determined.

In Case 3 pathological information became available so the cause of conduction block was confirmed.

Case Report 3

An 18-year-old man with muscular dystrophy had developed a severe extension deformity of the neck. Surgical correction of the deformity by fixation of the cervical spine using Luque's rods and bone grafting was done. As surgery was at the neck, an epidural electrode could not be placed so the scalp SEP from the posterior tibial nerves was monitored during surgery. During placement of sublaminar Luque's wires, the scalp SEP changed to low amplitude, long latency and then disappeared over 30 minutes (Figure 3). At that time blood pressure fluctuated widely. Despite various procedures, the SEP had not returned by the end of surgery. Twelve hours later, the patient died. Autopsy revealed a small intramedullary haemorrhage 4 mm × 10 mm with adjacent hypoxic change in the dorsal horns at C8-T1 level[8].

Case 3 is the only publication of the pathological change accompanying a positive SEP monitoring event. Almost certainly there was cord compression by Luque's wires and death probably resulted from disturbance to the sympathetic nervous system output. Once again, monitoring observed events but could not influenced them.

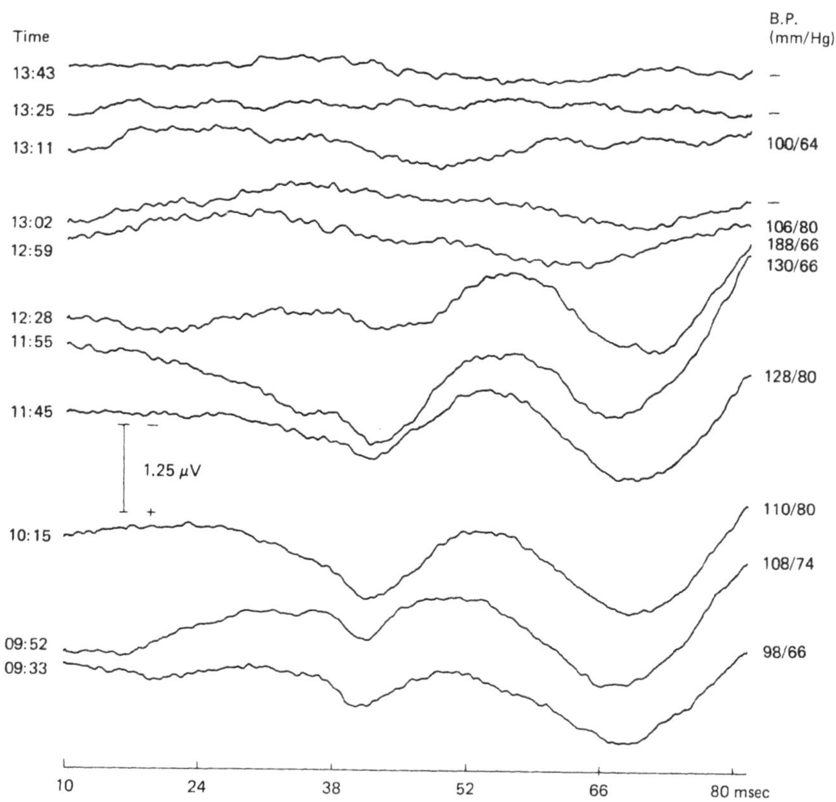

Figure 3 **Case 3** Scalp recorded somatosensory evoked potentials following stimulation of the posterior tibial nerve are shown for this 18-year-old male undergoing fixation of the cervical spine. The time sequence reads from bottom upwards with the time on the left of the traces and blood pressure on the right. The latency of P40 became prolonged before the potential disappeared. Note the blood pressure fluctuations. (Reprinted with the permission of Annals of Neurology[8])

In these three cases there was damage to a healthy spinal cord at surgery. In Case 1, corrective action promptly resulted in restoration of function and recovery was complete. Monitoring was helpful in this case. In Case 2, the damage was observed but nothing could be done about it. Had monitoring, therefore, been a waste of time? Supposing there had been no monitoring? There would have been no extensive laminectomy and search for compression. Postoperatively, one would have had to consider reasons for the weak leg which would have led to extensive investigations and perhaps even surgical re-exploration. With the knowledge of the events revealed by monitoring the patient and his doctors were spared this difficulty.

In only one of these three chosen examples did spinal cord monitoring perform its expected duty. It warned of damage, corrective action was taken and prevented a disaster. Can one expect more of monitoring? There is more one can do and learn as suggested by following the scheme below (Figure 4).

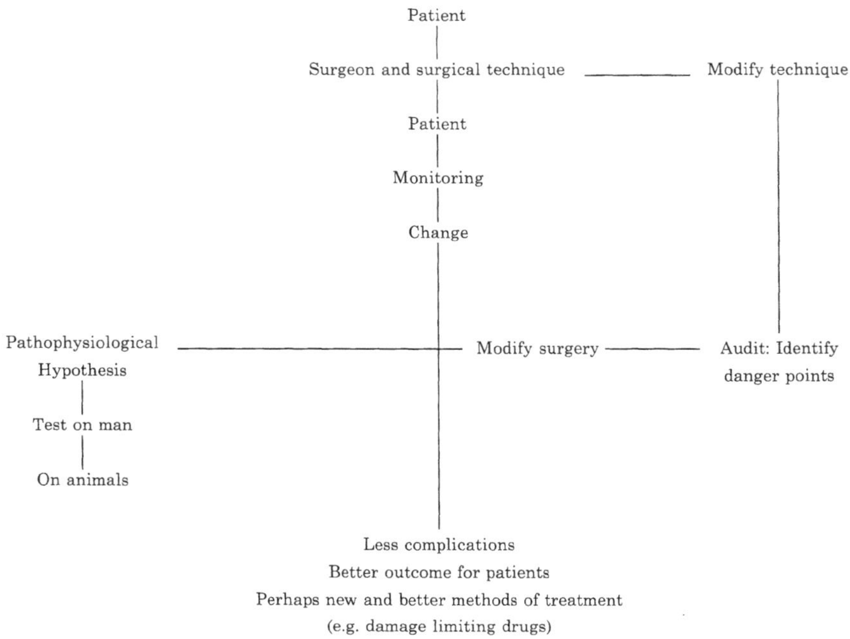

Figure 4 A scheme for monitoring

Down the centre of the scheme the usual monitoring process is represented with feedback to the surgeon if the monitored signal changes. To the right is represented the audit of monitoring; to do this effectively requires large numbers of patients, perhaps from several centres. The first and obvious audit questions concern limitation of complications or simply 'Does it work?'; are there false negatives, etc? One can also recognize danger points in surgery and suggest modifications to the technique to avoid these. On the left of the scheme are the more scientific activities. Monitoring should lead to questions about the pathophysiological processes from which one can generate new and testable hypotheses. In order to test some of the hypotheses, the technique of monitoring may be modified and some hypotheses may have to be tested on animal models. As an example, the hypothesis has been proposed that descending pathways, the cortico-spinal tract in particular, may be more sensitive to damage and may, therefore, be a more logical pathway to monitor. The technology for doing this is available and is extensively discussed in this volume. We already have enough evidence to suggest that this activity, interesting though it may be, for scientific reasons, will contribute very little to monitoring. The SEP method gives so vanishingly few false negatives that a trial to find a more sensitive method of monitoring would take years to accomplish simply because of the large number of cases necessary to prove the point. Perhaps our time could be better spent.

Audit of a large number of patients shows clearly the 'danger points' during surgery. In our experience and that of others, Luque's wires are dangerous things. The message from monitoring teams to surgeons must be to find safe alternatives to inter-lamina wiring. This would reduce spinal cord damage very significantly.

There must be continuing scientific enquiry into the mechanism of axonal damage. Once this is understood we will be able to take other sensible steps to limit damage. It may be possible to find drugs which will help to prevent injury or assist the recovery process.

CONCLUSION

Spinal cord monitoring is of value in preventing neurological deficit at the time of surgery, as in Case 1. At other times it warns of problems which we cannot immediately remedy, Cases 2 and 3. Retrospective consideration of cases similar to 2 and 3 leads to improvements of techniques and greater awareness of dangers. The lessons might be to reduce dislocations gently without allowing sudden movement; to be very careful with or, better still, avoid Luque's wires; to be more aware that loss of potentials may mean a severe cord lesion and the possibility of vasomotor instability in cervical cord and high thoracic lesions.

Monitoring in more detail with multi-channel recording will help to define the processes underlying cord damage. Cellular mechanisms need to be worked out by a combination of clinical and laboratory studies.

References

1. Gasser HS, Graham HT. Potentials produced in the spinal cord by stimulation of dorsal roots. American J Physiol. 1933;103:303–320.
2. Bernhard CG. The cord dorsum potentials in relation to peripheral source of afferent stimulation. Cold Spring Harbour Symp Quant Biol. 1952;17:221–232.
3. Shimoji K, Higashi H, Kano T. Epidural recording of spinal electrogram in man. Electroenceph Clin Neurophysiol. 1971;32:236–239.
4. Shimoji K, Matsuki M, Ito Y, Masurko K, Maruyama M, Iwane T, Aida S. Interactions of human cord dorsum potential. J Applied Physiol. 1976;40:79–84.
5. El Negamy E, Sedgwick EM. Properties of a spinal somatosensory evoked potential recorded in man. J Neurol Neurosurg Psych. 1978;41:762–768.
6. Pool JL. Electrospinogram (ESG): Spinal cord action potentials recorded from a paraplegic patient. J Neurosurg. 1945;3:192–198.
7. Sawa H. Spontaneous electrical activities obtained from human spinal cord. Foila Psychiat Neurol Jap. 1947;2:165–179.
8. Shukla R, Docherty TB, Jackson RK, Weller RO, Sedgwick, EM. Loss of evoked potentials during spinal surgery due to spinal cord hemorrhage. Annals Neurol. 1988;24:272–275.

7
A multicentre survey of spinal cord monitoring outcome

M. R. NUWER and L. G. CARLSON

INTRODUCTION

Somatosensory evoked potential (EP) monitoring has been used for over 20 years to help assure safety of the spinal cord during certain procedures. The technique has grown in popularity during the past 12 years. In some medical communities, the technique has grown to be widely used and well accepted, being used in most procedures in which the spinal cord is at risk for significant damage. Such monitoring has especially been applied to procedures to stabilize and correct scoliosis and similar disorders of the spinal column.

For any new medical procedure, there comes a time to assess the accuracy of the test as it is used in common clinical practice. For EP spinal cord monitoring, such assessments of accuracy are hindered by the relative infrequency of neurological complications. Indeed, the Scoliosis Research Society Survey[1] previously found an overall risk of 0.72% for neurologic complications resulting from procedures to treat scoliosis. To assess EP spinal cord monitoring, my colleagues and I have undertaken to survey the Scoliosis Research Society again. This survey has been carried out and recently the results have been presented elsewhere[2]. Further analysis of survey results is still in progress.

The survey collected information about the rates of neurologic complications as well as the occurrence of EP changes during monitoring. It also assessed the criteria used for 'change' and various technical aspects of monitoring. Both the surgeons and the monitoring physicians or technologists were included in this survey process. A summary of the published survey results is presented here, along with some results from the additional analysis done since that time.

THE SURVEY PROCESS

The initial phase of this survey was carried out at the time of the Scoliosis Research Society (SRS) meeting in Amsterdam, The Netherlands, held in

September, 1989. This meeting was held jointly with the European Spinal Deformities Society (ESDS). The survey document was subsequently circulated by mail to all members of SRS, whether or not they had been present at the Amsterdam meeting. This initial survey document requested information on the surgeons' years of experience with EP spinal cord monitoring, asking for the number of years of such experience, the average number of monitored cases per year, and the average number of all OR procedures conducted by that surgeon with or without EP monitoring. This survey document, referred to as Phase I, also inquired regarding the occurrence of any neurologic postoperative deficits, separated by whether or not the evoked potentials had changed during those cases. The occurrence of evoked potential changes without neurologic deficits was also surveyed. Physicians were asked to enumerate the instances in which any of these occurrences has transpired, and were asked subsequently for specific clinical information regarding each such occurrence. Finally, the surgeons were asked for the name and address of the monitoring person who assists them.

In a follow-up survey, referred to as Phase 2, the monitoring physician or technologist was contacted. The monitorist was asked independently for an enumeration of the instances in which neurologic deficits had occurred during monitoring, with or without evoked potential changes. Information about the occurrence of evoked potential changes without neurologic deficits was also solicited. The monitorist was asked further about the kind of equipment used, the filter settings, stimulating and recording electrodes employed, criteria for 'change', the use of inhalation anaesthetics and other technical factors.

Because only a minority of the surgeons' monitorists had responded during Phase 2, an additional Phase 3 survey was subsequently carried out. This continued to pursue collecting the same information as had been solicited for Phase 2, but supplemented that data base with additional such cases. In addition, some of the clinical details of neurologic deficits were further defined through direct communications with some of the surgeons. Phase 3 is considered separate from Phase 2 partly because of the manner in which this work has been published. Phases 1 and 2 were published originally[2]. Since the time of that publication, the additional Phase 3 information has been obtained and publication of the extended results are planned.

Statistical analyses were done using SYSTAT[3]. Comparative statistics used chi-square testing, sometimes requiring collapsing of individual continuous variables into two or six categorical ranges. Many distributions were highly skewed, thereby precluding appropriate use of analysis of variance techniques.

SURVEY CLINICAL RESULTS

In all, 243 surgeons responded to this survey. Of these, 184 were included in the data base. The remaining 59 surgeons were not included in the data base because they had no experience with EP spinal cord monitoring. In the data base, 153/184 (83%) of the surgeons were from the USA, representing 24 of the 50 states. The non-USA surgeons represented 22 additional nations on six continents. Of the respondents, 80% were members of the Scoliosis Research Society.

The surgeons had between one and 18 years of monitoring experience, median six years, representing a total of 1235 man-years of experience of monitoring. Within the scope of those monitoring years, these surgeons conducted 25 to 3000 operating room procedures, median 500 procedures. An aggregate total of 114 164 surgical procedures were performed by these 184 surgeons during the years in which they used monitoring. That total includes both the monitored and the unmonitored cases. Restricting ourselves to just those cases in which EP monitoring was carried out, individual surgeons reported experience with 10–1900 monitoring cases, with a median of 233 monitored cases and an overall estimated 59 303 monitored cases.

During the same time, these surgeons conducted wake-up tests on some patients. The range of wake-up tests performed was 0 to 1800 cases for each surgeon, with a mean of 26 cases and an aggregate total of 19 973 wake-up cases among all the surgeons. Wake-up tests were used more often than EPs by 10% of the respondent surgeons. Wake-up tests were used in every case by an additional 11% of the respondents. Half the surgeons (50%) used the wake-up tests only when EPs changed. No wake-up tests at all were employed by 18% of the surgeons. The ratio of wake-up tests to evoked potentials was also calculated. When the calculation was performed for each surgeon, the median value of this ratio was 13%. That is to say, half of the surgeons performed wake-up tests on more than one-eighth of their cases in which EP monitoring was used, the other half performing wake-up testing on fewer than one-eighth of their monitored patients.

The use of inhalation anaesthetic agents, such as isoflurane, has been of concern to those who perform EP spinal cord monitoring. Such agents can substantially attenuate the scalp potentials as recorded in some techniques. In this data base, 50% of the respondents generally avoided the use of such inhalation anaesthetic agents.

Among the patients monitored, 60% had scoliosis, 7.5% fractures, 6.5% kyphosis and 5.5% spondylolisthesis. The variety of other situations monitored included spinal tumours, degenerative disease, cervical decompression and fusion, and various vascular procedures.

SURVEY TECHNICAL RESULTS

Technical data have now been obtained from 90 respondents. The technical data were obtained from the physicians or technologists who provided the monitoring services.

Stimulation techniques are described in Table 1. The direct spinal cord stimulation is included in the category of 'other' stimulation sites. Of the groups who stimulate the lower extremities simultaneous stimulation of both was reported by 11% of the respondents, alternating stimulation of the right and left legs by 87% and use of one leg only by 2% of the respondents.

Recording techniques are shown in Table 2. The preponderance of scalp and neck surface recordings corresponds to the high proportion of US

Table 1 Stimulation techniques

Sites stimulated	*(As % of respondents)*
Posterior tibial	87
Peroneal	20
Popliteal	6
Upper extremity site	31
Other	25

Note that some respondents used more than one site

Table 2 Recording techniques

Kinds of recording paradigm		*(As % of respondents)*
Epidural		15%
Spinal ligamental		2%
Scalp or neck only:		
Low filter 1–15 Hz		41%
Low filter 20–30 Hz		43%
Line filter		
Generally off		75%
Often on		25%
Number of recording electrodes		
Range	1–16	
Median value	4	
Trials per EP		
Median value	300	

monitoring teams in this data base. This preponderance of surface recordings has important implications for the interpretation of the survey outcome. The poor representation of epidural and spinal ligament techniques in this data base suggests that the outcome data here should be applied only to the surface recording techniques, as too few of the epidural and ligamental centres were surveyed in this process. As a result, some of the outcome data is further differentiated, as noted below, to denote the results only for the US portion of this survey which used almost exclusively the surface techniques.

Additional recording technique information was also obtained. The 50 or 60 Hz line filter was left off by 75% of the technical respondents. The number of recording electrodes varied from 1 to 16, with a median of four recording electrodes employed. A median of 300 trials were used to create each evoked potential tracing used for monitoring purposes.

The criteria for change are shown in Table 3. Most respondents used both latency and amplitude criteria. Most respondents used a 50% change from baseline as the criterion for determining an amplitude change. Half the respondents used a 7–10% increased latency as the criterion for determining a latency change. Also, 95% of respondents indicated that the short latency (less than 50 ms) peaks were used for monitoring purposes, rather than middle latency (50–100 ms) peaks.

Table 3 Peaks and criteria for change

Peaks monitored at scalp or neck:	*(As % of respondents)*
Short latency	95%
Middle latency	5%
Criterion for change	
Amplitude only	3%
Latency only	7%
Both	90%
Criterion for amplitude change	
<50%	20%
50%	75%
>50%	5%
Criterion for latency change	
<5%	30%
7–10%	51%
15%	11%
>15%	8%

Table 4 Surgical outcome

	Analysis by cases (By no.)				*Analysis of rates (By %)*			*Respondents reporting 'None'*
	Range	*Median*	*Mean*	*Sum*	*Range*	*Median*	*Mean*	
False negative	0–9	0	0.27–0.46	51–98	0–10	0	0.165	76%
True positive	0–20	0	1.1–1.5	200–268	0–10	0	0.452	54%
False positive	0–75	1	3.3–5.2	609–943		0.75	1.590	38%

SURGICAL OUTCOME RESULTS

The surgical outcome results are shown in Table 4 for all respondents. This analysis is broken down both by the number of cases reported by a surgeon, and also by the rate at which such problems were encountered by each surgeon.

In Table 4, the entries for mean and sum are displayed as a range for the analysis by cases. Such entries need an explanation. In the course of the survey process, some surgeons expressed their results in a somewhat contingent or uncertain manner. On some occasions, they reported '1 or 2' instances of an adverse outcome, such as a true positive problem. On other occasions, the adverse outcome results reported by the surgeon differed slightly from those reported by the technologist. For both of these circumstances, we analysed both a minimum and a maximum value for each surgeon–monitorist team. The lower value reported reflects the lower values of the reporting, whereas the higher value reported reflects the maximum possible value as per the team's report. In some cases, the discrepancy is apparently due to some very marginal neurologic problems, where the team had difficulty deciding whether they should count that outcome as an adverse outcome or not. For the analysis of rates, the maximum value was used. As such, the analysis of rates may over-report the neurologic deficits in cases where even the surgical team was uncertain about whether it should be counted.

For our purposes here, a 'false negative' monitoring case was defined as one in which a new postoperative neurologic deficit occurred despite an absence of evoked potential changes during surgery. Neurological deficits include even those which were minor or transient, so long as the surgeon–monitoring team reported such cases during this survey process. 'True positive' monitoring cases were defined as those in which the evoked potentials did change significantly during the surgical procedure and new neurologic deficits did occur postoperatively despite the EP warning. This includes those cases where the surgeon did intervene, and potentially reduced the severity of the neurologic complication, i.e. potentially altering a severe neurologic outcome into a mild new neurologic deficit. Such changes are not amenable to analysis in this kind of survey process, so that the rate of such alterations in outcome remains invisible to us here except as described further below. 'False positive' monitoring events were defined as those in which the evoked potential did change during the case, but no significant new neurologic deficit was apparent postoperatively. Of course, some of these may represent technical problems in the monitoring. Others may represent false positives due to abrupt anaesthetic administration, thereby altering EPs. Still other false positive events may represent other true physiologic occurrences, some of which were enigmatic. Finally, other 'false positive' events may represent those in which the patient was at true high risk for a postoperative neurologic deficit, but the EP warning allowed the surgeon to intervene and thereby prevent an actual neurologic deficit. Unfortunately, the survey process does not allow us to differentiate among these various possible causes of false positive events.

Note that there is a non-linearity in analysis of mean rates. One cannot simply sum mean rates of changes. The best analysis of the neurologic deficit rate in this survey is obtained by summing the maximal value of false negative and true positive instances, and then dividing by the total number of monitored cases. The maximal value of neurologic deficits here is 366 cases, among 59 303 EP monitored spinal procedures. This yields an overall rate of 0.62% neurologic deficits. If the more conservative survey outcome results are used, there were at least 251 adverse neurologic deficits, corresponding to a rate of 0.42% neurologic deficits in this survey population.

The survey revealed an improvement in severe or persistent neurologic deficits compared with the results of the earlier MacEwen et al. survey[1]. The rate of *major* neurological deficits fell from 0.61% in the MacEwen et al. survey to 0.24% of monitored cases in the present survey. The rate of persistent neurological deficits has also declined, from 0.46% to 0.31%.

THE EFFECTS OF EXPERIENCE

As analysis of these survey results proceeded, it became clear that the data showed a substantial effect of monitoring-team experience. Surgeon–

Table 5 Geographical effects

USA users had:	*USA*	*Other*
1. Less use of epidural techniques	7%	60%
1A. More recording channels	5.4 (means)	3.7
1B. Less use of inhalation anaesthetic	44%	80%
2. Fewer inexperienced users (< 100 monitored cases)	18%	42%
2A. Fewer problem outcomes:		
False negative rate	0.13 (means)	0.41
True postive rate	0.42	0.63
False positive rate	1.51	2.10

monitorist teams had the fewest adverse neurologic outcomes when the team was most experienced in carrying out these procedures. Two groups were defined, one with 100 cases or less of experience with monitoring, the other with more than 100 cases of experience. The effect of experience was statistically significant for both the false positive and false negative rates, with the true positive rate showing a trend that did not quite reach significance. Further analysis of the effects of experience are still in progress. Experience can influence outcome in several ways. Technical problems become sorted out and the team members gain confidence in the EP tool, allowing prompter surgical intervention when EPs do change. Also, the surgeon may learn to change his own technique when presented with EP feedback over many cases.

GEOGRAPHICAL EFFECTS

In Table 5 the surgical outcomes are separated by geography. The majority of all respondents in this survey were from the USA. It made sense, therefore, to separate the respondents into two categories, those from the USA and those from other nations. In one way, this would allow a more homogeneous subset of respondents to be analysed within the USA. It was noted that the respondents from other nations used rather heterogeneous recording techniques. Even the epidural techniques used in Japan are different from those used in the United Kingdom, these two nations representing the largest number of respondents in the non-USA portion of this survey.

Corresponding to the greater use of epidural techniques in the non-USA group, there was also a tendency toward fewer recording channels and greater use of inhalation anaesthetic agents. Unexpectedly, the proportion of relatively inexperienced monitoring teams (fewer than 100 monitored cases total) was substantially higher among the non-USA respondents in this survey. We believe that this is the principal reason why the neurologic deficits and problems in the non-USA group were higher than that in the USA portion of the respondents. The relative rates of such adverse outcomes are shown in Table 5.

The principal conclusion of this geographical analysis is to suggest that further analysis of this data is warranted. The further analysis should be restricted to the more homogeneous group of USA users, who would

Table 6 Outcome dependence on recording technique: comparison of two surface recording techniques

	Neurologic deficit rate			
	None	Less than 1%	1% or more	Total
Open filter	11	16	11	38
Restricted filter	19	16	5	40
Total	30	32	16	78
			(Difference significant, $p < 0.05$)	
Mean rate of neurologic deficits among cases:				
With open filter, e.g. 1 Hz				0.63%
With restricted filter, e.g. 30 Hz				0.47%

represent a group of surgeon–monitorist teams who use a set of techniques and operate in a medical culture which is more homogeneous than can be obtained by looking at worldwide survey results. These results should not be interpreted as implying that the non-USA results are worse than those in the USA, since the geographical differences are primarily due to differences in the proportion of experienced teams.

DEPENDENCE ON RECORDING TECHNIQUES

Several types of surface recording technique are in common use. The two most common approaches are those suggested originally by Nash and Brown[4–6] using a 1 Hz low filter; and that proposed initially by Nuwer and Dawson[7–9] which espoused the use of a 30 Hz low filter. Surgical outcomes were assessed for a subgroup of respondents who could be categorized into one of these two techniques of open filters (1 Hz low filter) or restricted filter (30 Hz low filter). The rate of false-positive monitoring cases was not significantly different between the two techniques of surface recording. The rate of neurologic deficits was somewhat different, as shown in Table 6. Chi-square testing revealed statistical significance, with less neurologic deficits seen when employing the restricted filter technique.

OTHER SURVEY RELATIONSHIPS

Several other interesting or important relationships were revealed by this survey, as reported in the initial survey publication[2]. A surgeon's false positive, false negative and true positive rates are all positively correlated, with a correlation coefficient of $r = 0.3–0.7$. This may correspond to the effect of inexperience described above. Also, the rate of neurologic deficits was higher for those cases in which kyphosis was present, as opposed to those for scoliosis, spondylolisthesis or a fracture.

The use of both amplitude and latency criteria together was associated with lower false negative rates. As noted above, most users do employ both monitoring criteria together. This seems to be an appropriate strategy. However, low thresholds for reporting amplitude change (amplitude change criterion of less than 50%) was associated with a higher false positive rate, but was not associated with a lower false negative rate. This suggests that the monitoring teams should be suspicious of changes in the 30–40% range, but that the 50% criterion for amplitude change is appropriate for clinical use. It was also noted that the wake-up test was unrelated to the occurrence of false-negative cases. This is of particular importance, since it validates the clinical strategy of using evoked potentials to carry out the main portion of monitoring, rather than relying on the wake-up test for every surgical case despite monitoring.

COMMENTARY

Despite its popularity in many medical communities, lingering doubts continue to surface about the efficacy of EP spinal cord monitoring (e.g. Aminoff[10]). Such concerns have been raised especially by reports of false negative monitoring cases, where EPs remained normal but the patient suffered new neurologic deficits[11–18]. As reviewed elsewhere[19], many such reports actually make sense physiologically in that the wrong pathway was monitored, or the onset of clinical deficits was delayed, or the EPs really did change. The present survey tries to look at false negative and other problems, taking a more broad overview of the rates of various problems among a large series of cases.

The survey results seem reassuring. The number of neurologic deficits is small, indeed smaller than the previously surveyed rate. Some improvement in the deficits rate may be from other improvements in surgical technique, but this is balanced by the trends toward operating on higher-risk patients and surgically instituting greater degrees of correction and rotation.

We have assessed problems with false positive cases in the past[20]. In this survey there is a persistent rate of such false alarms, but the rate seems reasonably small overall. Many false positive events may be real, physiological changes rather than due to technical or anaesthetic problems.

These survey results seem to validate the usefulness of EP spinal cord monitoring during scoliosis procedures. It does appear to be a technique able to identify complications. It is associated with a reduction in rates of neurologic deficits, not only in comparison to historical rates but also when comparing experienced monitoring teams to inexperienced ones. Rates of paraplegia and paraparesis, the feared major complications, are particularly lower in this monitoring survey compared to their historical rates.

We hope that the encouraging clinical outcome, and the technical factors described here, will be helpful for those clinical teams carrying out monitoring now as well as those contemplating beginning such a program.

References

1. MacEwen GD, Bunnell WP, Sriram K. Acute neurological complications in the treatment of scoliosis: a report of the scoliosis research society. J Bone Joint Surg [Am]. 1975;57A:404–408.
2. Dawson EG, Sherman JE, Kanim LEA, Nuwer MR. Spinal cord monitoring: results of the

Scoliosis Research Society and the European Spinal Deformity Society Survey. Spine. 1991;16:S361–S364.
3. Wilkinson L. SYSTAT: The system for statistics. Evanston, IL: SYSTAT Inc; 1989.
4. Nash CL Jr, Lorig RA, Schatzinger LA, Brown RH. Spinal cord monitoring during operative treatment of the spine. Clin Orthop. 1977a;126:100–105.
5. Nash CL Jr, Schatziner LH, Brown RH, Brodkey J. The unstable thoracic compression fracture: its problems and the use of spinal cord monitoring in the evaluation of treatment. Spine. 1977b;2:261–265.
6. Engler GL, Spielholz NI, Bernhard WN, Danziger F, Merkin H, Wolff T. Somatosensory evoked potentials during Harrington instrumentation for scoliosis. J Bone Joint Surg [Am]. 1978;60A:528–532.
7. Nuwer MR, Dawson E. Intraoperative evoked potential monitoring of the spinal cord: enhanced stability of cortical readings. Electroencephalogr Clin Neurophysiol 1984a;59: 318–327.
8. Nuwer MR, Dawson E. Somatosensory evoked potential monitoring: measurement of variability. In: Nodar RH and Barber C, eds, Evoked Potentials II. Boston: Butterworth; 1984b:510–513.
9. Nuwer MR, Dawson E. Intraoperative evoked potential of the spinal cord: a restricted filter, scalp method during Harrington instrumentation for scoliosis. Clin Orthop. 1984c;183: 42–50.
10. Aminoff MJ. Intraoperative monitoring by evoked potentials for spinal cord surgery: the cons. Electroencephal Clin Neurophysiol. 1989;73:378–80.
11. Tamaki T, Noguchi T, Takano H, Tsuji H, Nakagawa T, Imai K, Inoue S. Spinal cord monitoring as a clinical utilization of the spinal evoked potentials. Clin Orthop. 1984;184: 58–64.
12. Wilber RG, Thompson GH, Shaffer JW, Brown RH, Nash CL Jr. Postoperative neurological deficits in segmental spinal instrumentation. A study using spinal cord monitoring. J Bone Joint Surg [Am], 1984;65A:1178–1187.
13. Lesser RP, Raudzens P, Lüders H, Nuwer MR, Goldie WD, Morris III HH, Dinner DS, Klem G, Hahn JF, Shetter AG, Ginsburg HH, Gurd AR. Postoperative neurological deficits may occur despite unchanged intraoperative somatosensory evoked potentials. Ann Neurol. 1986;19:22–25.
14. Ginzberg HH, Shetter AG, Raudzens PA. Postoperative paraplegia with preserved intraoperative somatosensory evoked potentials. J Neurosurgery. 1985;63:296–300.
15. Molaie M. False negative intraoperative somatosensory evoked potentials with simultaneous bilateral stimulation. Clin Electroencephalogr. 1986;17:6–9.
16. Harper CM, Daube JR, Litchy WJ, Klassen RA. Lumbar radiculopathy after spinal fusion for scoliosis. Muscle and Nerve. 1988;11:386–391.
17. Johnston CE, Happel LT Norris R, Burke SW, King AG, Roberts JM. Delayed paraplegia complicating sublaminar spinal instrumentation. J Bone Joint Surg [Am]. 1986;68A: 556–563.
18. Ben-David B, Haller G, Taylor P. Anterior spinal fusion complicated by paraplegia: a case report of a false-negative somatosensory-evoked potential. Spine. 1987;12:536–539.
19. Nuwer MR. Evoked potential monitoring in the operating room. New York: Raven Press; 1986.
20. More RC, Nuwer MR, Dawson EG. Cortical evoked potential monitoring during spinal surgery: sensitivity, specificity, reliability, and criteria for alarm. J Spinal Disorders. 1988;1:75–80.

8
Round Table Debate on International Spinal Cord Monitoring Practice

In the wake of Dr Nuwer's presentation of the results of the Scoliosis Research Society survey of spinal cord monitoring mainly in the United States, eminent delegates from Great Britain, Japan and Belgium were asked to give a brief synopsis of the prevailing practice and attitude in their own countries.

Present status of spinal cord monitoring in Japan

T. TAMAKI

In Japan, spinal cord monitoring utilizing the spinal cord evoked potential, which is evoked by direct stimulation to the spinal cord, was developed in 1972. Since then, this method has been employed at many facilities. In 1990, the Department of Health and Welfare of the Japanese government finally certified the observation of the spinal cord evoked potential during surgical intervention to the spine, spinal cord and the aorta, as a method of patient's benefit which is covered by national health insurance. Before this authorization, intraoperative spinal cord monitoring was regarded as an experimental and technical adjunct of surgical procedures to the spine and the spinal cord. Despite the economic negligence of the Japanese government concerning spinal cord monitoring, widespread clinical use and basic research have been made of the spinal cord evoked potentials to utilize them for functional spinal cord monitoring.

To understand the present status of spinal cord monitoring in Japan and the degree of reliance placed on it, questionnaires were sent to the departments of orthopaedic surgery of 80 medical schools in Japan, and 68 replies were collected in the spring of 1990.

Fifty-one facilities had already employed intra-operative spinal cord monitoring and 17 were planning to use it in the near future. Ten had no experience and no plan to introduce the practice.

Practically used methods, as listed in Table 1, consisted of about 10 techniques. It seems that no best single method had been developed, although it should be understood that the investigators were selecting appropriate methods to cover the spinal cord segments which were at risk during the operation. Among these, the most frequently used method was the spinal cord evoked potential generated by direct stimulation of the spinal cord (SCEP). The cortical somatosensory evoked potential (SEP) was used routinely at seven institutions. At another 18 institutions, SEP was employed on selected cases whose high cervical spinal cord had to be monitored. Conductive and segmental spinal somatosensory evoked potentials after stimulating peripheral nerve trunks (SSEP) were used routinely at 7 and 8 institutions respectively, although 27–28 institutions used it on selected cases. When this inquiry was performed in 1989, motor evoked potentials (MEP)

Table 1 Utilized methods

	Institution	
	routine	*selective*
1. Wake-up test	4	16
2. Cortical SEP	7	18
3. Conductive spinal SEP	8	27
4. Segmental spinal SEP	7	29
5. Spinal-spinal SCEP	22	25
6. Spinal-muscle CMEP	1	7
7. Cortico-spinal MEP	2	6
8. Cortico-muscle MEP		1
9. Cortical mag. stim. MEP		1
10. Multimodality	8	12

recorded from the spinal cord after stimulating the brain and compound motor evoked potential (CMEP) recorded from the peripheral muscles were routinely used at only two and one institutions respectively.

So-called multimodality monitoring in which multiple methods were performed simultaneously was utilized routinely at eight institutions and another 12 used combined methods on selected cases. No unanimity was found, although when this inquiry was performed many institutions were mainly using evoked potentials recorded from the vicinity of the spinal cord while not so many were using motor system monitoring.

The wake-up test was used routinely at four institutions and selectively at 16 units. The majority of them were utilizing the wake-up test as an adjunct to a neurophysiological monitoring technique. A total of 4562 clinical cases were monitored before the end of 1989 at 51 institutions. They could be enumerated as follows: spinal cord myelopathy 1640 cases, scoliosis 1180, spinal cord tumour 151, diseases at the cranio-cervical junction 147, aorta aneurysm 140, others 700.

Fifteen institutions had clinical experience of more than 100 cases. Twenty-two false-negative cases were reported from 12 institutions. Five of them were cases of spinal cord tumour, three abdominal aorta aneurysm, nine scoliosis and five others. Eight scoliotic patients were judged to be excluded as their postoperative neurological complaints were derived from injured segmental nerve roots of thoracic level which were not covered by the spinal cord monitoring.

On the other hand, in cases of spinal cord tumour and abdominal aorta aneurysm, selective injuries of the spinal cord neural tissues were not detected by the methods used on these cases. Further detailed analysis should be performed to understand the mechanism of such false negative findings.

The overall reliability of the methods used was also investigated. Forty four institutions (88%) judged their method of spinal cord monitoring to be reliable, three institutions (6%) regarded it as unreliable, and another three institutions neglected to answer properly. Interestingly, these six institutions belonged to the group whose success rate of intraoperative spinal cord monitoring was less than 50%, hence the immature technology of their monitoring team was thought to be responsible for the poor reliability.

Responding to the question about the most and least reliable methods, 33 institutions answered that the most reliable method was the spinal cord evoked potential generated by the direct stimulation to the spinal cord (SCEP). Thirteen institutions evaluated the spinal somatosensory evoked potential (SSEP) as reliable. In contrast, only two institutions considered the somatosensory cortical evoked potential (SEP) to be reliable while 12 institutions responded that the SEP was not reliable.

The most characteristic feature of monitoring in Japan may be the personnel who actually carry out the technique. At 43 institutions (84%) orthopaedic surgeons who had neurophysiological knowledge were exclusively involved; at six departments, an orthopaedic surgeon or an anaesthesiologist and technician were responsible for it. At only two institutions, well-trained technicians were performing the monitoring. This fact may provide the reason why so many neurophysiological methods were arbitrarily selected. Another factor differentiating Japanese practice from that in other countries is that Japanese orthopaedic surgeons treat spinal cord tumours and another space-occupying lesions of the spinal canal and are therefore able to evaluate the efficacy of monitoring methods for themselves.

During the past two years, multimodality monitoring using MEP and SCEP or another potentials has been frequently performed at many facilities. Furthermore, many experimental and clinical reports relating to the compound muscle action potentials after stimulating the spinal cord have been presented. These developments may lead to further advancement of monitoring of complicated spinal cord functions intraoperatively.

Spinal cord monitoring in the UK

R. K. JACKSON and N. J. SMITH

In 1989 the Association of British Clinical Neurophysiologists (ABCN) expressed concern about the standards of spinal cord monitoring in the UK. There was also uncertainty about how much monitoring was actually going on during scoliosis and who was doing it. As a result, in 1990 a Joint Working Party was set up by the British Scoliosis Society (BSS) and the ABCN to evaluate the current state of cord monitoring in the UK and to produce an agreed 'Code of Practice'.

One of the more striking findings of the survey was that 40% of scoliosis surgeons were not using continuous cord monitoring at all, but were still relying on a wake-up test during the operation. Another interesting finding was that only 50% of cord monitoring was being done by neurophysiologists, while 25% was being done by medical physicists and 25% by surgeons or their anaesthetists. In nearly all cases sensory evoked potentials were being used and the majority of centres preferred epidural to surface electrodes for scoliosis surgery.

An audit of fifteen hundred operations for correction of spinal deformity showed that there were fewer false positives when monitoring was performed by neurophysiologists. Otherwise there was very little to choose between the various monitoring teams, except that there was a noticeable lack of medical supervision when monitoring was performed by medical physics departments. It was reassuring that there had been no false negatives and that the occurrence of true positives was less than 2%.

Although it might be our long-term objective to have all monitoring for scoliosis surgery done by neurophysiology departments, this is certainly not a realistic prospect in the short term, since too few neurophysiologists are available particularly in specialist orthopaedic hospitals. The Working Party also recognised the fact that most scoliosis surgeons feel perfectly comfortable with their current arrangements. It was therefore recommended that the existing arrangements be continued, provided that acceptable monitoring standards were maintained and that adequate training was provided for monitoring staff.

The monitoring standards we would like to see are set out in the 'Code of Practice' recommended by the Joint Working Party. We felt it was equally important to address the question of technician training and that this was most properly in the remit of the ABCN.

With regard to responsibilities, it was the Working Party's recommendation that:

1. the acquisition and interpretation of tracings be the responsibility of the monitoring team;
2. clinical decisions during the operation be the responsibility of the operating surgeon.

The question of when cord monitoring should be used during scoliosis surgery was excluded from the terms of reference of the Working Party. It was felt by the British Scoliosis Society that as long as a substantial number of orthopaedic surgeons do not feel the need to use spinal cord monitoring every time they operate on the spine, or indeed every time they correct spinal deformity, we are not presently in a position to say that spinal cord monitoring should be mandatory.

Spinal cord monitoring for spinal trauma or neurosurgical operations on the spine was not covered by the ABCN/BSS survey, but the Code of Practice would be applicable to these fields as well as scoliosis surgery.

General perception of the usefulness and failings of spinal cord monitoring. The U.C.L. experience

J. M. GUERIT

INTRODUCTION

We have been monitoring somatosensory evoked potentials (SEP) for about five years in vascular and orthopaedic surgery. There is a general agreement among vascular surgeons that spinal cord monitoring in descending aorta surgery improves patient outcome, and this is largely confirmed in our statistics. The opinion of orthopaedists is more variable: some do not believe in monitoring, some consider it mandatory, some find it simply useful. Spinal cord monitoring is not a medico-legal obligation in Belgium and is not specifically covered by health insurance. Owing to its poor economic acceptability, the criteria of actual clinical yield of the technique should be studied. Two criteria will be considered: the pathophysiological adequacy of the technique and the initial agreement between the surgeon, the neurophysiologist, and the anaesthesiologist.

THE PATHOPHYSIOLOGICAL ADEQUACY OF THE TECHNIQUE

The technique used should be adapted both to the anatomy of the neural structure at risk and to the pathophysiology of the damage to be avoided. Moreover, it should remain compatible with the anaesthetic conditions. With regard to anatomy, the differential sensitivity of motor and somatosensory EPs to lesions of the ascending and descending spinal pathways is well documented in this volume. Two factors of the pathophysiology of spinal lesions should be considered: the neural structures actually at risk and the time course of the lesions to be avoided. These factors are differentially involved in vascular and orthopaedic surgery, which most probably partially explains the variable perception of the clinical usefulness of the technique by vascular and orthopaedic surgeons.

In vascular surgery, spinal ischaemia is the only event to be prevented, and we know that high degrees of ischaemia, sufficient to result in clinical

Table 2 The initial agreement between the surgeon, the anaesthesiologist, and the neurophysiologist

The surgeon
Agrees to wait for the completion of a pathophysiological process
Agrees to adapt his procedure to the output of the neurophysiologist
Agrees to learn and to be open to clinical neurophysiology
The anaesthesiologist
Agrees to adapt his techniques to the recording of SEPs
Agrees to keep the neurophysiologist continuously informed
The neurophysiologist
Agrees to question his own technique continuously
Agrees to try to improve interactively
Agrees to adapt his technique to the needs of the anaesthesiologist
Agrees to devote a great deal of time to surgical monitoring

sequellae, most probably give rise to significant SEP alterations. Moreover, the time course of SEP alterations consecutive to spinal ischaemia is rather slow, and, when abnormalities are found, they can still be reversed even after 10 or 20 min. Thus, SEPs are particularly suited to the prevention of spinal cord ischaemia. In orthopaedic surgery, several mechanisms can explain the lesions (direct injury, compression, ischaemic damage). Little is yet known about the differential involvement of these mechanisms in many orthopaedic situations. Thus, the site of the lesion to be avoided is not necessarily known in advance and, in particular, it can be very restricted in focal injuries of the spinal cord. This implies the a priori probability of missing a lesion to be higher in orthopaedic than in vascular surgery. Moreover, the time course of the spinal damage varies as a function of the mechanism involved. Thus, the detection of some lesions can be incompatible with the time needed to obtain a reliable motor or somatosensory EP.

THE INITIAL AGREEMENT BETWEEN THE SURGEON, THE NEUROPHYSIOLOGIST AND THE ANAESTHESIOLOGIST

In our opinion, a clear agreement between the surgeon, the neurophysiologist, and the anaesthesiologist it is an absolute prerequisite for neurophysiological monitoring to be useful. If it cannot be reached, we prefer not to begin monitoring. To provide the surgeon with an unjustified safety element seems to us much more hazardous than not providing him with any safety at all. Both a general and a particular level should be considered.

The *general* level concerns the decision taken by a surgeon, an anaesthesiologist, and a neurophysiologist to work together in some operations. The rules of the game are given in Table 2 and should be clearly accepted in advance. Three points should be emphasized. First, it is senseless to perform neurophysiological monitoring if the surgeon does not agree to adapt his attitude to the data provided by the neurophysiologist; in particular, he should be clearly informed and should accept the necessity of waiting whenever any decision is to be taken on the basis of the occurrence of a slow pathophysiological process. Second, it is evident that the neurophysiological method and the anaesthesia should be compatible with each other. However, both the anaesthesiologist and the neurophysiologist should consider the possibility of adjusting their techniques. For instance, our anaesthesiologists try to avoid gases whenever possible if cortical SEPs are to be performed, but they prefer to use them in aortic coarctation repair in small children. Thus, we decided to adapt our own technique, that is, to replace the recording of cortical SEPs by that of the lemniscal far-field, which can be very easily obtained in well-curarized children, even under gas anaesthesia. Third, it is our opinion that neurophysiological monitoring should be performed by a neurophysiologist. As a matter of fact, we advocate the use of sophisticated techniques, as we are convinced that their specificity is higher. This implies that a rather complex analysis has to be performed during the operation. Such analysis presumes sufficient training in neurophysiology and is often incompatible with the multitude of tasks the anaesthesiologist has to take upon himself during the operation.

At a *particular* level, each operation should be individually designed, and one should try to define all the events that are most likely to occur, in order to develop specific responses. The rules of the dialogue between the surgeon and the neurophysiologist should also be clearly defined. In practice, a choice must be made between immediately warning the surgeon of all neurophysiological alterations or waiting until the alteration is confirmed or judged significant. The advantage of the former system is easier detection of the factor involved, if this factor exists, but it also increases the risk of false alarms, which, in turn, increases both the operation time and the stress on the surgeon. Therefore, we inform the surgeon of all events, but the surgeon, in the absence of any evident cause, continues and waits for confirmation or invalidation of the first warning.

Somatosensory and spinal cord evoked potential studies in humans

9
Spinal cord evoked potentials in cervical and thoracic myelopathy

H. BABA, N. KAWAHARA, S. NAGATA, Y. KIKUCHI, K. TOMITA and S. IMURA

INTRODUCTION

Spinal cord evoked potentials (SCEP) recorded from the epidural space reflect directly spinal cord function and have been used for the diagnosis of the decompressive areas. We have used the spinal cord stimulation and recording technique since the recorded evoked responses are stable in configuration and amplitude[1]. In particular, it is informative for level-specific diagnosis of diseases with multilevel vertebral involvement. It is important to elucidate any SCEP abnormality and use in the evaluation of postoperative neurologic improvement. A restrospective study was made on 75 cases who underwent SCEP examinations.

MATERIALS AND METHODS

A study was made on 75 cases, undergoing surgical decompression and fusion for cervical and/or thoracic myelopathy. SCEP were tested in all patients before surgery, and 16 were examined postoperatively to monitor the neurologic improvement. There were 52 patients with cervical myelopathy; 42 with cervical spondylotic myelopathy (CSM), and 10 with ossification of posterior longitudinal ligament (OPLL). Eight patients out of 42 with CSM had cervical disc herniation at a single level. Twenty-three patients had thoracic myelopathy, 20 of whom had OPLL and/or ossification of the ligamentum flavum involving more than two vertebrae. The remaining three had vascular disease of the thoracic spinal cord.

SCEP were obtained by direct stimulation of the spinal cord at the lower levels of the lesions[2]. In this study, the main concern was ascending SCEP changes at and around the lesions. A five-polar epidural recording electrode (Unique Medical, Tokyo) was used to reduce testing time. Stimulus intensity was set just above the patient's sensory threshold, this usually being forty volts. A reference electrode was placed in the subcutaneous tissue between the stimulating and recording electrodes. Responses occurring within 10 ms following the triggering of the stimulus were amplified by a DISA 15CO4 (DANTEC DISA, Skovlunde, Denmark) with a band pass of 10 Hz to 2 kHz, and averaged 32 to 64 times with a DISA N1500 EMG System. The initial negative spike (N_1) was stable under all conditions and mainly analysed in the present study.

Abnormal SCEP were categorized as normal, moderate to severe amplitude reduction, disappearance, and positive-going waves. In each case, SCEP measured from the intact levels were accepted as the control response for each individual. The relationship among SCEP data, Japanese Orthopaedic Association Scores (JOA scores), and postoperative neurologic improvement rate[2] was investigated.

RESULTS

In CSM, 31 out of 34 patients showed abnormal SCEP, positive-going waves and severe amplitude decrease of more than 50% of control, at mostly narrowed levels. Twenty-nine patients showed abnormal SCEP at mostly unstable levels of the cervical spine. All eight patients with myelopathy due to disc herniation showed abnormal SCEP at the level of disc derangement. In CSM including disc herniation, more than 91% of the abnormality rate was observed at narrowed levels, this being compatible with spondylotic changes causing myelopathy. Only three of 32 patients with CSM showed no abnormality in SCEP at the levels of spondylosis. In 10 cases with OPLL at multiple levels, eight patients showed abnormal SCEP at most severely compressed sites, and the other two, at discrete sites of OPLL indicating clinical instability.

Figure 1 summarizes the relationship of SCEP abnormality and JOA scores in cervical myelopathy. Thirty patients presenting positive-going waves at levels affected by myelopathy showed smaller scores, but these waves were not specific for the severity of myelopathy. The scores for patients with and without positive-going waves were essentially the same.

In sixteen patients, perioperative and postoperative SCEP up to 10 days following surgery were monitored by a recording electrode placed at the site of operation. In these patients, SCEP abnormality and neurologic improvement rates at more than six months following surgery were compared. Four of 16 patients showed improved SCEP after decompression during surgery, and the average neurologic improvement rate was 85%, this being quite an excellent result. Eight patients with improved SCEP within one week postoperatively had an average improvement rate of 71%, and for the other four showing no improved SCEP even 10 days after surgery, it was 42% (Figure 2).

In thoracic myelopathy, abnormal waveforms of SCEP were recognized in all patients. Fourteen patients with positive-going waves had a mean preoperative JOA score of 3.5 ± 2.1 points, and eight with marked amplitude attenuation, 6.8 ± 1.7 points. These scores differ significantly (Wilcoxon's

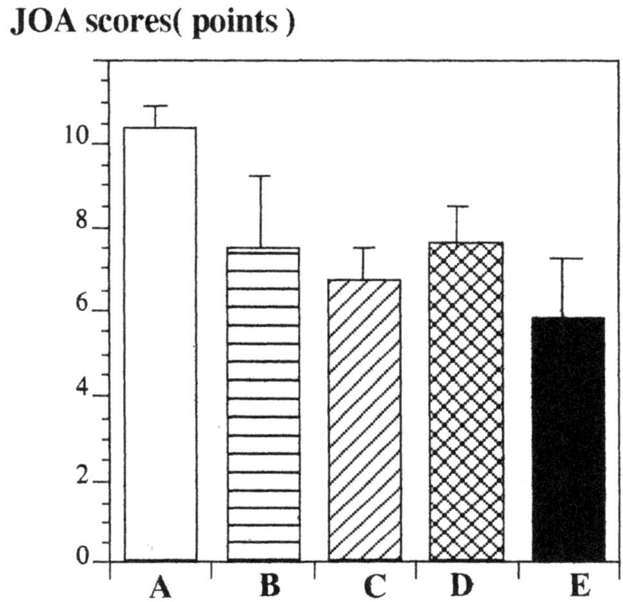

Figure 1 Abnormality of spinal cord evoked potentials and JOA scores. The maximum number of points was 11. A: Normal SCEP (n = 6), B: Slight amplitude decrease (n = 8), C: Marked amplitude decrease or disappearance (n = 8), D: Positivity in SCEP without abnormality in cranial sites of lesion (n = 20), E: Positivity with cranial abnormality (n = 10)

test, $p < 0.05$). The above former group showed a postoperative improvement rate of 46%, and the latter, 52%, the difference not being significant.

DISCUSSION

Spinal cord evoked potentials are widely used in spinal surgery[3–5]. The decompressive manoeuvre and corrective procedure are apparently well monitored by SCEP. The preoperative diagnosis of responsible levels is also facilitated by SCEP monitoring. These potentials being mainly reflective of long tract sensory pathways in the spinal cord[6] they have definite limits in the diagnosis of myelopathy. In compressive myelopathy, abnormal SCEP are frequently recorded at injured spinal cord levels. Although SCEP would appear useless for motor tract monitoring, they are still useful for level-specific diagnosis of compressive myelopathy.

Abnormal SCEP indicate the extent of spinal cord conduction disturbance. In cervical myelopathy, more than 90% of the patients showed abnormal SCEP at the level of most significant spondylosis. Some patients showed abnormality at the levels of symptomatic instability such as retrolisthesis on extension film. In clinical situations, the final decision of the area to be decompressed is made on the radiologic as well as neurophysiologic data. In cases of more than three vertebral levels, preoperative SCEP test for

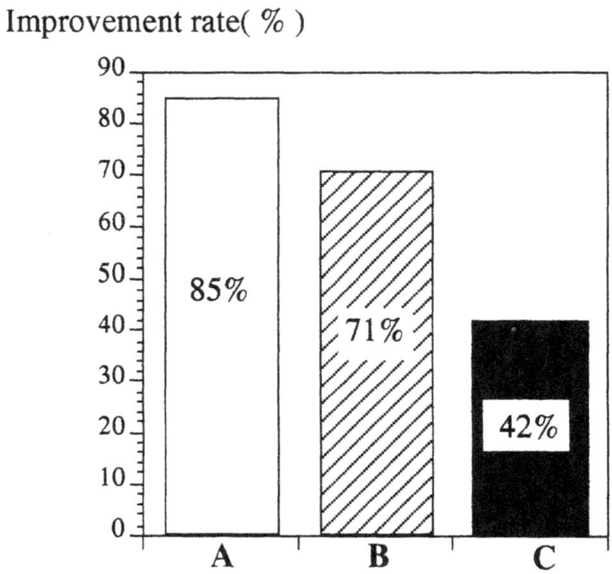

Figure 2 Neurologic improvement and recovery of spinal cord evoked potentials. A: SCEP recovered during surgery (n = 4), B: SCEP recovered within 1 week (n = 8), C: No recovery in SCEP within 10 days postoperatively (n = 4)

localization is particularly useful. For anterior decompression of relatively limited areas, areas for decompression and fusion should be carefully examined by every means of diagnosis. Due to the limitations of decompressive surgery, good results may be difficult to achieve.

Controversy remains over the standards of amplitude abnormality in SCEP. The present authors[6], in an experimental study of ascending SCEP, found the spinal cord to lose neurophysiological viability when the amplitude of ascending SCEP at compressed sites exceeded 50% of the control value. We also noted 30% amplitude attenuation to be possibly critical for the spinal cord. In compressive myelopathy with injured spinal cord function, a 30% amplitude decrease may truly be critical for the spinal cord. This must be kept in mind in SCEP monitoring. In the cervical spine, early recovery of SCEP during or shortly after surgery was found to correlate with favourable results. Although the severity of disease is correlated with preoperative SCEP, postoperative neurologic improvement cannot be assessed only by preoperative SCEP. This study shows SCEP to be useful for diagnosis in cases of multi-level involvement, but not for prognostic evaluation.

References

1. Baba H, Tomita K, Umeda S. Clinical study of spinal cord evoked potentials. In: Ducker TB, Brown RH, eds, Neurophysiology and standards of spinal cord monitoring. New York: Springer-Verlag; 1988: 216–221.

2. Kikuchi Y, Baba H, Kawahara N. Experience of diagnosis of thoracic myelopathy using spinal cord evoked potentials. In: Shimoji K, Kurokawa T, Tamaki T, Willis WD Jr, eds. Spinal cord monitoring and electrodiagnosis. Berlin Heidelberg: Springer-Verlag; 1991: 461–471.
3. Schramm J, Krause R, Shigeno T, Brock M. Experimental investigation on the spinal cord evoked injury potentials. J Neurosurg, 1983;59:485–492.
4. Nash Jr CL, Brown RH. Current concepts review: Spinal cord monitoring. J Bone Joint Surg. 1989;71A:627–630.
5. Baba H, Tomita K, Kawahara N. Spinal cord monitoring: Diagnosis and prognostic evaluation of cervical myelopathy. J Jpn Orthop Assn, 1992b; 66:S219.
6. Baba H, Nagata S, Kawahara N, Tomita K, Imura S. Spinal somatosensory evoked potentials in compression and distraction injury. Presented at 65th Annual Meeting of American Academy of Orthopaedic Surgeons, Feb 20–25, 1992a, Washington DC.

10
Spinal cord evoked potentials in spinal cord intermittent claudication

H. BABA, N. KAWAHARA, K. TOMITA, K. NAKAHASHI and S. IMURA

INTRODUCTION

Intermittent claudication is commonly observed in lumbar spinal stenosis and obstructive arterial disease. However, this condition due to compressive lesions to the spinal cord is very rare. Spinal cord intermittent claudication presents with exaggerated spasticity during walking. Reduced spinal cord blood flow during walking may be the cause of this symptom but no article has described the pathogenesis of compressive lesions to the spinal cord, by which intermittent claudication may be caused. Two cases of thoracic ossification of the ligamentum flavum (OLF) with spinal cord intermittent claudication are presented.

CASE REPORTS

Case 1

In 1986, a 72-year-old male noted vague dullness around the buttocks and posterior aspect of the thighs on walking for about 20 minutes. Shortly after resting, the symptoms completely disappeared. In 1988, he noted severe dullness in the lower extremities after walking for about 50 metres. Because of no improvement even with medication from his doctor, he was admitted to our Medical Center in October, 1989. There was nothing remarkable in his medical history. Physical examination showed no significant neurologic signs in the lower extremities except slight sensory disturbance at the L_5 and the S_1 dermatones experienced for more than 20 years. Bowel and bladder function were normal. There was no muscular atrophy in the lower extremities. Before walking, deep tendon reflexes (DTRs) in the lower extremities were normoactive and no pathological reflexes were seen.

However, after walking for approximately 50 metres, DTRs in the lower extremities became hyperactive, and he felt the desire to urinate as well as abnormal sensation at the sacral area. Very shortly after resting, his symptoms completely disappeared. Radiographic examination of the lumbar spine showed only spondylotic changes but in the thoracic spine there was OLF at the T_{9-10}, T_{10-11} and T_{11-12} levels by which the spinal cord was compressed posteriorly (Figure 1). After walking, spinal cord evoked potentials (SCEP) recorded from the posterior epidural space showed the disappearance of the third negative component at the T_{10} level. The first negative spike showed an amplitude reduction at the T_{10-11} and the T_{11} levels (Figure 2). From these radiographic and SCEP findings, OLF at the T_{10-11} level was judged principally responsible and that at the T_{9-10} level subclinically responsible for spastic cramps in walking. Laminectomy of T_9 and T_{10} completely relieved muscular spasticity during longer walking periods. At a follow-up visit in October, 1991, two years after surgery, he was without any muscular discomfort and well except for sensory disturbance of the feet.

Case 2

A 19-year-old achondroplastic man became unable to walk for more than approximately 240 metres. He complained of severe muscle cramps in the thighs and could walk no more. Before admission on 17 August 1989, he had never noted muscular discomfort in the lower extremities except minimal sensory disturbance at the posterior aspect of the legs. Neurologically, he had slightly decreased tactile sensation below the L1 level, but pain and thermal sensation were essentially normal in this area. DTRs in the lower extremities were normoactive and no pathological reflexes were observed. Bowel function was normal, but there was slight urinary incontinence. In the lower extremities, especially in the quadriceps muscle, the patient experienced severe muscle cramps after walking followed by abnormal dullness in the lower extremities. When muscle cramps appeared, there were markedly hyperactive DTRs in the lower extremities. An abnormal tingling sensation at the sacral area made a longer walk impossible. However, progressive muscular spasticity and abnormal sensation during walking resolved with a short rest. Along with spinal canal stenosis, spinal cord compression at the T_{4-5} level by OLF was found. SCEP showed positive-going waves at the T_4 level and latency delay at the T_{4-5} levels (Figure 3). Limited laminectomy of T_4 and T_5 resulted in excellent improvement. At two weeks after surgery, the patient could walk without aid. At a follow-up visit in March, 1992, the patient was well and had returned to his original light work.

DISCUSSION

Spinal cord intermittent claudication is considered to occur due to reduced spinal cord blood flow during walking. Classical reports refer to the underlying pathogenesis of vascular diseases such as arteriosclerosis and arterio-venous malformation[1–4]. In such diseases, compromise of the blood

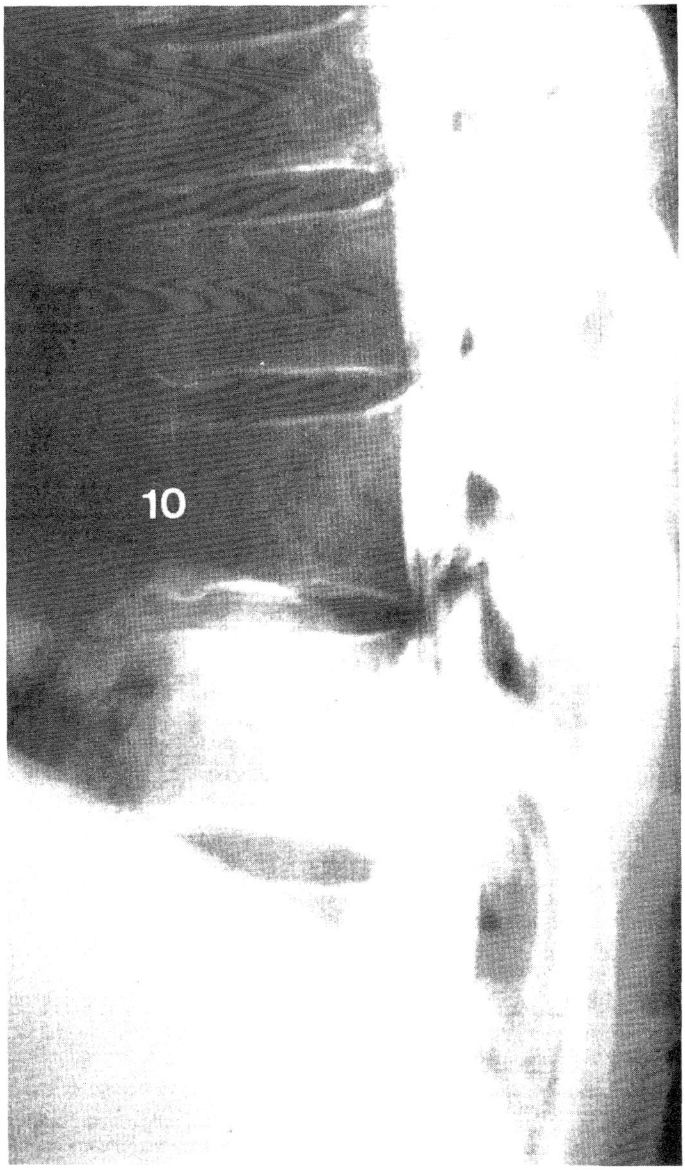

Figure 1 Myelogram of case 1 showing posterior compression of the spinal cord

supply during walking leads to a relatively ischaemic state of the spinal cord. However, verification of this pathogenesis is difficult in clinical situations.

Patients with this disorder usually do not show neurologic signs at rest. After walking some distance, they show signs of spasticity especially in the

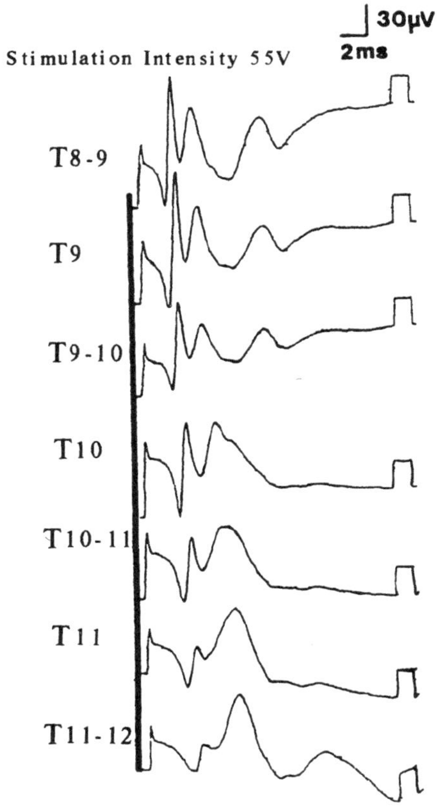

Figure 2 Spinal cord evoked potentials of case 1 showing disappearance of the third negative spike at the T_{10} level

muscular tonus of the lower extremities. Most patients feel muscular dullness or weakened muscular tonus without significant sensory involvement as observed in cauda equina claudication due to lumbar spinal stenosis. Shortly after resting, the muscular symptoms disappear. Clinical symptoms are characterized by long tract pyramidal signs in varying degrees. Hyperactive DTRs, pathological reflex such as a Babinski reflex and occasionally urinary incontinence are other features of this disorder. Of sensory modalities, vibration sense is the most and earliest compromised and position sense the latest involved. However, in the cases presented, there was no distinct sensory disturbance associated with claudication.

Dejerine[1], Reichert et al.[5], and Jelinger et al.[6] reported symptoms of spinal cord intermittent claudication arising from transient ischaemia of the spinal cord. As far as the authors know, there is no report of ossified ligamentum flavum compressing the thoracic spinal cord, causing intermittent spinal cord dysfunction. Narrowing of the spinal canal may be critical for blood flow to the spinal cord. Even without arterial disease, any compressive lesions of the

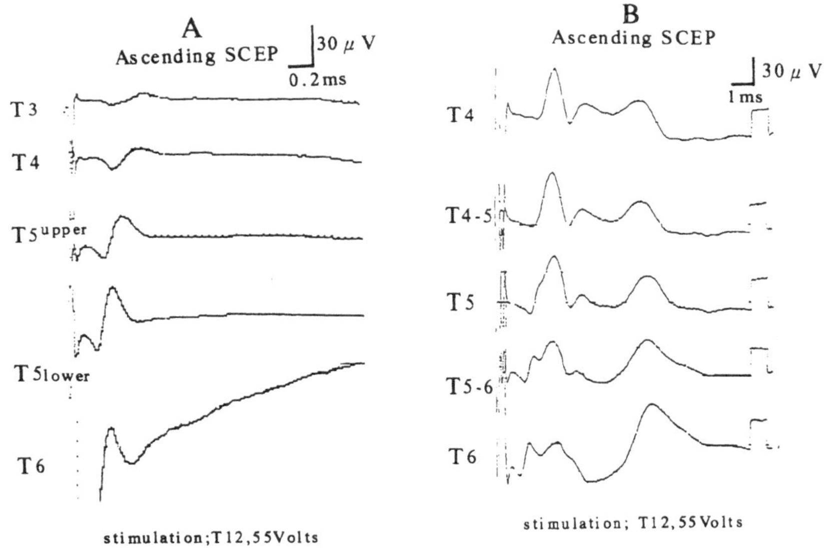

Figure 3 Preoperative spinal cord evoked potentials of case 2 (A) showing positive-going waves at the T_4 level, but which were normalized 12 days postoperatively (B)

spinal cord may compromise spinal cord blood flow. Local circulation may be disturbed by OLF but verification of decreased circulation by OLF in clinical situations seems impossible. Abnormal SCEP provide distinct evidence of spinal cord dysfunction. The disappearance of the negative spike or positive-going waves indicate conduction disturbance. Most patients with thoracic myelopathy due to compressive lesions show abnormal SCEP at the lesions responsible for myelopathy with approximately 90% sensitivity[7]. Across compressive lesions shown radiologically, no abnormality is detected in SCEP when the lesions are asymptomatic for the spinal cord. In this study, osseous compressive lesions for OLF were the cause of transient thoracic myelopathy. In one case, SCEP showing positive-going waves became normal after surgery demonstrating the reversibility of both symptoms and electrophysiologic conduction disturbance.

Spinal cord intermittent claudication is a diagnosis, based on observation of the clinical symptoms. We stress that osseous compression of the spinal cord, although the pathogenesis remains unknown, may be a cause of spinal cord intermittent claudication.

References

1. Dejerine J. Sur la claudication intermittente de la moelle epiniele. Rev Neurol. 1906;14: 341–350.
2. Dupre ME. Claudication intermittente par arterio-sclerose de la moelle. Rev Neurol. 1915;24:498–500.

3. Demanet JC, Dustin Jr P, Perier O. Ramollissement de la moelle par embolisation de cristaux de cholesterides. Rev Belg Path. 1960;27:347–354.
4. Garcin R, Godlewski S, Lapresle J, Fardeau M. Syndromes vasculaires aigus probables de la partie inferieure de la moelle. Rev Neurol. 1959;100:212–229.
5. Reichert FR, Rytand DA, Bruck EL. Arteriosclerosis of the lumbar segmental arteries producing ischaemia of the spinal cord and consequent claudication of the thighs. Am J Med Sci. 1934;187:794–806.
6. Jelinger K. Neumayer E. Claudication of the spinal cord and cauda equina. In: Vinken PJ, Bruyn GW, eds. Handbook of clinical neurology. Amsterdam: Elsevier–North-Holland, 1972;12:507–547.
7. Kikuchi Y, Baba Y, Kawahara H, Nagata S, Tomita K, Nomura S, Yugami H. Experience of diagnosis of thoracic myelopathy using spinal cord evoked potentials. In: Shimoji K, Kurokawa T, Tamaki T, Willis WD Jr, eds. Spinal cord monitoring and electrocardiagnosis. Berlin Heidelberg: Springer-Verlag; 1991:461–471.

11
Somatosensory cerebral potentials evoked by paraspinal stimulation: an approach to spinal cord monitoring

C. CALLEGARINI, G. CIUCCI, D. MISEROCCHI, R. QUATRALE, G. SERRA and G. G. REBUCCI

INTRODUCTION

In recent years several papers have been published on the usefulness of evoked potentials in the diagnosis of root entrapment in the lumbar or cervical region. These methods have improved the diagnosis of radiculopathies. A variety of recording techniques has been presented and many authors described the use of dermatomal stimulation[1-7]. The usefulness of somatosensory potentials evoked by stimulation of mixed nerves (such as peroneal or tibial nerves) in spinal cord pathologies is widely described in the literature.

A common technique is the electrospinogram recording but the spinal potentials evoked over the thoraco-lumbar or cervical vertebral bodies by stimulation of a lower limb nerve, although feasible in infants and young children, are often problematical in adults for two reasons:

1. the potentials recorded percutaneously are frequently small and often contaminated with excessive noise;
2. when the subjects under examination are very stout or muscular the signal to noise ratio of direct percutaneous recording is lower.

The invasive methods (i.e. by using epidural needle electrodes) are clearly not routinely applicable.

We think that dermatomal stimulation of the thoracic region can be useful in the diagnosis of myelopathies and also allows a reasonable estimation of the level of a lesion within the spinal cord. The afferent spinal cord conduction in humans can be studied by paraspinal stimulation as appropriate[8]: this method can play a useful role in the electrophysiologic investigation of cord disease especially if not visible by neuroimaging or in pinpointing accurately the region that must be studied by neuroimaging.

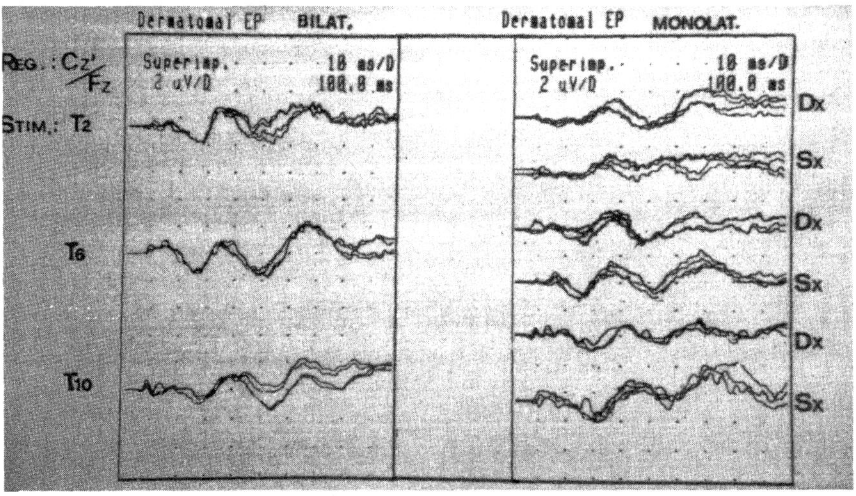

Figure 1 Somatosensory evoked potentials following paraspinal stimulation at different levels (T2, T6, T10) in a normal subject. Bilateral stimulation is on the left side; unilateral stimulation is on the right (Dx: right unilateral stimulation; Sx: left unilateral stimulation)

MATERIALS AND METHODS

Fourteen healthy subjects (6 males and 8 females) aged between 21 and 71 years were studied. Height was between 141 and 182 centimetres. The recordings were made with the subject in supine and comfortable position. The skin temperature was kept at least at 30 degrees centrigrade and the room temperature at about 23 degrees centrigrade. Disc electrodes were placed on the scalp at Cz' and Fz reference for recording. We stimulated the paravertebral skin with two pairs of disc electrodes, three centimetres from the midline, at three different levels (T2, T6 and T10). The cathode was directed medially. A ground electrode was placed on the seventh cervical spine.

The stimulus was given at a frequency of 1 Hz. Its intensity was 4 times above the threshold of sensation; the pulse duration was 0.2 milliseconds. The amplifier band pass was between 1 Hz and 500 Hz. Unilateral stimulation was performed in all subjects. In 10 subjects unilateral stimulation on left and right side was made, in order to study the acceptable variability between the two sides in a normal population. In all subjects the SEP stimulation of the tibial nerve was made. The data were obtained using a Dantec Counterpoint.

RESULTS

As we can see in Figure 1, cortical SEP by unilateral and bilateral paravertebral stimulation consisted of two positive and two negative peaks (P1, N1, P2, N2). In Table 1, we present the values of amplitude and latency peaks of evoked potentials following bilateral and unilateral paraspinal

Table 1 Latencies and amplitude values of somatosensory evoked potentials following paraspinal stimulation (T2, T6, T10)

		Bilateral stimulation						
		Onset	*P1*	*N1*	*P2*	*N2*	*N1 pre*	*N1 post*
T2	Mean	17.21	23.84	35.23	45.72	61.06	2.04	1.98
	SD	1.88	2.62	4.33	5.44	3.85	1.33	1.02
T6	Mean	18.86	25.89	37.01	49.07	64.78	2.35	1.92
	SD	2.08	2.20	2.84	3.92	5.17	0.98	0.82
T10	Mean	21.56	28.52	39.22	50.74	66.19	2.82	2.31
	SD	2.13	2.35	4.14	4.93	7.21	1.82	0.77
		Unilateral stimulation						
		Onset	*P1*	*N1*	*P2*	*N2*	*N1 pre*	*N1 post*
T2	Mean	17.51	24.11	34.45	45.04	61.12	1.29	1.11
	SD	1.61	2.44	3.95	3.93	5.77	0.57	0.57
T6	Mean	19.75	25.89	35.84	46.93	65.23	1.52	1.60
	SD	2.46	2.03	3.08	3.37	4.06	0.73	0.52
T10	Mean	21.65	27.21	38.24	50.02	64.66	1.46	1.31
	SD	2.28	1.79	3.08	4.30	6.44	0.67	0.58

stimulation. The cortical SEP amplitude by bilateral stimulation was always higher than the unilateral one, but the morphology was similar (Figure 1). The P1–N1 complex was always the most sharply distinguishable. The late components were less repeatable and their latency values had a higher standard deviation (Table 1). Using this technique and with these normal values, we have studied some clinical cases affected by spinal cord pathology. We present two of our clinical cases.

Case 1

The first case is a 36-year-old-man in whom a non-Hodgkin lymphoma over the diaphragm was diagnosed in August 1989. Since then he has received both chemotherapy and radiotherapy over the mediastinal region. In May 1991 he developed over a few days, a spastic paraparesis with ataxia, a lack of the superficial sensitivity below T5–T6, and vibratory anaesthesia below the iliac crest.

MRI of the dorsal cord (Figure 2) showed alterations consistent with the effects of massive radiation treatment administered: reduced density of the vertebral bodies with a spinal cord diffusely enlarged, oedematous and with a hyperintense band in the mid-dorsal section: a severe myelopathy. From the neurophysiological point of view it was difficult to examine the patient using the standard technique of SEP by stimulation of tibial nerve at the ankle: as shown in Figure 3 apart from, a good lumbar signal, the early cortical potentials are completely altered in their structure. This morphology prevented us from giving a correct evaluation of the seat of the damage along the central somatosensory pathway. On the contrary, the responses to paraspinal stimulation show a progressive attenuation of the signal obtained passing from the cephalic sectors to the most caudal one along the

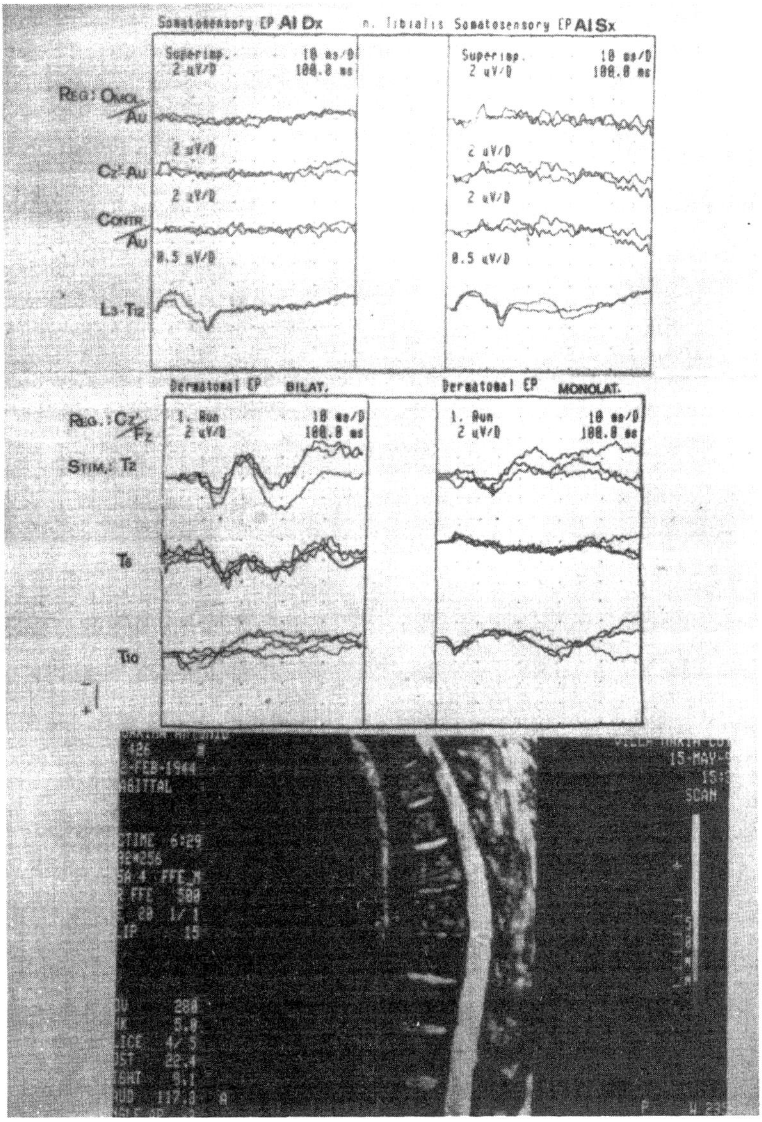

Figure 2 Case no. 1 (B. A., 36 years old). From the top downwards; somatosensory evoked potentials by tibial nerve; paraspinal stimulation at different levels (T2, T6, T10); MRI of dorsal cord

spinal cord: the degree of information we get from this type of stimulation is certainly higher than that given by the standard techniques, providing the neurophysiological evidence of damage in the lower-intermediate sectors of the spinal cord.

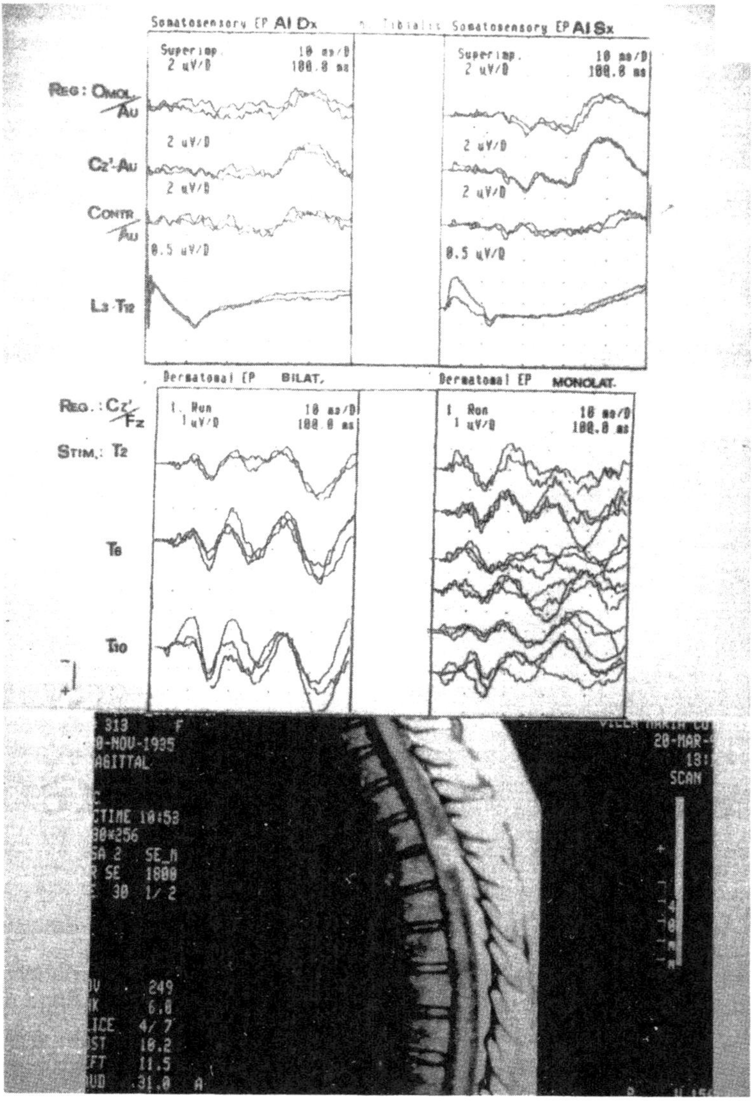

Figure 3 Case no. 2 (B.G., 55 years old). From the top downwards: somatosensory evoked potentials by tibial nerve; bilateral (left side) and unilateral (the upper trace is the right unilateral stimulation, the lower trace is the left unilateral stimulation) paraspinal stimulation; MRI of dorsal cord

Case 2

A 55-year-old lady came to us on account of altered sensitivity to cold in the left leg which had been present for six months. Initial investigation included an X-ray and a CT scan of the lower spine, which were negative. Symptoms progressed until she had an irritating sensation in the left leg as if it were swollen or anaesthetized. On examination the following abnormal signs were present:

- hyperreflexia in the lower limbs, especially on the right;
- loss of pain and thermal sensations on left side with a T8–T9 level;
- reduced vibratory sensation on the right side from the iliac crest downwards.

As we can see in Figure 3, the SEP to lower limb stimulation shows an increase of the central conduction time and a certain degree of morphologic alteration of the structure. Examining the signal obtained by paravertebral stimulation, a good signal was recorded at all levels with bilateral stimulation, but the response to unilateral stimulation shows a progressive loss of symmetry of the evoked signal towards the most caudal sectors: on the right, stimulating the paravertebral skin at a level T10, the response is abnormal compared to that from the left. MRI of dorsal cord shows a tumour, which is probably intradural but extra-medullary arising from the right, almost completely occluding the spinal canal at T5 level and displacing the cord. We think this clinical case is extremely interesting as it seems to confirm the usefulness of the technique of thoracic DSEPs by unilateral stimulation which has not previously been described in the literature. The potential evoked by posterior tibial nerve stimulation shows only impairment of conduction along the central somatosensory pathway, without defining the level. The thoracic DSEP shows a medullary pathology, it identifies the level, and manifests its lateralization with unilateral stimulation.

DISCUSSION

The approach of clinical neurophysiology to the study of the spinal cord is often unsatisfactory, especially when we have to deal with segmental lesions. With the techniques which are generally used in the routine clinic the neurophysiologist is able to document quite well a possible block of conduction along the central afferent pathways, but less well the level where this block is located. A good neurologic examination, together with a suitable knowledge of the segmental architecture of spinal cord is superior to instrumental investigation. From the instrumental point of view, the recording of the spinal response (electrospinogram) during the acquisition of the SEP by tibial nerve cannot often be used since it is burdened with artifacts, especially when it is done with patients who scarcely cooperate, who have a strong muscular constitution or are obese. For neurophysiologists it is therefore important to follow a method which can go beyond the limits of the spinal electrogram, which can be easily performed and might help in the diagnosis of medullar pathologies.

The technique of the potentials evoked by thoracic dermatomal activation may meet our requirements and our studies show it is a feasible method which supplies repeatable, quantifiable and standardized data. Besides it allows us to obtain signals from the scalp even with unilateral stimulation and for this reason it is possible to study 'lateral' pathologies, which is absolutely impossible with the recording of the spinal electrogram.

Of course different questions are still unsettled and they should be faced during the progress of research: for example it will be necessary to define with a higher precision the nature of the structures we are stimulating, to establish with a better quantitative accuracy what is the error due to displacement between medullary segments and vertebral bodies; again the degree of overlapping or thoracic dermatones.

Even though there are these limits, we think that the thoracic DSEP may help to show any medullary damage with a good degree of approximation, especially when the clinician has a well founded suspicion about them and the neuroimaging is of no help.

However, the use of this technique on a broad scale is a goal which is premature at present: the extension of data in the normal population and a careful validation of the method in pathological conditions are still necessary for this method and for its useful, correct application.

References

1. Scarff T, Dalmann D, Toleikis JR, Bunch WH. Dermatomal somatosensory evoked potentials in the diagnosis of lumbar root entrapment. Surg Forum. 1981;32:489–491.
2. Aminoff MJ, Goodin DS, Barbaro NM, Weinstein PR, Rosemblum ML. Dermatomal somatosensory evoked potentials in unilateral radiculopathy. Ann Neurol. 1985;17:171–176.
3. Katifi HA, Sedgwick EM. Somatosensory evoked potentials from posterior tibial nerve and lumbosacral dermatones. Electroenceph Clin Neurophysiol. 1986;65:249–259.
4. Katifi HA, Sedgwick EM. Evaluation of dermatomal somatosensory evoked potentials in the diagnosis of lumbo-sacral root compression. J Neurol Neurosurg Psychiatry. 1987;50:1204–1210.
5. Pop PHM, Oepkes CT, Notermans SLH, Stegelman DF. Dermatomal somatosensory evoked potentials of the lumbar and cervical roots. Eur Arch Neurol Sci. 1988;238:22–27.
6. Ciucci G, Callegarini C. Potenziali evocati dermatomerici. In: Arrigo A, Carreras M, De Grandis D, Serra G, eds. Elementi di elettromigrafia e neurofisiologia clinica. Roma: Marrapese; 177–197.
7. Slimp JC, Rubner DE, Snowden ML, Stolov, WC. Dermatomal somatosensory evoked potentials: cervical thoracic and lumbosacral levels. Electroenceph Clin Neurophysiol. 1992;84:55–70.
8. Goodridge A, Eisen A, Hoirch M. Paraspinal stimulation to elicit somatosensory evoked potentials: an approach to physiological localization of spinal lesions. Electroenceph Clin Neurophysiol. 1987;68:268–276.

12
The effect of changes in epidural temperature on spinal evoked potentials during scoliosis surgery

J. E. FITZGERALD, C. J. GRIFFITHS, K. W. MITCHELL,
M. A. LEONARD, M. GIBSON and C. K. McKNIGHT

INTRODUCTION

Intraoperative monitoring of spinal evoked potentials (SpEPs) during corrective surgery on scoliosis patients is now a well accepted practice in the UK[1] and is performed so that serious cord trauma occurring during the course of surgery may be rapidly identified.

Although the details of monitoring techniques vary between centres, generally the technique involves a transcutaneous stimulation of the common peroneal or posterior tibial nerve and measurement of the somatosensory potential rostral to the operative site, either by means of an electrode positioned in the epidural space or by scalp electrodes placed over the somatosensory cortex[2]. Recordings are made during the course of the operation and any significant changes are assessed in relation to the surgical procedure.

In general a reduction in waveform amplitude of at least 50% or complete loss of one component of the waveform is regarded as significant and highly indicative of spinal cord impairment[2]. Increases in latency of some or all of the components may also indicate a problem in spinal conduction[2,3].

However, Jones et al.[2,4] commonly observed a gradual increase in SpEP latency in the absence of any apparent cord damage during the course of the surgical procedure. In a previous report by our group[5], a study of 17 patients showed a mean increase in latency of 0.34 ms per hour, with a maximum in excess of 1 ms over 2 hours. Possible reasons for such changes could include: temperature, low blood pressure, anaesthesia, neurological fatigue/adaptation due to repeated stimulation, and changes in electrolyte balance.

The effect of temperature on conduction velocity in peripheral nerves is well known[6], but the effect on the spinal tract pathways has not been documented. It is however likely that their temperature dependence is similar to that of peripheral nerves. The exposed spine and introduction of metalwork at room temperature could make it particularly susceptible to cooling during surgery and lead to a possible decrease in conduction velocity of the section of the cord within the exposed spine. This hypothesis was investigated by measuring epidural temperature in relationship to changes in latency of spinal evoked potentials. The possibility of a more generalized cooling of the patient, affecting the whole conduction pathway, was also investigated.

METHODS

Monitoring of SpEPs and epidural temperature using a thermocouple to give an estimate of cordal temperature in the exposed spine, was performed in 26 patients undergoing corrective surgery for scoliosis using Cotrel Dubousset instrumentation. Of these, eight were also monitored for core temperature using a nasopharyngeal temperature probe (S&W 8050 – oesophageal probe) and S&W system 8000 monitor.

Patients were premedicated by diazepam (0.2 mg/kg) approximately one hour before induction of anaesthesia. Anaesthesia was induced with thiopentone (5 mg/kg), fentanyl (2.5 μg/kg), suxamethonium (1.5 mg/kg), and morphine (0.2 mg/kg). Anaesthesia was maintained with a morphine infusion (0.03 mg/kg/hour) and isoflurane (1–1.5%) with nitrous oxide and oxygen (60%/40%). Systolic blood pressure was maintained at 75–85 mmHg with labetalol, with a mean blood pressure of approximately 60 mmHg. End tidal CO_2 was maintained at 4 kPa. No competitive relaxants were used, to allow accurate localization of peripheral nerves and to demonstrate effective stimulation during the operation.

After induction of anaesthesia, the common peroneal nerve was localized near the head of the fibula of both legs, using a bipolar electrode positioned to elicit maximum dorsi-flexion of the foot when stimulated with a 0.1 ms current pulse at 20 Hz. Silver/silver chloride electrodes were then attached to these sites, and an isolated ground electrode strapped to each leg proximal to the stimulating electrodes. The current giving maximal dorsi-flexion of the foot was used as the stimulus during data acquisition. A platinum tipped recording electrode (diameter 1.3 mm) was placed in the epidural space rostral to the operative site and a stainless steel reference hook electrode inserted in muscle at the same vertebral level.

For data acquisition, typically 256–1024 epochs were recorded in each average using a Medelec Mistral, with the amplifier filters set to 100–3000 Hz. A repeat measurement of right and left leg stimulation was performed at regular intervals throughout the operation and at crucial stages. Figure 1 illustrates a typical recording, and the labelling of the waveform used.

Epidural temperature was monitored every 10–15 minutes using a copper-constantan thermocouple in a thermally sealed polythene tube (0.5 mm diameter) inserted alongside the recording electrode. The L1 and L2 latency change and temperature change from the start of monitoring was calculated.

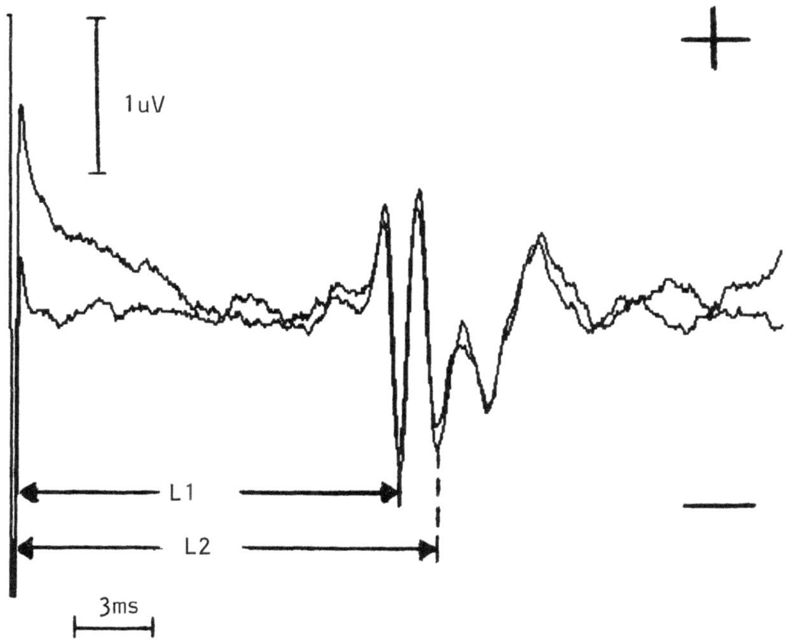

Figure 1 Illustration of a typical somatosensory evoked response recorded with an epidural electrode placed in the mid-thoracic region. Two independent recordings superimposed

RESULTS

Figure 2a shows data for a single patient where epidural temperature was observed to fall from 35.0°C to 34.0°C in association with an increase in L1 latency of 0.8 ms over the period of measurement. The L2 latency showed a similar pattern. In a similar way, for another patient (Figure 2b), the temperature was observed to fall and the latency to increase for the first half of the measurement period, but then both the temperature and latency stabilized until the end of the recording period.

The change in L1 latency data for the whole group are shown in Figure 3a and in Figure 3b the same latency data are plotted against epidural temperature changes. In general terms 16 of the 26 patients showed an overt increase in L1 latency (>0.3 ms) associated with a decrease in epidural temperature over the period of measurement. In nine patients no measurable temperature drop, and no overt L1 latency changes were observed over the measurement period. One patient showed no overt increase in latency despite an apparent initial decrease followed by an increase in epidural temperature. In all cases, L1 and L2 latency changes showed a similar pattern. Linear regression analysis performed on the data of Figure 3b showed an overall increase of 0.5 ms per °C drop, with a correlation coefficient of 0.65.

For the patient whose results are shown in Figure 2a, it can be seen that epidural temperature change and core temperature changes track each other, but with an offset of approximately 0.5°C. This was a general trend for the

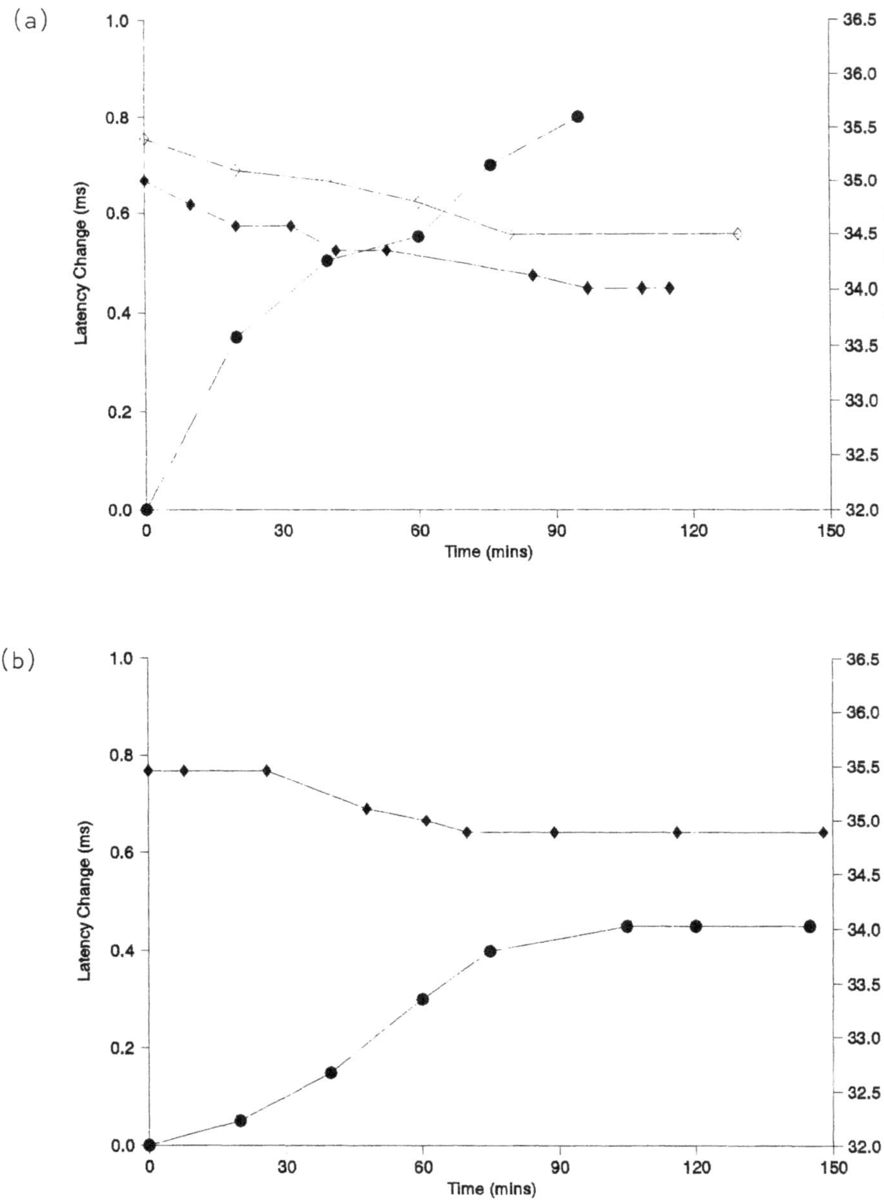

Figure 2 L1 latency change (●) (ms) (mean of right and left leg), epidural (◆) and core (◇) temperature change (°C) against time (mins) for two individual patients

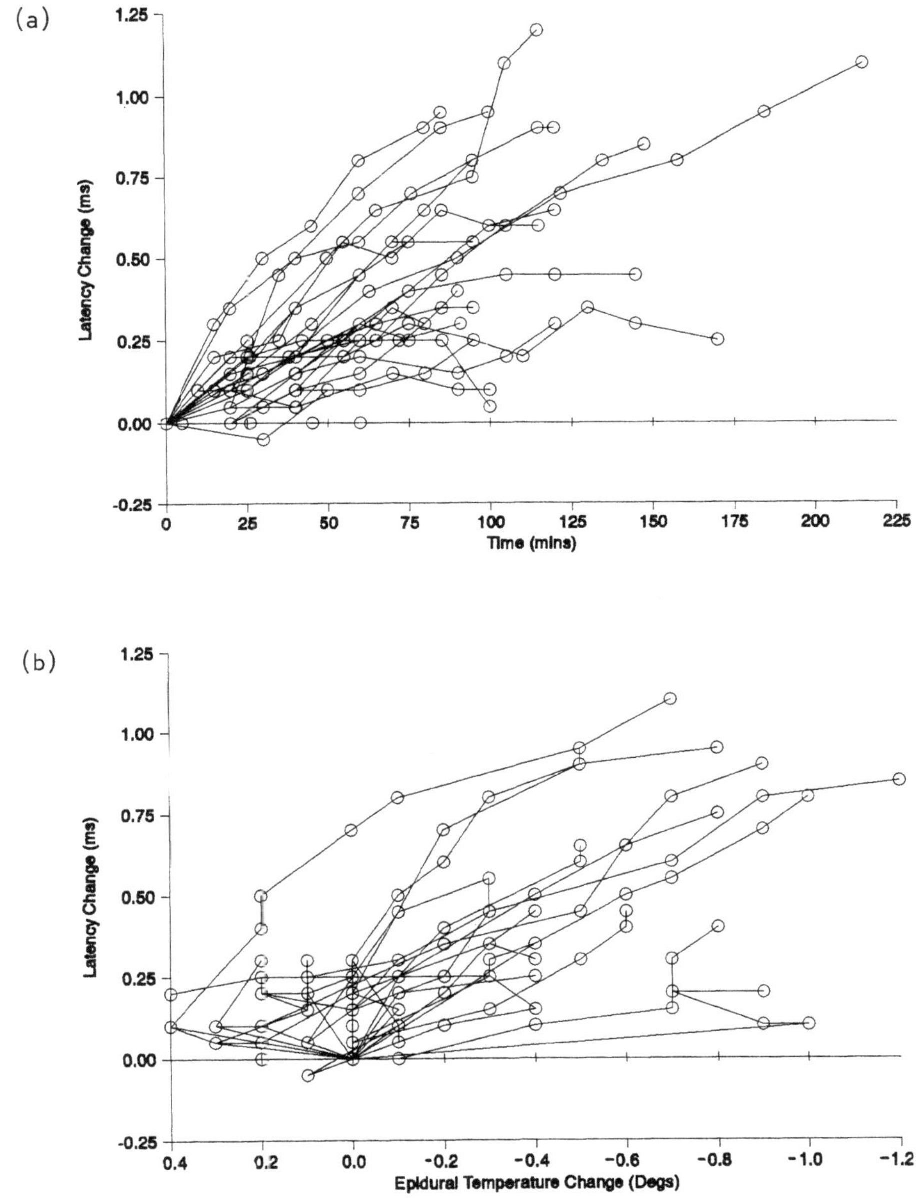

Figure 3 L1 latency change (ms) (mean of right and left leg) against (a) time (mins) and (b) epidural temperature change (°C) for 28 patients. (O–O individual patient data)

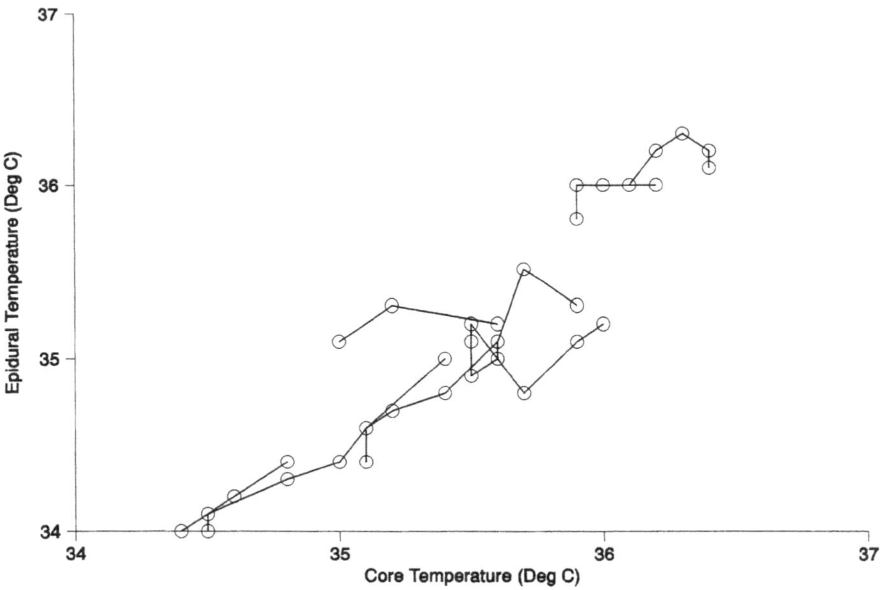

Figure 4 Epidural temperature (°C) against core temperature (°C) for eight patients

eight patients whose core temperature was monitored, as demonstrated in Figure 4 which illustrates epidural temperature plotted against core temperature for this group. A linear regression (gradient 1.15, intercept 5.7) gives an approximate difference of 0.4°C over the range of interest with a correlation coefficient of 0.93.

DISCUSSION

Of the possible factors that may influence waveform latency, the temperature effect was investigated as its effect on peripheral nerve conduction velocity is well documented[6].

The results imply a relationship between a drop in epidural temperature and an increase in latency. The data illustrated in Figure 3b demonstrate a mean increase in latency with epidural temperature of 0.5 ms per °C drop. It is interesting to consider this finding in the context of how different parts of the pathway are cooled. Initially it was anticipated that cooling of the conduction pathway would occur predominantly in the region of the exposed spine, as indicated by the changes in epidural temperature. However, the data of Figure 4 demonstrate a strong association between core temperature and epidural temperature for the group data with the latter on average 0.4°C lower over the range of interest. The association is also true for most of the individual patients in this figure. This suggests that a stable thermal gradient between core temperature, epidural temperature and room temperature had usually established by the time measurements commenced. Thus, it seems likely that cooling of the whole pathway contributed to latency changes and epidural temperature changes were representative of this.

If this is the case, a simple model allows the expected changes in latency for a given overall temperature change to be calculated. Assuming that the conduction velocity of central nerves have the same temperature dependency as for peripheral nerves, that is a reduction in conduction velocity of $2\,ms^{-1}$ per °C drop[6] and assuming overall conduction velocities[4] of $50\,ms^{-1}$–$60\,ms^{-1}$, the expected increase in latency for a waveform, whose L1 latency was of the order of 15 ms, may be deduced to be approximately 0.5–0.6 ms per °C drop.

The close agreement of the prediction of this simple model with the mean temperature dependence of latency (0.5 ms per °C) for the data in Figure 3b plus the additional evidence of Figure 4, suggest that temperature changes can account for much of the latency change often seen during monitoring and that these changes are mainly the result of overall temperature changes affecting the whole of the conduction pathway.

However there is wide variability in Figure 3b which also requires consideration. The low correlation coefficient (0.65) is partly due to the data being heavily weighted around zero (nine patients did not show significant temperature or latency changes), but there is also large variability between patients. A number of factors may contribute to this. Although the finding of Figure 4, for most patients, shows a strong correlation between epidural and core temperature, some patients do not. When the epidural and core temperatures do not change similarly, the length of the exposed spine is likely to contribute to the variability as the epidural temperature measurement will be representative of a greater or lesser proportion of the whole pathway. Other physical factors that could also contribute are: patient size, room temperature fluctuations, location of the epidural temperature probe and effects of the inserted metalwork. The epidural temperature probe, placed at the end of the operating site, often only a few centimetres from the heart, could register a higher temperature than it would if placed in a more exposed region of the spine.

Other factors unrelated to epidural temperature can not be excluded, e.g. blood pressure, anaesthesia, adaptation or fatigue of the neuronal pathways due to repeated stimulation, or changes in electrolytic balance. However, it is thought unlikely that blood pressure could account for the trend in latency increase observed throughout the operation as a constant blood pressure of 60 mmHg was maintained throughout the procedures. Patients were also maintained at constant levels of anaesthesia throughout the operative procedure, but the possibility that the anaesthetic agent redistributed from one body compartment to another contributing to such observed latency changes cannot be ruled out. Additionally adaptation or fatigue of the neuronal pathway due to repeated stimulation and its concomitant effect on SpEPs cannot be excluded. This is currently being investigated in a supplementary study.

In conclusion it can be said that the results show a strong association between an increase in latency and a drop in temperature of the conduction pathway, supporting the hypothesis that there is a causal relationship between the two. Epidural temperature changes reflect changes in the overall pathway. In the context of spinal cord monitoring the study highlights the need to distinguish between sudden changes in latency that may indicate cord trauma and those which appear progressively over the course of the operation. The latter, if interpreted accordingly, are of trivial consequence and reinforce the need to make regular and systematic measurements throughout the surgical procedure with particular emphasis on those elicited immediately prior to and following distraction.

References

1. Jackson RK. The current state of monitoring in Britain. J Bone Joint Surg. 1992;74B (suppl.1):84.
2. Jones SJ, Edgar MA, Ransford AO, Thomas NP. A system for the electrophysiological monitoring of the spinal cord during operations for scoliosis. J Bone Joint Surg. 1983; 65-B:134–139.
3. Bradshaw K, Webb JK, Fraser AM. Clinical evaluation of spinal cord monitoring in scoliosis surgery. Spine. 1984;9:636–643.
4. Jones SJ, Edgar MA, Ransford AO. Sensory nerve conduction in the human spinal cord: epidural recordings made during scoliosis surgery. J Neurol Neurosurg Psychiatry. 1982;45:446–451.
5. FitzGerald JE, Griffiths CJ, Mitchell KW, Leonard MA, Gibson M, McKnight CK. Spinal evoked potential changes observed in the absence of cord trauma during scoliosis surgery. J Bone Joint Surg. 1992;74B(suppl.1):87.
6. Buchthal F, Rosenfalck A. Evoked action potentials and conduction velocity in human sensory nerves. Brain Res. 1966;3:1–122.

13
Threatened spinal cord damage due to Luque's wires

A. J. GALE, E. M. SEDGWICK and R. JACKSON

INTRODUCTION

During a six-year period spinal cord monitoring has been done on a total of 121 cases during spinal fixation. In 105 cases (104 of which were idiopathic scoliosis) a post-operative neurological assessment was recorded and these were analysed for intra-operative monitoring events. A monitoring event is defined as a fall in the epidurally recorded potential to 50% or less of the original amplitude which could not be attributed to technical causes and about which the surgeon was informed.

METHODS

Epidural recordings were taken from a platinum tipped wire electrode introduced into the epidural space at the rostral end of the incision and passed a further 2–4 segments rostrally. The reference electrode was on the nearby muscle. Averages of the potential following alternate right and left posterior tibial nerve stimulation were recorded. A fall in amplitude of either potential of 50% or more was notified to the surgeon and a note made of the procedure being undertaken at that time. Monitoring ceased as the wound was finally closed.

RESULTS

A positive event, a decrement exceeding 50%, occurred in 16 cases. A formal audit of the monitoring procedure is shown in Table 1[1]. The definitions of the terms used in Table 1 are:

Sensitivity: Probability that test or procedure result will be positive when the disease is present.

Table 1 Binary table – audit of intraoperative epidural recording

Test result	*Neurological Deficit* Present	Absent
Decrement >50% Test positive	4	12
Decrement <50% Test negative	0	89
Total	4	101
Grand total	105	

Sensitivity 100%
Specificity 88%
Positive predictive value 25%
Negative predictive value 100%

Specificity: Probability that the test or procedure result will be negative when the disease is not present.

True positive rate: Sensitivity (percent).

False positive rate: 100 minus specificity (percent) = probability that test or procedure result will be positive when the disease is not present.

False negative rate: 100 minus sensitivity (percent) = probability that test or procedure result will be negative when the disease is present.

Negative predictive value: Probability that the disease is not present when the test or procedure is negative.

The negative predictive value was 100% – there were no false negatives. The positive predictive value was only 25% – that is, only four of the 16 cases with a monitored 'event' had evidence of subsequent neurological deficit.

Table 2 shows the reduction of amplitude of the epidural potential taking the initial potential as 100% and the final potential at wound closure. The table includes the 12 'false positive' patients only; those who had a fall in amplitude to 50% or less and had no post-operative neurological deficit. In five cases, the initial fall in amplitude remained unchanged for the remainder of the operative period. In the remainder, the amplitude improved to 40% or better, but in only one did it fully recover to the original value. Wake-up tests were performed in only three of these patients – we have abandoned this procedure except for occasional checks.

Table 3 shows four patients who had neurological deficits post-operatively. Only in Case 4 did the deficit persist and it recovered fully in 6–8 months. The other patients showed temporary sensory symptoms lasting up to 10 days. The fall in amplitude was to zero, 25% or 30%, and two of the patients with temporary sensory symptoms recovered to 80% and 70% of the initial amplitude.

Tables 2 and 3 show the surgical activity at the time of the monitoring event. In 13 out of the 16 cases, sublaminar wiring was being done, either passing the wires or tightening them (Table 4). Wiring was done in all 16 cases. In no case was there an anaesthetic explanation for the fall in potential.

Table 2 False positive epidural events. These patients had a fall of $>50\%$ in the epidural potential but no neurological sequelae

Case	*Age and sex*	*Lowest epidural potential amplitude Initial = 100%*	*Surgical activity at time of drop*	*Surgical intervention*	*Amplitude at closure*
1	16♂	50%	Passing wires	Nil	50%
2	9♀	0%	Tightening wires	Nil	100%
3	14♂	40%	Tightening wires	Rectangle removed realigned and re-wired	65%
4	20♂	50%	Passing wires	Nil	75%
5	16♀	50%	Tightening wires	Wire slackened off	70%
6	16♀	50%	Inserting rod and tightening wires	Nil	70%
7	12♀	35%	Compression	Nil	35%
8	18♀	50%	Distraction	Relaxation	50%
9	19♂	0%	Tightening wires	Nil	40%
10	51♀	50%	Passing wires	Nil	80%
11	11♀	40%	Tightening wires	Nil	40%
12	14♀	40%	Passing wires	Nil	40%

Table 3 True positive epidural events. Patients with a fall of $>50\%$ in the epidural potential who had post-operative neurological sequelae

Case	*Age and sex*	*Lowest epidural potential amplitude Initial = 100%*	*Surgical activity at time of drop*	*Surgical intervention*	*Epidural potential amplitude at closure*	*Post-op neurosurgery*
1	19♂	0%	Tightening wires	Rectangle removed – re-contoured	80%	Partial sensory loss L1–L5 left leg
2	13♀	30%	During wiring	Nil	70%	Paraesthesia front both legs for 6 days
3	17♀	25%	Tightening wires	Loosened wires	25%	Hyperaesthesia front left thigh over 10 day period
4	34♂	0%	2 minutes post reduction of dislocation	Laminectomy at fracture site	0%	Residual weakness right foot Intermittent parathesia right thigh

Table 4

Surgical activity at time of monitored event	
Passing sublaminar wires	5
Tightening wires	8
Distraction	2
Compression	1
Hypotension, etc	0
Total events	16

Blood pressure remained stable in all. Case 4 who was left with weakness is described in an earlier chapter (Sedgwick, this volume).

SUMMARY

In four patients, decrements of 70% or more of the original amplitude of the epidural potential were associated with neurological deficits which were temporary in three, but more serious in one. It may be significant that only this case had an absent potential at the end of surgery. The majority of monitoring events, 13/16, were associated with Luque's wires. It has to be recommended that surgeons consider carefully the use of sublaminar wires and try to find an alternative.

References

1. Gruner PF, Mayewski RT, Mushlin AI, Greenland P. Selection and interpretation of diagnostic tests and procedures. Annals of Internal Medicine. 1981;94:553–600.

14
Comparison of common peroneal and posterior tibial nerve stimulation for routine spinal cord monitoring during surgical correction of scoliosis

C. J. GRIFFITHS, J. E. FITZGERALD, K. W. MITCHELL, M. GIBSON and M. A. LEONARD

INTRODUCTION

Spinal cord monitoring is routinely practised in over 50% of UK centres performing surgical correction of scoliosis[1]. There is much in common in the technique used by different centres, most using the somatosensory evoked potential technique, but one notable difference is the site of stimulation used. Some use the common peroneal (CP) nerve[2] and others the posterior tibial (PT) nerve at the level of the knee[3].

The relative merits of the two sites for routine monitoring were assessed by applying stimulation electrodes to both sides, stimulating the common peroneal and posterior tibial nerves in turn and comparing the evoked epidural potential (EPs).

METHOD

Seventeen patients undergoing routine somatosensory spinal cord monitoring during surgical correction of scoliosis were studied. Stimulation electrodes were applied to the leg over the CP and PT nerves using the following technique.

Common peroneal nerve

The site was located by feel where the nerve passes over the head of the fibula. The optimal response to stimulation (20 Hz, 0.1 ms pulses) was checked with an exploratory probe (Medelec 16893) and disposable electrodes (Medicotest A-10-VS; 6 mm diam) were applied to the site, cathode cephalic. The currents for threshold and maximal motor responses were found and the latter used for monitoring.

Posterior tibial nerve

The site was located visually (a few mm above the popliteal fossa, lateral to the midline). The response to stimulation at this site was confirmed using the exploratory probe and electrodes (Medicotest Q-10-A; diameter 19 mm) were applied about 30 mm apart, cathode cephalic. The currents for threshold response and approximate maximal response (less well defined for this nerve and stimulus duration sometimes increased to 0.3 ms) were found and the latter used for monitoring.

Monitoring

The spinal SSEPs were recorded using a monopolar epidural electrode (Medelec 53523), usually positioned between T3 to T6, and an indifferent hook electrode in nearby muscle, with an amplifier bandwidth of 100 Hz–3 kHz. The averaged EP for 100–1000 stimuli (Medelec Mistral system) was recorded twice at each of the four sites (CP and PT for right and left legs). The set of recordings was repeated every 15–20 minutes and at critical times during surgery. In some patients, stimulation currents below and above the established monitoring currents were used to assess the effect on EP amplitude.

RESULTS

For the whole group of patients, the means ($\pm$ standard deviations) of the threshold stimulation currents were 14 $\pm$ 10 mA for the CP nerve; 23 $\pm$ 14 mA for the PT nerve. For the maximal currents, as used for monitoring, they were 27 $\pm$ 16 mA for the CP nerve; 49 $\pm$ 17 mA for the PT nerve. (Note: To stimulate at levels greater than 50 mA – the stimulator's maximum output – for 0.1 ms, the duration was increased to 0.3 ms. For the above calculations, where applicable, an equivalent current at 0.1 ms was estimated on the basis that stimulation energy is proportional to the product of the current squared and time. For example, 50 mA for 0.1 ms is approximately equivalent to 30 mA for 0.3 ms.)

In those patients for whom currents above and below the established monitoring current were used to assess the effect on EP amplitude, no additional increase in amplitude was seen for the CP nerve. In most cases this was also true for the PT nerve, but in two patients, increasing the current did give a further increase in EP amplitude of later components.

Figure 1 illustrates EPs from a typical patient for stimulation of the CP and PT nerves of their right leg. The morphology of two negative peaks followed by a third, slower negative peak is typical of EPs found at the upper

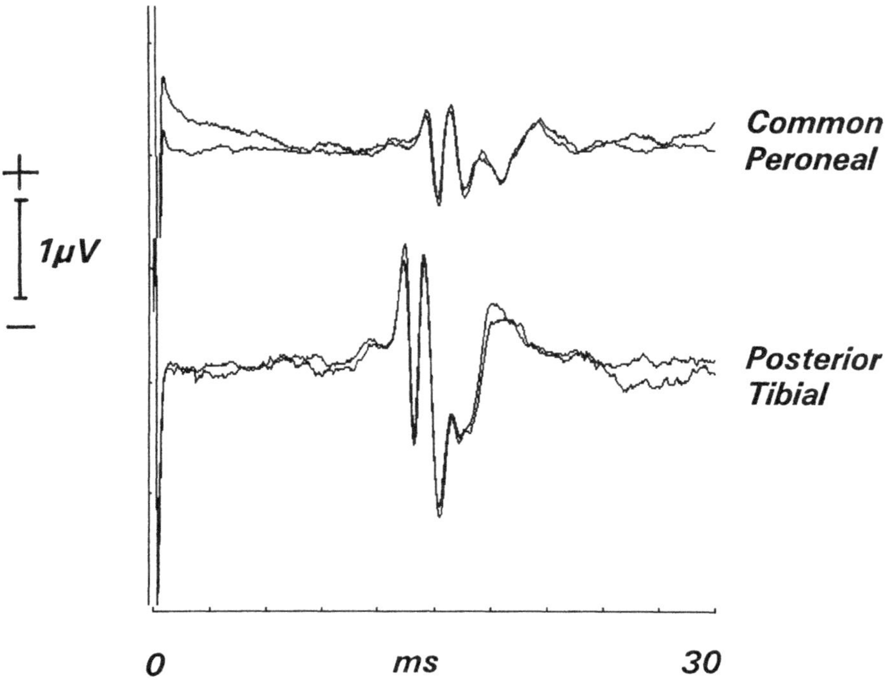

Figure 1 Typical EPs to stimulation of the common peroneal (CP) nerve (upper traces) and the posterior tibial (PT) nerve (lower traces). These waveforms are for the right leg of one patient

thoracic level and is very similar for the two stimulation sites. However, the amplitude is larger and the latency shorter for PT as compared to CP.

Figure 2 shows the measurements made on the waveforms for each set of recordings and includes a table of the mean ($\pm$ standard deviation) amplitude and latency measurements for CP and PT stimulation across all patients, combining right and left leg data. The increased latency of almost 2 ms for CP stimulation is consistent with the increased length of the pathway for this site.

Within each patient, the A1 and A2 amplitude measurements (see Figure 2) for PT stimulation were divided by those corresponding to CP stimulation. This was done separately for the right and left legs and the four values obtained were averaged to give an overall PT/CP ratio for that patient. Figure 3 shows the ratios so obtained for all patients, a clear difference being observed in favour of PT potentials, the mean ratio being 2.5, standard deviation (s.d.) ± 0.7.

DISCUSSION

The study, and experience gained while performing it, revealed clear differences between the two sites in practice. PT nerve stimulation was

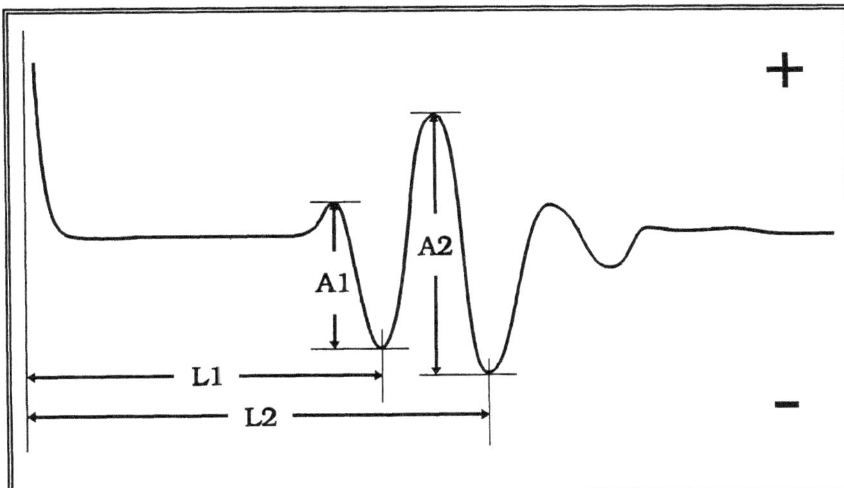

	A1 (μV)	A2 (μV)	L1 (ms)	L2 (ms)
CP	-0.95 (±0.52)	-0.69 (±0.33)	15.8 (±1.9)	17.4 (±2.0)
PT	-2.08 (±1.56)	-1.64 (±0.72)	14.4 (±1.8)	16.1 (±1.9)

Figure 2 Idealized upper thoracic EP waveform showing the amplitude (A1 and A2) and latency (L1 and L2) measurements made for each recording. For each patient, the mean of these measurements was calculated for stimulation of the CP nerve and the PT nerve (combining right and left leg data) and the overall mean and standard deviation for the 17 patients are shown below

advantageous in producing considerably larger EPs and accurate localization of the nerve was not essential to obtain them. However, the disadvantages were that the maximal and threshold motor responses to stimulation were less well defined and localization of the site using an exploratory probe was less precise.

CP nerve stimulation, on the other hand, revealed well defined maximal and threshold responses and localization of the nerve by feel and exploratory probe was more precise. Against this, a serious disadvantage was that the EPs were of much lower amplitude which sometimes made obtaining good recordings difficult.

One possible cause of the difference in amplitude measurement was the larger size of electrode used at the PT site. However, this seemed unlikely as increasing the stimulation current of the CP nerve above the level established for monitoring using the technique described did not increase further the EP amplitude in any of the patients assessed in this way. This, and the accuracy with which the CP nerve could be localized, made it unlikely that a larger electrode would allow more effective neuronal stimulation. The smaller electrode gave a higher current density for the same total current and was used to be consistent with our earlier practice. The larger electrode used for PT stimulation gave a greater area of stimulation which was taken

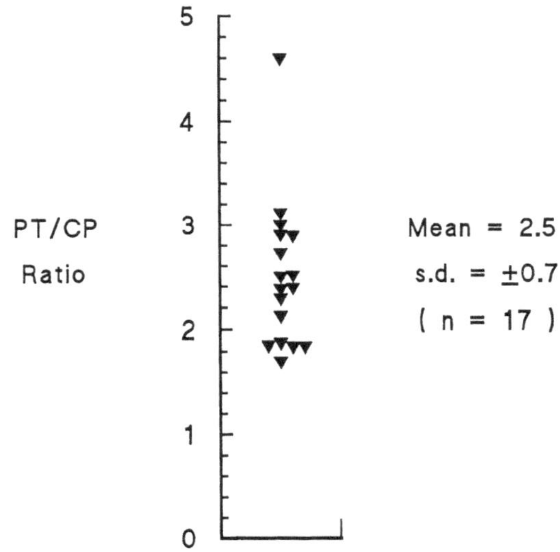

Figure 3 Ratio of amplitude for PT nerve stimulation to amplitude for CP nerve stimulation. Each point is the mean for four estimates for each patient: A1 and A2 for both left and right leg

to compensate for the lack of precision in localizing the nerve but required a higher total current to give an equivalent current density. Further investigation of the effect of size, shape and position of the PT stimulation electrodes on EP amplitude is warranted.

CONCLUSION

Obtaining a response to posterior tibial stimulation which is more than double the amplitude of common peroneal stimulation allows potentials of comparable signal to noise ratio to be obtained in less than a quarter of the time. Additionally, where noise and artefacts occur, signal quality is invariably better which is sufficient justification for preferring posterior tibial nerve stimulation, despite its less well defined localization and motor responses.

References

1. Jackson RK. The current state of cord monitoring in Britain. In: Proceedings of the combined meeting of British Society and Nordic Scoliosis Society, Newcastle, June 1991. J Bone Joint Surg. 1992;74B(suppl.1):84..
2. Bradshaw K, Webb K, Fraser AM. Clinical evaluation of spinal cord monitoring in scoliosis surgery. Spine. 1984;9:636–643.
3. Jones S, Edgar MA, Ransford AO. Sensory nerve conduction in the human spinal cord: epidural recordings made during scoliosis surgery. J Neurol Neurosurg Psychiatry. 1982;45:446–451.

15
An explanation of the high 'false-positive' rate of lower-limb SEPs in descending aorta surgery

J. M. GUÉRIT, P. Y. ETIENNE and R. DION

INTRODUCTION

The incidence of paraparesis or paraplegia following surgical repair of aneurysms of the thoracic or thoraco-abdominal aorta has been reported to be in the range of 2% to 25%[1]. These complications are likely to occur in two circumstances: first, too long a period of aortic cross-clamping associated with a low distal perfusion pressure (<60 mmHg); second, the exclusion of critical intercostal or lumbar arteries. The first mechanism can be prevented by short cross-clamping time or an efficient atrio-femoral shunt. The second mechanism can theoretically be prevented by reimplantation on the graft of critical arteries. However, this procedure often has to be performed in a blind fashion and can sometimes uselessly increase the cross-clamping duration, thereby increasing the risk of the occurrence of the former mechanism. Thus, adapting the surgical procedure to on-line monitoring of the spinal cord is likely to improve neurological outcome[2].

The use of lower-limb SEPs as a spinal-cord monitoring tool in these operations has been criticized for three main reasons: the insensitivity of SEPs to the central-cord syndrome, the inability to demonstrate any statistical influence of SEP monitoring on patient outcome, and too high a rate of false positives, that is, SEP alterations not followed by any neurological deficit[1,3]. As regards the first criticism, well-documented false-negative reports are extremely rare, if any. To explain this finding, Nuwer[4] hypothesised that, although SEPs can theoretically remain stable in the presence of spinal cord ischaemia, this degree of ischaemia would not be so severe as to result in a significant postoperative neurological deficit. Conversely, a high degree of ischaemia, sufficient to result in clinical sequelae, would give rise to significant SEP alterations. As regards the second criticism, it has never been clearly stated to what extent surgical teams arguing against any statistical benefit of monitoring actually took neurophysiological data into account in order to adapt their surgical strategy. As regards the third criticism, the aim of the present paper is to demonstrate that many of the so-called 'false positives' are not true 'false positives' but correspond to other real events not primarily involving the spinal cord.

MATERIAL AND METHODS

Twenty-seven patients were studied. The aetiologies of their aortic lesions were traumatic rupture of the aortic isthmus (n = 8), thoracic aneurysms (n = 11), thoraco-abdominal aneurysms (n = 7), and abdominal aneurysm (n = 1). Whenever possible, an atrio-femoral shunt was installed before aortic cross-clamping in order both to prevent distal ischaemia and to maintain SEP monitoring for signs of segmental spinal ischaemia. The shunt was actually used in 21/27 patients. The first step of the procedure consisted in cross-clamping the aorta between the left subclavian artery and the diaphragm. The proximal suture was then performed, and no vessels were reimplanted in the absence of SEP alterations within the first 20 minutes after cross-clamping. In patients undergoing repair of thoraco-abdominal aneurysms, the proximal clamp was removed and replaced at various levels of the abdominal aorta. Re-implantation of intercostal or lumbar arteries was decided upon the basis of appearance of SEP signs of spinal cord ischaemia.

The right and the left posterior tibial nerves (PTNs) were alternately stimulated at the ankle (rate: 3.1/sec; stimulus duration: 200 μs; stimulus intensity: 39 mAmp). Recordings were performed on four channels: popliteal fossa-patella (channel 1), L1 spinous process-L5 spinous process (channel 2), C4 spinous process-linked ears (channel 3), Cz-linked ears (channel 4). Both the normal curves and the analysed parameters are shown in Figure 1. The peripheral nerve-action potential (PNAP) and N24 corresponding to the afferent volley in the cauda equina were recorded in channels 1 and 2, respectively. A negativity corresponding to the afferent volley travelling

Figure 1 *Top*: example of post-induction SEPs. A negativity at the active electrode (first electrode of each derivation) is represented as an upward deflexion. FF: lemniscal far-field, Popl: popliteal fossa, Pat: patella, L1 (L5, C4): L1 (L5, C4) spinous process, LE: linked ears. Note a slight increase in P2 latency, most probably due to anaesthesia. *Bottom*: traumatic rupture of the aortic isthmus, relative contra-indication of anticoagulation and shunt (fracture of the pelvis). Three to four SEP recordings were made after induction of the anaesthesia and before thoracotomy. The mean latencies and amplitudes of peripheral nerve action potential (PNAP) (*per*), lumbar N24 (*lumbar*), cervical N30 (*cerv*), and parietal P1 (*par*) were calculated from these recordings and constituted our baseline. In this and subsequent figures, all latencies are represented in terms of the latency difference (△ *latency*) and amplitudes in terms of the amplitude percentage (*% amplitude*) with respect to the baseline. Aortic cross-clamping was performed 43 min after the beginning of the monitoring. A 3 ms increase of parietal P1 and cervical N30 latencies was observed after 12 min and 17 min, respectively. These were followed by a similar lumbar N24 latency increase after 22 min and by an increase of the PNAP latency after 27 min. The amplitudes of parietal and cervical potentials decreased more rapidly than the amplitude of lumbar potentials. The PNAP amplitude remained stable over the period of study. This pattern is compatible with a global lower-body ischaemia due to aortic cross-clamping that affects first the spinal and, later on, the peripheral nerve

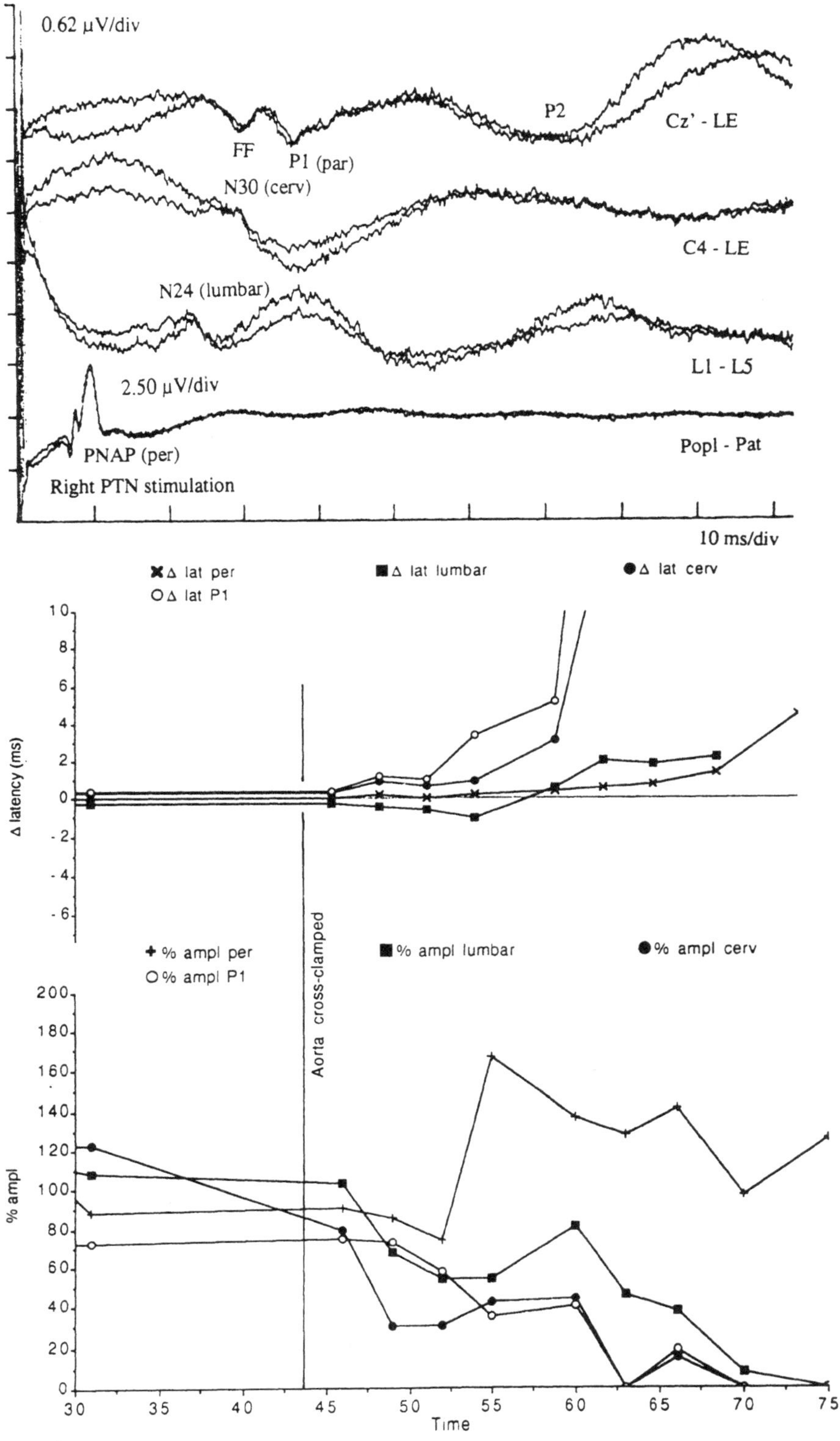

0.62 µV/div
FF
P1 (par)
P2
Cz' - LE
N30 (cerv)
C4 - LE
N24 (lumbar)
L1 - L5
2.50 µV/div
PNAP (per)
Popl - Pat
Right PTN stimulation
10 ms/div
Δ lat per
Δ lat lumbar
Δ lat cerv
Δ lat P1
Δ latency (ms)
Aorta cross-clamped
% ampl per
% ampl lumbar
% ampl cerv
% ampl P1
% ampl
Time

through the cervical cord was recorded in channel 3. The use of a linked-ear reference allowed us to record both the lemniscal P30 and the parietal 'W' complex in channel 4. Only the P1 component was considered in the parietal complex. A baseline was defined after induction by three or four recordings on each side. Two criteria were used in order to define SEP abnormalities after aortic cross-clamping: *static criteria*, that is, SEP amplitude and/or latency outside the baseline range, and *dynamic criteria*, that is, a progressive worsening of curves over two or three recordings.

RESULTS

SEP abnormalities

Five types of SEP abnormalities were defined on the basis of the differential channel-to-channel involvement and their relationship to peroperative events.

Type I (Figure 1)

A progressive increase of P1 latency was observed 3 min to 15 min (mean: 7.2 ± 3.2 min) after proximal aortic cross-clamping. This was followed after a few minutes by a similar deterioration of both the lumbar N24 and the PNAP. If the cross-clamping time was long enough and in the absence of shunt, P1 eventually disappeared after a delay ranging from 18 min to 29 min (mean: 23.7 ± 4.2 min). Clamp removal or shunt insertion gave rise to P1 reappearance after a delay ranging from 2 min to 12 min (mean: 5.4 ± 3.8 min). These alterations were observed in 11 patients irrespective of the PTN stimulated. These were only observed in the absence of an atrio-femoral shunt or after the shunt had been removed for technical reasons. These were consequently interpreted as reflecting distal ischaemia, spinal and peripheral, related to the non-perfusion of the lower body.

Type II (Figure 2)

These abnormalities were only observed after left PTN stimulation in patients in whom the atrio-femoral shunt was installed in the left common iliac artery. A progressive decrease of P1 amplitude was observed, simultaneously with a decrease of amplitude and an increase of latency of the PNAP, 15 min to 32 min (mean: 21.4 ± 8.4 min) after cannulation of the common iliac artery. Both potentials eventually disappeared after 33 min to 52 min (mean: 41.5 ± 9.0 min). The abnormalities recovered 3 min to 164 (mean: 52.5 ± 76.1 min) after shunt removal, and all patients awakened without neurological sequelae. We failed to find any evidence of SEP alteration specifically linked to aortic cross-clamping. These alterations were observed in six patients and were interpreted as reflecting peripheral nerve ischaemia to common iliac artery cross-clamping.

Type III (Figure 3)

These alterations were observed in four patients undergoing repair of thoraco-abdominal aneurysms. An atrio-femoral shunt had been installed in all of

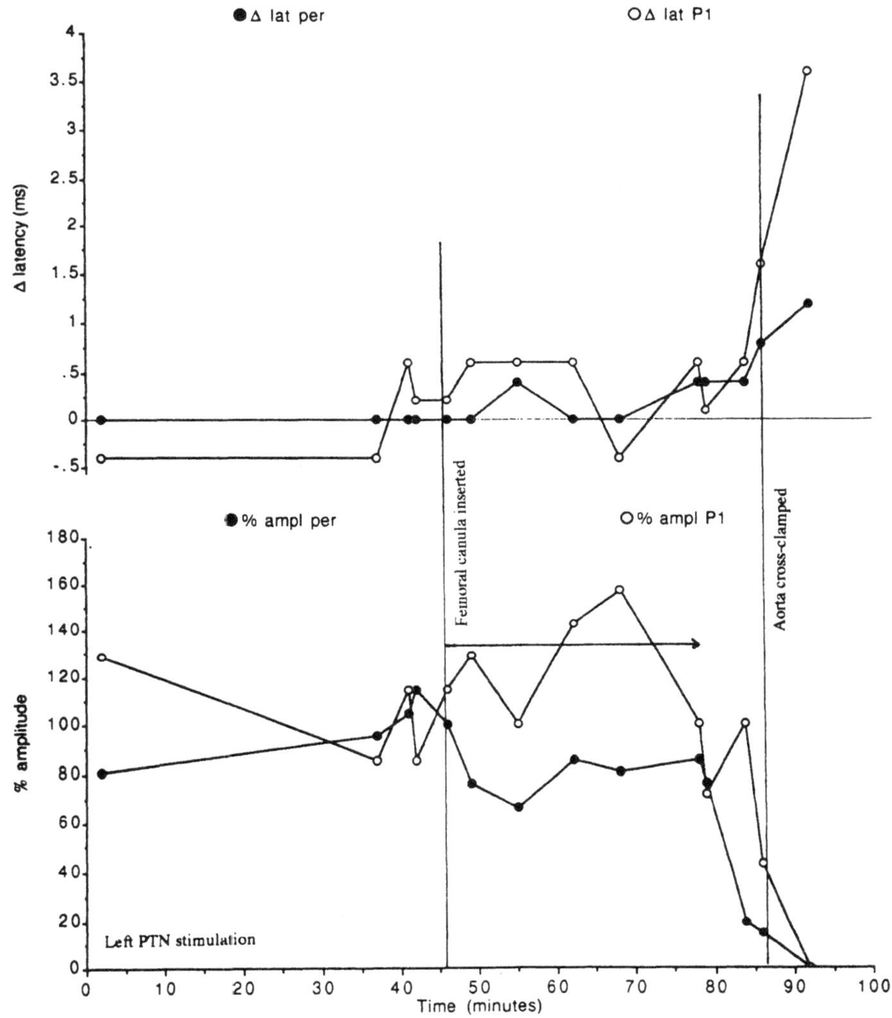

Figure 2 *Type II abnormalities* – Traumatic rupture of the aortic isthmus – same representation as in Figure 1. The femoral canula was inserted in the left common iliac artery that was cross-clamped 46 min after the beginning of the monitoring. The amplitudes of both the P1 and the PNAP simultaneously decreased 34 min after cannula insertion. This amplitude decrease was followed after 3 min by an increase of both the PNAP and P1 latencies. All these abnormalities were observed before aortic cross-clamping, so they cannot reflect spinal ischaemia. In this case, P1 abnormalities are actually explained by peripheral nerve ischaemia due to the exclusion of the left common iliac artery. This type of abnormality has always been observed after stimulation of the posterior tibial nerve ipsilateral to the common iliac artery into which the shunt was placed

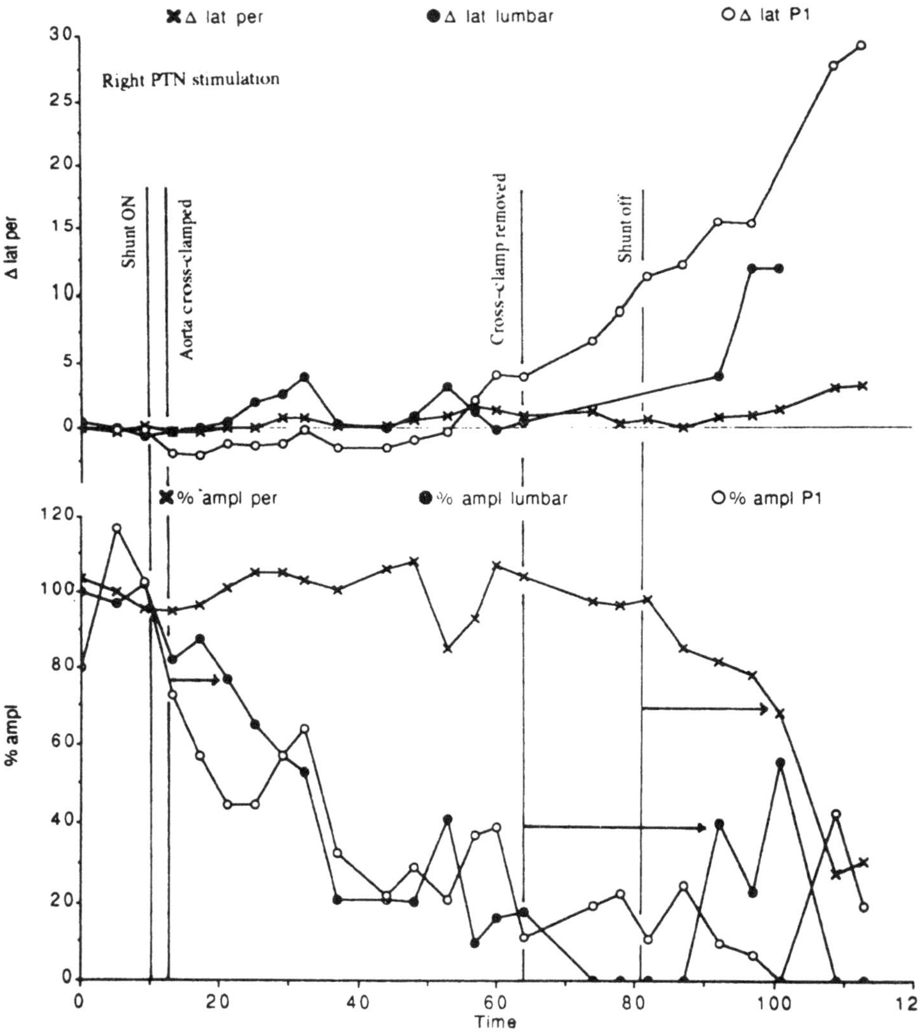

Figure 3 *Type III abnormalities* – Surgical repair of a thoraco-abdominal aneurysm. The right PTN was stimulated and no SEP alterations were found until shunt placement. For technical surgical reasons, the aorta had to be immediately cross-clamped between the left subclavian artery and the iliac bifurcation. Seven minutes after cross-clamping, an abrupt amplitude decrease of both the parietal P1 and the lumbar N24 was observed, while the PNAP amplitude remained stable over the whole period of cross-clamping. All latencies remained stable during the first 45 min after aortic cross-clamping, and a late P1 latency increase was observed. In this case, P1 alterations appeared to occur simultaneously with alterations of the lumber spinal centres but to be independent of any peripheral nerve involvement. Moreover, the absence of P1 latency increase argues against any predominant involvement of the spinal conduction pathways. This type of abnormality is interpreted as reflecting segmental spinal ischaemia consecutive to the exclusion of critical spinal feeding arteries

the patients. The first cross-clamping between the sub-clavian artery and diaphragm had not revealed any alteration. After the proximal suture, the distal clamp was placed more distally in order to re-anastomose the renal and the mesenteric arteries and to perform the distal suture of the graft. A decrease of amplitude of both the parietal and the lumbar potentials was observed after a delay ranging from 5 min to 12 min (mean: 8.5 ± 2.9 min). P1 latency remained stable in two patients and increased in two. Lumbar arteries were re-anastomosed in all four patients. P1 disappeared in two patients after 19 min and 25 min and reappeared 16 min and 2 min after clamp removal. In two patients, the clamp was removed before the P1 disappearance. We interpret these alterations as reflecting segmental spinal ischaemia linked to the exclusion of critical arteries.

Type IV (Figure 4)

This type of alteration was observed in one patient undergoing repair of a thoracic aneurysm. In order to improve the proximal suture, the surgeon cross-clamped the aorta between the left carotic and the left sub-clavian, thereby excluding the left carotid. A sudden decrease of P1 amplitude to right PTN stimulation was observed 4 min after clamping while no changes were noted in the PNAP or the lumbar and cervical activities, and the lemniscal P30. P1 disappeared 7 min after cross-clamping and reappeared 3 min after clamp removal. This alteration was interpreted as reflecting ischaemia in the left carotid artery.

Type V

In one patient, a sudden increase of P1 latency was observed without any change in the PNAP and the spinal activities after both right and left PTN stimulation (contrary to Type 4 changes that were only observed after right PTN stimulation). These changes were observed 2 min after the brisk occurrence of a 2 min long systemic hypotension (systolic blood pressure: 70 mmHg) and recovered 9 min after re-establishment of a normal blood pressure.

Patient outcome

Twenty-two patients awakened without neurological sequelae. Three patients undergoing repair of a thoraco-abdominal aneurysm died from multi-organ failure during or within a few hours after surgery. Paraparesis was noted in two patients. In one of these patients, the aorta had been cross-clamped between the diaphragm and the iliac artery for 119 min after Type III alterations had been observed. Magnetic resonance imaging revealed the presence of a lumbar centro-medullary ischaemia. Cardiac arrest occurred 20 days after surgery. The other patient was a 75-year-old man with a ruptured thoracic aneurysm who already presented distal paraparesis and sensory disturbances distal to the Th10 level before surgery. Preoperative SEPs were markedly disturbed with a major increase of the N24-P1 interpeak latency. No further SEP abnormalities were noted during surgery. Although a dense paraparesis was noted in awakening, lower-limb power progressively improved in the days following surgery, and the patient regained voluntary movements of the extremities one week after surgery. The patient died of cardiac arrest 10 days after surgery.

DISCUSSION

Most papers dealing with SEP monitoring in descending aorta surgery[1–3,5–7] only reported one-channel cortical recordings and were thus unable to differentiate SEP abnormalities consecutive to peripheral, spinal, or intracranial nervous dysfunction. Both the use of multichannel SEP and a careful integration of neurophysiological, anaesthetic, and surgical data allowed us to conclude that several factors could account for cortical SEP impairment. Five types of SEP alterations were identified.

Type I alterations were observed in all seven patients in whom no atrio-femoral shunt was used during the procedure or after it had been removed for surgical reasons. That the subcortical and cortical SEP alterations systematically preceded the PNAP alterations led us to consider these alterations as reflecting primarily ischaemia of the distal spinal cord, the peripheral nerve being involved later on. Type III alterations were observed after the exclusion of aortic segments of varying length in four patients in whom distal perfusion was maintained through an atrio-femoral shunt. These were very typically associated with the preservation of the PNAP in all three patients in whom it could be adequately recorded and were rapidly reversed by intercostal or lumbar re-perfusion in both patients in whom it could be promptly performed. We consider that Type III alterations reflect segmental spinal cord ischaemia consecutive to the exclusion of 'critical vessels'. Both Type I and Type III SEP alterations occur within the first 15 minutes after failure of correct perfusion of the spinal cord. Whenever any reversibility is likely, it manifested itself within the first 15 minutes after re-perfusion.

Type II abnormalities were observed in six cases and can be clearly distinguished from Type I and Type III on the basis of several criteria: first, we failed to find any evidence of any cause of spinal cord ischaemia that could be correlated with the abnormalities. Second, the alterations of the PNAP occurred simultaneously with the alterations of the subcortical and

Figure 4 *Type IV abnormalities* – Surgical repair of an aneurysm of the descending thoracic aorta. At the end of the operation, the surgeon cross-clamped the left carotid artery in order to revise his proximal suture. An abrupt decrease of P1 amplitude was observed 4 min after carotid cross-clamping. However, both the latency and the amplitude of the lemniscal far-field (FF) (*indicated by a cross superimposed on each trace*) remained stable throughout the procedure. We did not observe any change of the peripheral, lumbar, or cervical potentials (not shown in the figure). P1 rapidly recovered after clamp removal. These changes are interpreted as reflecting ischaemia restricted to the left carotid territory. The FF integrity implies that these alterations cannot be due to spinal or vertebro-basilar ischaemia. These alterations were observed after right PTN stimulation. Note that Type V abnormalities due to systemic hypotension were qualitatively and quantitatively similar to Type IV abnormalities but were observed after both left and right PTN stimulation

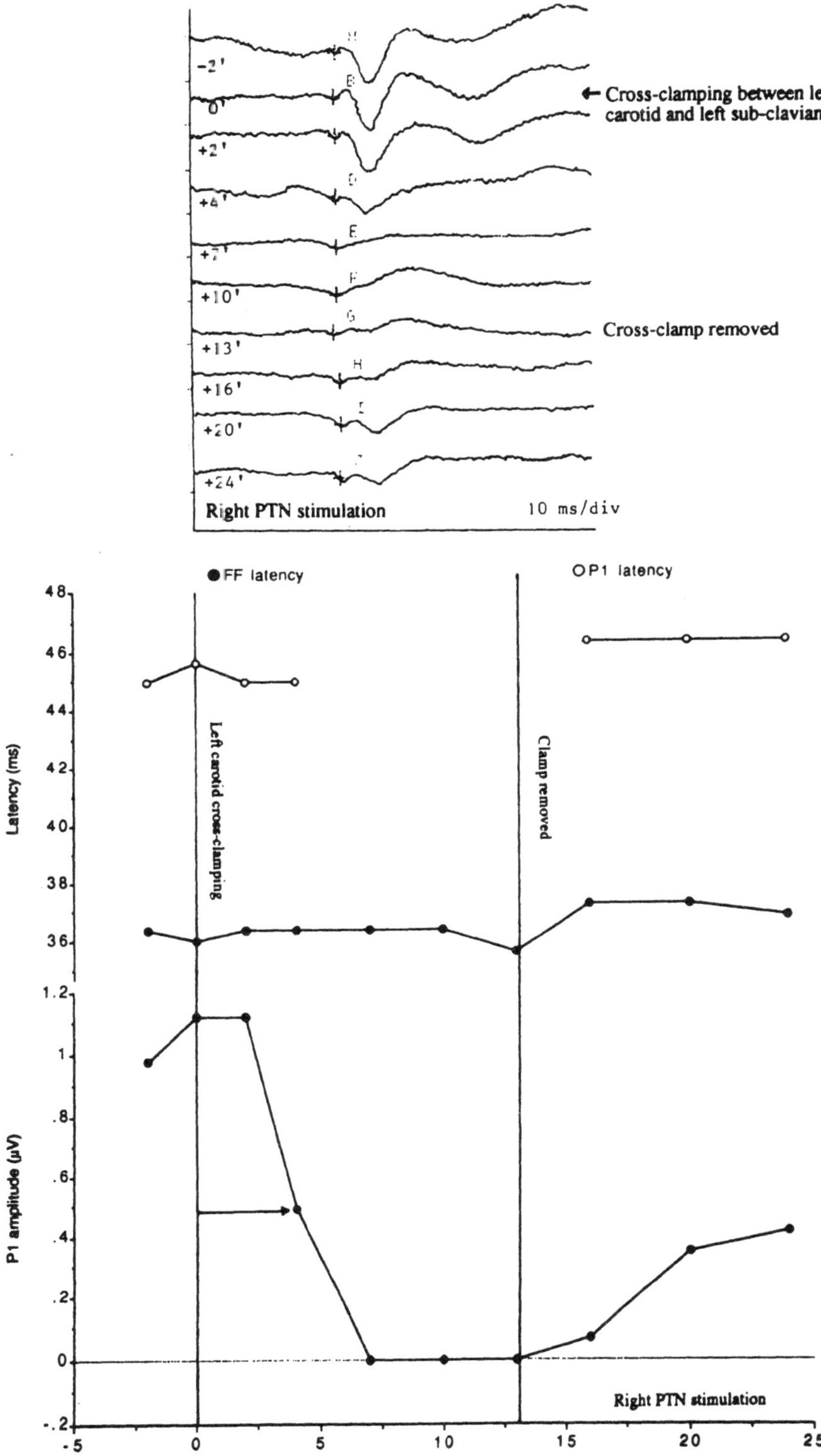

-2'
0'
+2'
+4'
+7'
+10'
+13'
+16'
+20'
+24'
← Cross-clamping between left carotid and left sub-clavian
Cross-clamp removed
Right PTN stimulation
10 ms/div
●FF latency
○P1 latency
Latency (ms)
P1 amplitude (µV)
Left carotid cross-clamping
Clamp removed
Right PTN stimulation

cortical activities. We conclude that these are actually consecutive to peripheral nerve ischaemia. This hypothesis is strengthened by the observation that Type II alterations were observed ipsilaterally to the femoral artery into which the atriofemoral shunt was placed. Taking the placement of the cannula as the primary event, Type II alterations occurred with a much longer latency than Type I and Type III alterations, and the delay of reversibility appeared also to be much longer. Type IV abnormalities were only observed in one patient and could be clearly related to ischaemia in the left carotid territory. Both their onset and offset latencies were very similar to those of the SEP alterations associated with carotid cross-clamping in carotid endarterectomies. Type V abnormalities were associated with systemic hypoperfusion and, unlike the Type IV alterations, were observed after stimulation of both the left and the right PTN. Neither Type IV nor Type V abnormalities were associated with any alteration of cervical or brain-stem activities, so the peripheral nerve and the spinal cord are excluded as the origin of the alterations.

Overall, 22 peroperative SEP abnormalities were observed in 18/27 patients (67%). In 15 cases, these were obviously related to distal (11 Type I alterations) or segmental (four Type III alterations) spinal ischaemia. These cases are clearly 'true positives'. However, we failed to find any evidence of a cause of spinal ischaemia in 7/22 cases (32%) (Types II, IV and V). These cases would most probably have been considered 'false positives' by many authors. Hence, our data clearly show that these abnormalities actually correspond to real events not primarily involving the spinal cord. We conclude that the rate of false positives in SEP monitoring of descending aorta surgery is over-estimated. A better estimate can be obtained by multichannel SEP recordings and careful integration of neurophysiological and surgical data.

Another important aspect of our study is that, although the reversal of SEP abnormalities seems to be a prerequisite for avoidance of paraplegia, it is not a sufficient condition: the duration of evoked potential loss is also important. This fact has already been pointed out in the dog by Laschinger et al.[8] and was examplified in one of our patients who developed postoperative paraplegia, although long duration (119 min) Type III abnormalities were reversed by re-implantation of critical vessels. It is our feeling, shared with other authors[2], that the SEP loss should never exceed 30 min duration. With this rule in mind, patients who become paraplegic in spite of ultimate recovery of transiently absent SEPs would not be considered false negatives[6, 7] but true positives.

References

1. McNulty S, Arkoosh V, Goldberg M. The relevance of somatosensory evoked potentials during thoracic aorta aneurysm repair. J Cardiothorac Vasc Anesth. 1991;5:262–265.
2. Cunningham JN, Laschinger JC, Spencer FC. Monitoring of somatosensory evoked potentials during surgical procedures on the thoracoabdominal aorta. IV. Clinical observations and results. J Thorac Cardiovasc Surg. 1987;94:272–285.
3. Crawford ES, Mizrahi EM, Hess KR et al. The impact of distal aortic perfusion and

somatosensory evoked potential monitoring on prevention of paraplegia after aortic aneurysm operation. J Thorac Cardiovasc Surg. 1988;95:357–367.
4. Nuwer MR. Evoked potential monitoring in the operating room. New York: Raven Press, 1986.
5. Maeda S, Miyamoto T, Murata H, Yamashita K. Prevention of spinal cord ischemia by monitoring spinal cord perfusion pressure and somatosensory evoked potentials. J Cardiovasc Surg. 1989;30:565–570.
6. Takaki O, Okumura F. Application and limitation of somatosensory evoked potential monitoring during thoracic aortic aneurysm surgery: a case report. Anesthesiology. 1985;63:700–703.
7. Zornow MH, Drummond JC. Intraoperative somatosensory evoked responses recorded during onset of the anterior spinal artery syndrome. J Clin Monit. 1989;5:243–245.
8. Laschinger JC, Cunningham JN, Cooper MM et al. Monitoring of somatosensory evoked potentials during surgical procedures on the thoracoabdominal aorta. I. Relationship of aortic cross-clamp duration, changes in somatosensory evoked potentials, and incidence of neurologic dysfunction. J Thorac Cardiovasc Surg. 1987;94:260–265.

16
Decomposition of spinal cord evoked potentials into propagating and stationary components

J. A. HALTER, I. ŠTĚTKÁŘOVÁ and M. R. DIMITRIJEVIĆ

INTRODUCTION

Spinal cord evoked potentials (SCEPs) recorded from the epidural space allow examination of both the propagating action potentials (PAPs)[2] and stationary action potentials (SAPs)[2]. PAPs reflect the integrity of the white matter spinal tracts while SAPs reflect the function of synaptic structures in the grey matter. At the segmental level at which the SAPs reach a maximal amplitude[3–5], the PAPs superimposed on the SAPs can be difficult to interpret since their amplitude is of the order of 10 times smaller than the SAPs. The PAPs can be made easier to interpret at the time of data acquisition if a highpass analog filter is used to attenuate the SAPs, which are primarily composed of lower frequency components. However, with this technique the SAPs are not recorded and therefore unavailable for examination. This study examines the utility of a digital filter as a tool to perform decomposition of the PAPs and SAPs after acquisition of SCEPs using wide bandwidth filter settings.

MATERIAL AND METHODS

Spinal cord evoked potentials were recorded from the epidural space in five patients (4 males, 1 female). Three patients sustained cervical spinal cord injury, two sustained head injury and one patient suffered from cerebral palsy. The average age was 31 years (14 to 48 yrs). All were in good health with no evidence of decubiti, bladder infections or other neurological disorders. SCEPs were recorded as an adjunct to evaluation of the therapeutic efficacy of spinal cord stimulation (SCS) for control of pain and spasticity. All patients gave informed consent for these recordings under a protocol which was approved by the Baylor College of Medicine Institutional Review

Board for Human Research.

SCEP recordings were obtained using a multielectrode spinal cord stimulation lead (Neuromed 1980 JF) placed percutaneously via a 16-gauge modified Tuohy spinal needle into the epidural space. Electrodes were 1 mm in diameter with 4 platinum-iridium contacts 4 mm in length, spaced at 10 mm centre-to-centre intervals with an overall length of 550 mm. All four electrodes were referenced to a surface Ag/AgCl Beckman electrode placed over the T6 processus spinosus (Figure 1A). Lateral and anterior–posterior (AP) X-rays were used to determine the position of the electrodes. In the final recording session, the lead was systematically removed. The tibial nerve was stimulated at supramaximal intensities at the popliteal fossa with a Grass S88 stimulator delivering 0.2 ms duration pulses. The data acquisition bandwidth was 8 Hz to 3.2 kHz. Analysis time was 40 ms with a sampling period of 72 μs. For each waveform 128 data sweeps were used. Stimulation was delivered during the quiescent phase of alternate ECG cycles.

After acquisition the waveforms were displayed and the component peaks measured (Figure 1B). These data were then processed with a scientific spreadsheet program (XESS*), allowing for computation of relative amplitudes and latencies. The digital filtering process was accomplished using the MATLAB† software package. Both finite impulse response (FIR) and infinite impulse response (IIR) filters of different cutoff frequencies were explored. An 8-pole Butterworth IIR filter with a cutoff frequency of 400 Hz was selected. This cutoff frequency was chosen based on subjective assessment of the degree of attenuation of the rising edge of the SAP. The IIR digital filtering algorithm allowed for zero-phase-shift computation, resulting in a 16-pole equivalent filter and exhibited less distortion at the temporal extremes of the data than did the FIR filter. Conduction velocities (CVs) were determined by taking the results of the latencies of the component peaks versus electrode position and using Mathematica‡ to perform a linear interpolation to the dataset.

RESULTS

Figure 2 demonstrates application of the digital filter to epidural SCEPs recorded from the T9 to L3 vertebral levels. In comparison to the original data (Figure 2A), note the complex of PAPs which result in the highpass case (Figure 2B) and the SAPs present in the lowpass data (Figure 2C). Application of the IIR filter to the SCEP data obtained from the other subjects produced similar results. Note the change in the waveform pattern as well as what appears to be a discontinuity of conduction at the inferior aspect of the T11 vertebral body. Figure 3 illustrates the latencies of major negative components versus the vertical position of the recording electrodes. In the original data (Figure 2A) there is a linear relation of latency to the

*Trademark, Applied Information Systems, Inc., Chapel Hill, North Carolina, USA.
†Trademark, The MathWorks, Inc., Natick, Massachusetts, USA.
‡Trademark, Wolfram Research, Inc., Champaign, Illinois, USA.

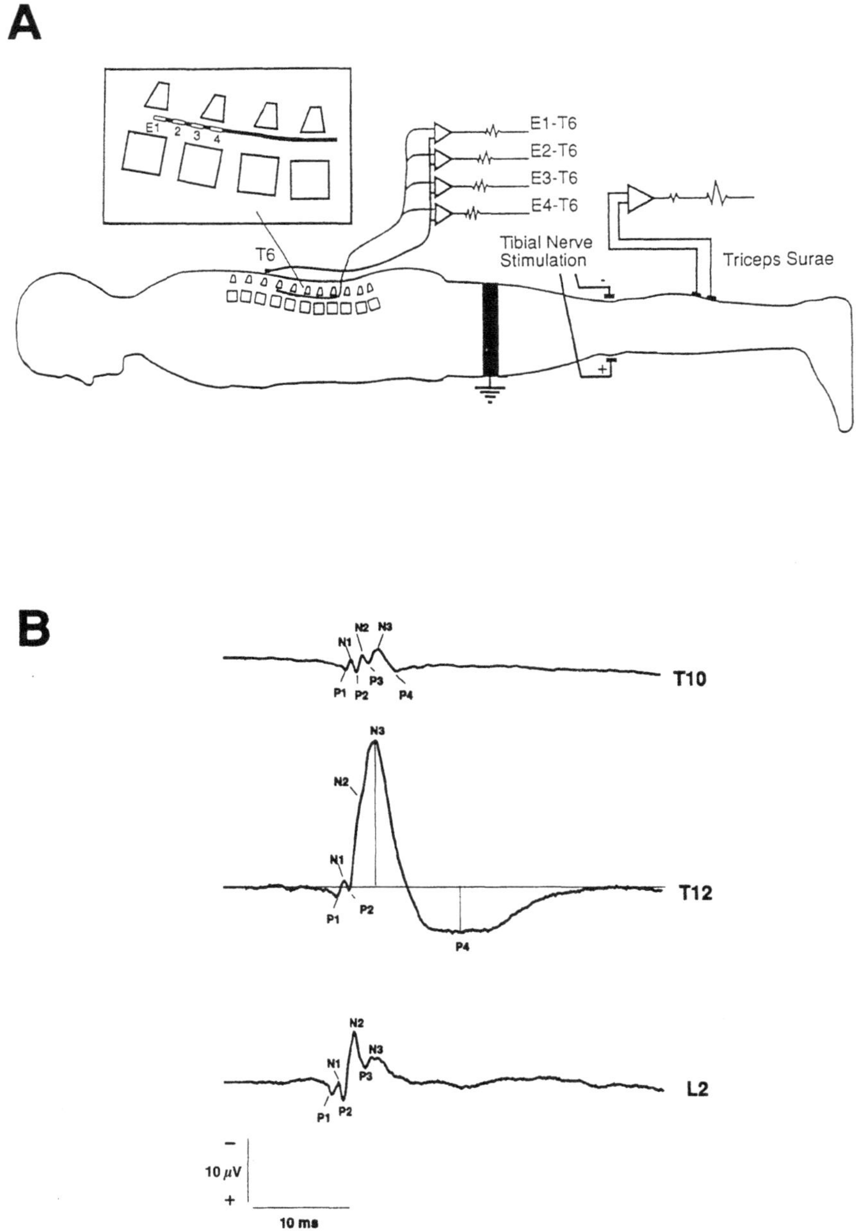

Figure 1 (A) Technique for recording epidural SCEPs. Stimulation was applied to the tibial nerve at the knee and a T6 surface reference electrode was used. (B) Illustration of major positive–negative SCEP peaks at the T10, T12 and L2 vertebral levels

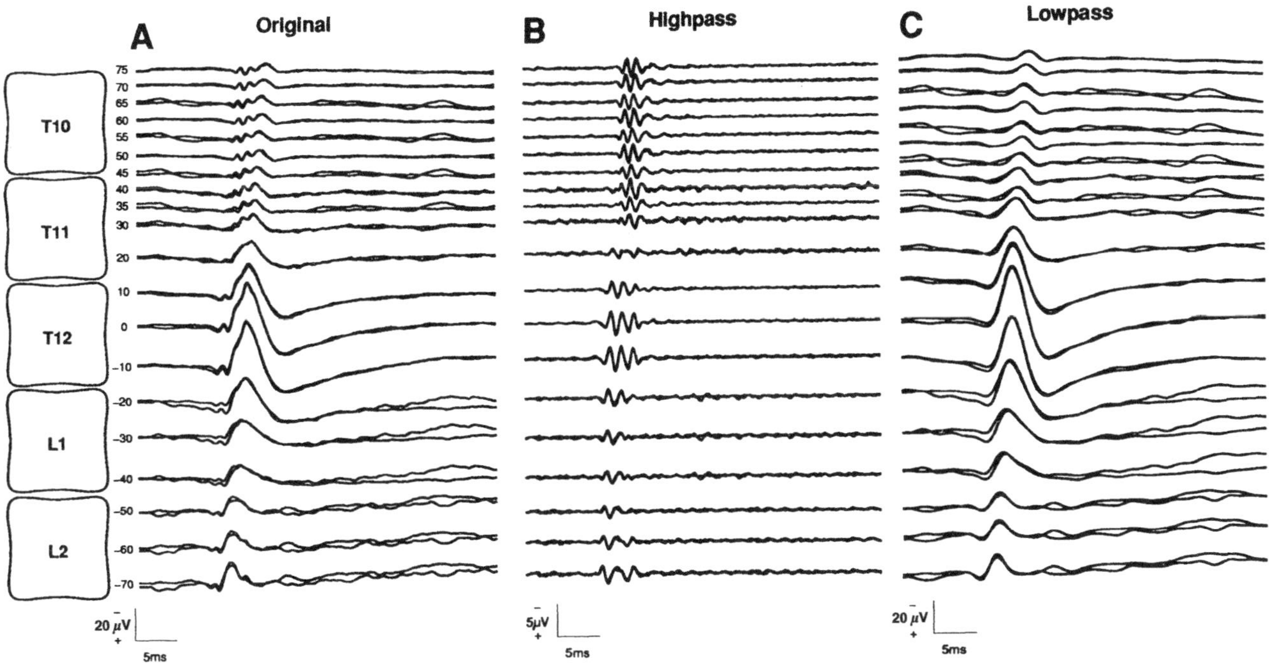

Figure 2 Application of the IIR digital filter to epidural SCEPs recorded at the T10–L2 vertebral levels. The vertical position of the recording site is shown in millimetres where zero is located at the centre of the T12 vertebral level. (A) Original data. (B) Highpass results. Note the change in character of the SCEPs at the T11 level. (C) Lowpass results. The subject was a 14-year-old male with cerebral palsy

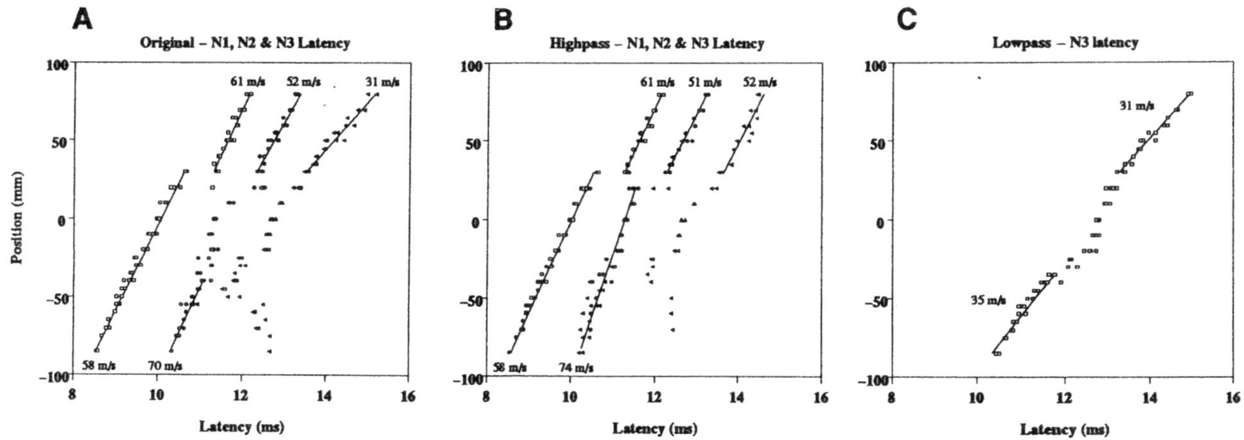

Figure 3 Graphs of the latencies versus position of the major negative SCEP components for: (A) original, (B) highpass and (C) lowpass SCEP data

position for each of the two major components present below the T12 level and the three major components present above the medial aspect of T11. The CV of the first negative component at regions below T12 (Figure 3A) is 58 m/s, followed by a negative component with an apparent CV of 70 m/s. Above the medial aspect of T11, the three negative components (N1, N2 and N3) had CVs of 61 m/s, 52 m/s and 31 m/s, respectively. In the highpass case (Figure 2B and 3B) a similar pattern is seen, with the N2 component present below the T11 vertebral level now clearly visible through the T12 vertebral level. The CVs of the components are comparable, with the exception of the N2 component present above the medial T11 level, which is 52 m/s versus 31 m/s in the original case. In the lowpass case (Figure 2C and 3C), the single negative component (N3) present shows a linear relation between latency and position below the T12 level (35 m/s) and above the medial T11 vertebral level (31 m/s), with a relatively constant latency in the region between.

DISCUSSION

In this study, the decomposition of the PAPs and SAPs is predicated on the assumption that the spectral content of the PAPs consists primarily of high frequency components and the SAPs of low frequency components. The spectral content of the PAP is dependent upon the conduction velocity, where higher CVs yield action potentials of shorter duration and therefore higher frequency components. The higher frequency components of the SAP spectral content are dependent on the rising edge of the initial depolarization (N3-wave) with most of the N3 and P4 wave energy concentrated in the low frequency components. SAPs can also be generated by PAPs arriving at different segmental levels at progressively increasing times, resulting in SAPs which can appear similar to PAPs.

Examination of Figure 2 shows that the highpass SCEPs allow easy identification of the peaks of the PAPs which are of smaller amplitude than the SAPs. At the inferior aspect of the T11 vertebral body, where the lumbar enlargement ends, there is a marked delay of conduction of all major negative peaks, a change in waveform appearance and difference CVs present below and above this level. This phenomenon has been reported by Halonen et al.[1] who found discontinuity of conduction between the T11 and T10 vertebral levels and hypothesized that activation of the lowest threshold fastest conducting peripheral afferents is followed by a delay in conduction just above the root entry level and repetitive volleys in rostral postsynaptic fibres. Figure 3A is similar in appearance to that presented by Halonen et al. with the exception of the presence of a second negative PAP component present below the L1 level. There is also some evidence of descending PAPs, which Halonen et al. attributed to possible activity in efferent fibres. After application of the highpass filter (Figure 3B) the second negative component appears to conduct to the level of discontinuity of conduction at the inferior aspect of T11. This second PAP now appears to be a possible source of activation of the second PAP seen above the medial aspect of the T11 vertebral level. There is also evidence of a third PAP with a CV of 52 m/s which may have been obscured by a progressively delayed SAP whose latency is presented in Figure 3C.

In summary, the advantages of using this digital filtering decomposition technique are in avoiding the loss of information at the time of data acquisition, relative ease of interpretation of the PAPs in the region where the SAPs are of much greater amplitude and the ability to selectively examine high and low frequency components. It would be straightforward to incorporate this method into existing computer-based evoked potential algorithms which would allow for realtime decomposition of the SCEPs. Further study can help establish the optimal parameters required to achieve maximal separation of the PAP and SAP components with digital signal processing methods.

Acknowledgements

Support for this work was provided by The Vivian L. Smith Foundation for Restorative Neurology, Houston, Texas.

References

1. Halonen JP, Jones SJ, Edgar MA, Ransford AO. Conduction properties of epidurally recorded spinal cord potentials following lower limb stimulation in man. Electroenceph Clin Electrophysiol. 1989;74:161–174.
2. Maruyama Y, Shimoji K, Shimizu H, Kuribayashi H, Fujioka H. Human spinal cord potentials evoked by different sources of stimulation and conduction velocities along the cord. J Neurophysiology. 1982;48:1098–1107.
3. Campbell JA, Miles JB. Intraspinal SEPs recorded from the vicinity of the dorsal root entry zone. In: Ducker TB, Brown RH, eds. Neurophysiology and standards of spinal cord monitoring. New York, Springer-Verlag. 1988:34–39.
4. Hallstrom YT, Lindblom U, Meyerson BA. Distribution of lumbar spinal evoked potentials and their correlation with stimulation-induced paresthesiae. Electroenceph Clin Neurophysiol. 1991;80:126–139.
5. Jeanmonod D, Sindou M, Mauguière F. The human cervical and lumbo-sacral evoked electrospinogram. Data from intra-operative spinal cord surface recordings. Electroenceph Clin Neurophysiol. 1991;80:477–489.

17
An atlas of human spinal cord evoked potentials

J. A. HALTER, I. ŠTĚTKÁŘOVÁ and M. R. DIMITRIJEVIĆ

RATIONALE

The literature has been steadily growing in size and sophistication regarding the electrical activity generated by the spinal cord and the relation of these events to function within the spinal cord. This growth in understanding of spinal cord physiology is enabling more effective interpretation of human spinal cord evoked potentials (SCEPs) obtained in the healthy and diseased or traumatized condition. The clinical utility of examining SCEPs has been demonstrated in intra-operative monitoring[1–3] and substantial quantities of human SCEP data have been collected in multiple centres[1–6]. Since 1981, five International Symposia on Spinal Cord Monitoring and Electrodiagnosis[3] have been held and more clinicians and researchers are studying the human spinal cord in this manner. The establishment of a *standard reference* of human spinal cord evoked potentials would serve these clinicians and researchers. Richard Brown, PhD, has pointed out the wide disparity in the methods employed to acquire SCEPs as well as the terminology used to refer to SCEPs and has challenged those in the field to come to consensus[7]. As it was in the cases of electroencephalography and electromyography, establishment of standards for acquisition, analysis and interpretation of human SCEPs is a crucial step in the creation of a new technique which is intended to gain widespread acceptance as a viable clinical tool. For the researcher, a database of human SCEPs contributed by clinicians collecting SCEPs during intra-operative monitoring and other investigators collecting human SCEPs under well-controlled conditions, offers a unique resource from which a new understanding can be reached. For these reasons and as a means to enhance communication between the scientific and clinical communities, we feel that it is time to develop an *atlas of human spinal cord evoked potentials* (the Atlas).

The Atlas will serve as a compendium of human SCEP data recorded from the surface of the skin, epidural space[1] or intrathecal space[2]. SCEPs obtained from within the spinal canal will be subdivided into those obtained from the dorsal, ventral and lateral regions of the spinal cord and the activity in the white and grey matter elucidated.[5]. SCEPs obtained from multiple stimulation conditions (peripheral nerve, spinal cord and cortical stimulation) will be included. The character and distribution of the SCEPs in the healthy, diseased and traumatized condition[4] will be documented and the ability to develop statistical measures of the SCEP properties will be enhanced. This will also serve those interested in the range of variability of the human SCEPs and their alteration in cases of compression of the spinal cord or pharmacological modification of spinal cord function.

Another factor motivating development of the Atlas is that an era has arrived where global telecommunication is a reality and the means is at hand for rapid interchange of data regardless of physical location. Efforts are underway to apply these rapidly evolving communications and computer technologies to facilitate research in the basic and clinical neurosciences[6]. In anticipation of the need to exchange information between computer-based neurophysiological instrumentation, a standard has been established for data interchange[8]. As these technologies become more ubiquitous, we envisage a version of the Atlas in electronic form. This would allow for rapid evolution of the information presented in the Atlas, ultimately providing a rapid index into the database of human SCEPs which could be accessed from the operating or examination room via the same instrumentation used to acquire the SCEPs.

FORM AND CONTENT

The initial draft of the Atlas is in the form of a handbook based on recording techniques and human SCEP data collected in our laboratories over a 10-year period which will serve as a framework for an expanded version which will merge human SCEPs collected from multiple centres. The present content of the Atlas is as follows:

1. Origin and Properties of SCEPs
2. Methods
3. Spinal Cord Evoked Potentials at Different Vertebral Levels
    - 3.1 Cervical
    - 3.2 Thoracic
    - 3.3 Lumbo-sacral
4. Spinal Cord Evoked Potentials in Disease and Trauma
5. Bibliography

Section 1 provides a brief summary of the origin and properties of spinal cord evoked potentials drawn from the literature, a comprehensive list of which is presented in the bibliography. Section 2 consists of a description of the methods employed to record the SCEPs[5] and discusses potential pitfalls and trouble-shooting procedures. Section 3 presents human SCEP data obtained from the dorsal, ventral and lateral aspects of different vertebral levels. Figure 1 presents a representative entry in the Atlas showing the

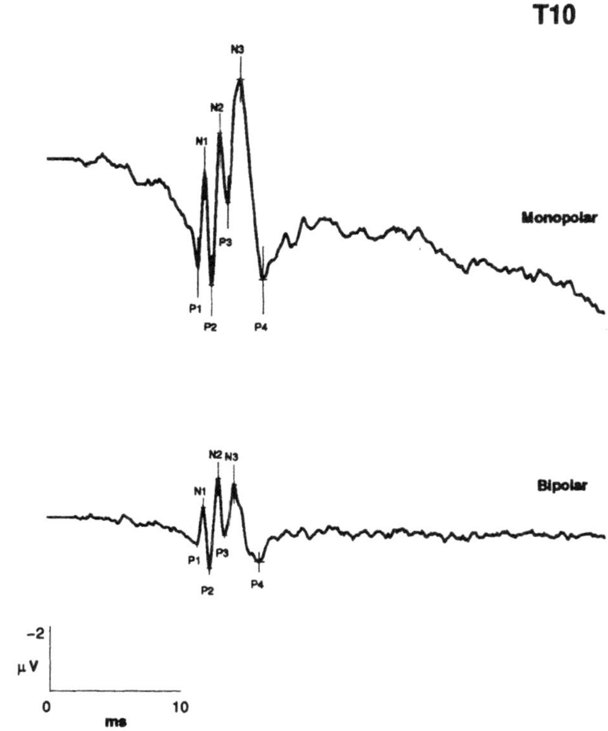

Figure 1 Representative entry in the Atlas illustrating the character of the monopolar and bipolar (10 mm electrode separation) SCEPs recorded from the dorsal aspect of the epidural space at the medial T10 vertebral level. Supramaximal stimulation was applied to the tibial nerve at the popliteal fossa. The recording conditions were identical to those stated in a companion chapter[5]. The table contains the mean values and standard deviations for the amplitudes and latencies of the component peaks

Component peaks for T10 SCEPs

	Monopolar		*Bipolar*	
Peak	*Latency* *ms*	*Amplitude* *μV*	*Latency* *ms*	*Amplitude* *μV*
P1	11.3 ± 0.1	1.7 ± 0.9	11.2 ± 0.1	0.1 ± 0.1
N1	11.8 ± 0.1	−0.4 ± 0.8	11.7 ± 0.2	−1.1 ± 0.3
P2	12.3 ± 0.2	2.1 ± 0.9	12.2 ± 0.2	0.7 ± 0.2
N2	12.9 ± 0.2	−1.4 ± 0.6	12.8 ± 0.2	−1.7 ± 0.5
P3	13.4 ± 0.2	0.2 ± 0.8	13.4 ± 0.2	0.1 ± 0.1
N3	14.4 ± 0.3	−2.6 ± 0.7	14.0 ± 0.2	−1.0 ± 0.6
P4	16.1 ± 0.3	2.2 ± 1.0	15.2 ± 0.4	0.6 ± 0.3

detailed character of the SCEPs recorded from the dorsal aspect of the epidural space at the T10 vertebral level. Both monopolar and bipolar presentations of the SCEPs are included with a table containing statistics for the amplitudes and latencies of component peaks. Section 4 presents examples of the character of the SCEPs recorded in cases of pathology and/or trauma.

A computer-based version of the Atlas has been implemented using a software package entitled, The Virtual Notebook System* (VNS). The VNS version of the Atlas allows the user to display graphics and text information describing the SCEP data by selecting icons on the computer screen. Accessing SCEP for different vertebral levels is accomplished by identifying the level of interest. The VNS is built on a relational database and designed for distributed network-based access which allows multiple individuals in different locations to access and selectively alter the Atlas.

FUTURE DEVELOPMENTS

The initial draft of the Atlas will be extended by inclusion of additional human SCEP data from our database and that contributed by other laboratories. This will include human SCEPs recorded in the healthy and pathological condition recorded from the dorsal, ventral and lateral aspects of the spinal canal using the major stimulus paradigms. Greater varieties of stimulus conditions will be incorporated such as spinal cord stimulation at different vertebral levels and aspects of the spinal canal as well as cortical stimulation. SCEP data will be included in which manipulation of the spinal cord will be accomplished by systemic, epidural or intrathecal application of pharmaceutical agents. The computer-based version of the Atlas will also be extended and will include the SCEP standards for data interchange and analysis as they are established. We envision new editions of the Atlas developing as more clinical and research workers participate.

Acknowledgements

Support for this work was provided by The Vivian L. Smith Foundation for Restorative Neurology, Houston, Texas.

References

1. Halonen JP, Jones SJ, Edgar MA, Ransford AO. Conduction properties of epidurally recorded spinal cord potentials following lower limb stimulation in man. Electroenceph Clin Electrophysiol. 1989;74:161–174.
2. Jeanmonod D, Sindou M, Mauguière F. The human cervical and lumbo-sacral evoked electrospinogram. Data from intra-operative spinal cord surface recordings. Electroenceph Clin Neurophysiol. 1991;80:477–489.
3. Shimoji, K, Kurokawa T, Tamaki T, Willis WD eds. Spinal cord monitoring and electrodiagnosis. Berlin: Springer-Verlag; 1991.
4. Halter JA, Haftek I, Sarzynska M, Dimitrijevic MR. Spinal cord evoked injury potentials in

*The Virtual Notebook System, a tool for collaborative work. Baylor College of Medicine and the Forefront Group, Inc., Houston, Texas.

patients with acute spinal cord injury. J Neurotrauma. 1989;6:231–245.
5. Halter JA, Stetkarova I, Dimitrijevic MR. Decomposition of spinal cord evoked potentials into propagating and stationary components. In: Jones SJ et al., eds. Handbook of Spinal Cord Monitoring. Dordrecht: Kluwer; 1993:146–151.
6. Pechura CM, Martin JB eds. Mapping the brain and its functions: Integrating enabling technologies into neuroscience research. Washington DC: National Academy Press; 1991.
7. Brown RH, Nash Jr CL. Standardization of the evoked potentials recording. In: Ducker TB, Brown RH, eds. Neurophysiology and Standards of Spinal Cord Monitoring. Berlin: Springer-Verlag; 1988:1–10.
8. Standard specification for transferring digital neurophysiological data between independent computer systems. Document number E1467-92, ASTM, Philadelphia, PA, USA, 1992.

18
A multichannel planar epidural electrode for spinal cord monitoring

P. D. HEATH, N. W. VALENTINE and A. ALLEN

INTRODUCTION

Spinal cord monitoring, entailing the recording of spinal cord SSEPs from the epidural space, has generally proven to be a practical and effective technique[1], such that it has now entered routine clinical use during much major spinal surgery presently undertaken in the UK[2]. However, we have noted some variability of the recorded potential, particularly during the crucial period of spinal distraction, which potentially restricts the sensitivity of the technique in detecting cord dysfunction. Such instability of the recorded response may in part reflect the mechanical limitations of wire electrodes commonly used. Their lateral flexibility prevents accurate electrode location relative to the preferred midline position between the posterior columns and such electrodes are likely to be positionally unstable during spinal distraction.

METHODS AND MATERIALS

A new electrode (Figure 1) was designed in cooperation with Forward Circuits Ltd (Tamworth), using circuit board interconnector technology. The electrode has a ribbon appearance, 35 cm long by 5 mm wide, but only 0.23 mm thick. It has a multilaminate structure, with layers of pyralux enclosing ten etched longitudinal copper conducting strips. The recording tip has rounded edges to prevent tissue trauma and ten 1 mm by 2 mm gold electrode surfaces arranged in five parallel pairs. At the proximal end there is a 10-way connector.

This electrode is resistant to lateral distortion such that the position of the recording tip may easily be determined from that part of the electrode protruding from the surgical wound. The electrode lies flat on the dura effectively 'loosely adherent' and is therefore less likely to be displaced relative to the cord during spinal distraction. The multichannel arrangement permits the optimal electrode pair to be selected and significant lateral changes in

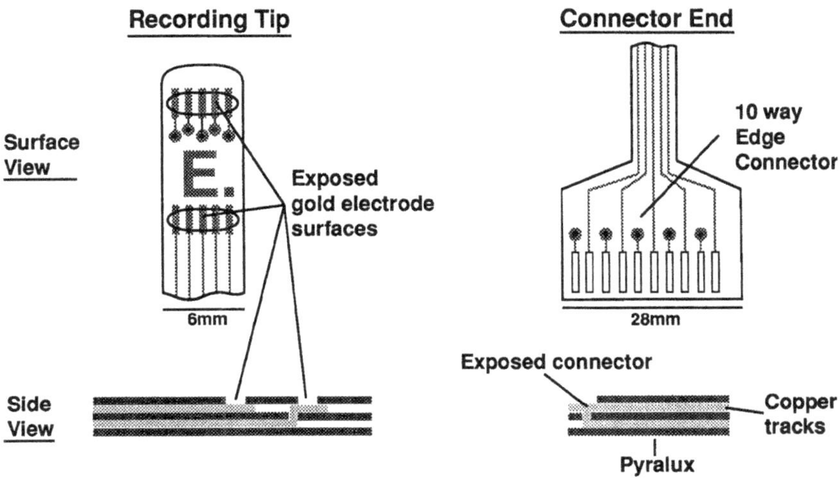

Figure 1 Diagram showing form and construction of the electrode (note: side view of connector end superimposes the exposed copper conductors which remain separated laterally)

electrode position relative to spinal cord potential generators to be detected. The electrode is however flexible longitudinally, so that it may be bent clear of the surgical field.

Evaluation was undertaken *in vitro* and in a feline model (with appropriate authorization) prior to entering clinical use.

RESULTS

In vitro studies, placing the electrode in an oscillating electrical field at frequencies varying between 100 Hz and 10 kHz, revealed no unwanted capacitance effects between adjacent channels. The electrode withstood repeated steam autoclaving at 130°C without alteration of shape or recording characteristics and may be reutilized.

In a feline experiment reproducible high amplitude potentials were recorded from the thoracic spinal cord to common peroneal nerve stimulation, with the maximal response over midline and near ipsilateral channels. Displacement of the electrode laterally resulted in a corresponding change of channel registering the maximal response. However this was readily distinguished from the effects of percussive trauma to the spinal cord which resulted in a profound global symmetrical reduction in signal.

The electrode has since been used successfully in six patients undergoing major spinal surgery. It is easily sited via an incision in the ligamentum flavum, the location of the recording tip relative to the midline is readily apparent and reproducible high amplitude recordings have been obtainable with a good signal to noise ratio. Recordings made in the vicinity of the midline suggest that displacements greater than 2 mm are usually required to alter signal amplitude significantly and the present electrode appears sufficiently stable positionally for such degrees of displacement to occur only rarely or when present to be obvious from the location of the distal electrode. Conversely reversible cord dysfunction has been readily identified from sequential recordings.

DISCUSSION

Utilizing circuit board interconnector technology we have designed and developed a multichannel planar electrode, which has more favourable recording characteristics than conventionally used wire electrodes, for intraoperative monitoring of epidural spinal cord somatosensory evoked potentials.

The lateral rigidity of the new electrode permits accurate positioning of the recording tip relative to the preferred recording site, usually between the posterior columns. The multichannel arrangement also provides information on any lateral variation in the recorded response. Hence any significant inadvertent electrode displacement is readily apparent both by change in angle of the proximal segment of the electrode protruding from the wound and, if severe, by change in the recording channel registering the maximal response. The situation may then be corrected by repositioning the electrode or selecting an alternative channel. In clinical practice potentials recorded during spinal distraction appear more stable than with wire electrodes, possibly due to the ribbon form having greater positional stability and relative adherance to the dura.

The new electrode is also convenient in that it may be bent back semi-permanently upon itself out of the field of surgery.

To date SSEP recordings of high amplitude and good signal to noise ratio have been obtained consistently, except where there has been evidence of spinal cord dysfunction.

Acknowledgements

We thank Forward Circuits Ltd (Tamworth) and Prof E Evans of the Department of Communication and Neuroscience, Keele for help in electrode design and evaluation and we wish particularly to thank Mr John Dove for his encouragement in this project.

PDH and NWV were supported in this project by the J Levy Scholarship from the British Scolioisis Society.

References

1. Jones SJ, Edgar MA, Ransford AO. Sensory nerve conduction in the human spinal cord: epidural recordings made during scoliosis surgery. J. Neurol Neurosurg Psychiatry. 1982;45:446–451.
2. Heath P, Smith NJ, Jackson RK, Webb PJ. Report of a joint Association of British Clinical neurophysiologists and British Scoliosis Society working party on spinal cord monitoring for scoliosis surgery. 1991.

19
Somatosensory evoked potentials (EPs) during anterior and posterior spinal surgery

L. HENDERSON, S. L. PAPASTEFANOU, A. HAMILTON, N. J. SMITH, S. CLARKE and J. K. WEBB

INTRODUCTION

The primary use for intraoperative evoked potential monitoring remains spinal surgery. Traditionally there have been four methods by which this can be done:

- somatosensory cortical evoked potentials;
- somatosensory epidural evoked potentials;
- spinal evoked potentials;
- motor evoked potentials.

The wake-up test is still claimed by some to be the gold standard.

Each method has its own advantages and disadvantages. This paper describes an unusual combined method of somatosensory evoked potentials recorded at surface cervical and cortical electrodes. The recognized advantages of this method are non-invasiveness, high sensitivity, versatility, preoperative and postoperative assessment by the same method[1]. It is its very sensitivity which also accounts for its major criticism, i.e. its responsiveness to physiological or certain pharmacological changes makes it potentially prone to low diagnostic specificity.

Controversy continues to surround the routine use of spinal cord monitoring for spinal surgery. Issues at stake include:

- Does it affect operating practice?
- Can it replace the wake-up test?
- Who should be doing the monitoring?
- Who should be monitored?

This study set out to review the Nottingham experience with these relevant clinical questions in mind.

METHODS

Patients had baseline recordings preoperatively, intraoperatively, and post-operatively if there was any new neurological deficit postoperatively.

Standard cortical electrode placements were used for recording the response from the posterior tibial nerve and median nerves according to the International 10–20 recording system. Spinal surface electrodes were located over the spinous processes of C2 and C7. Recordings were also made from electrodes at Erb's point and the popliteal fossa to test stimulus production and normal conduction times. Sensory evoked potentials are of a very small order of amplitude and have to be enhanced by 10^3 to 10^4 by a differential amplifier and then fed to an averager with capacity to edit the tracings and reject trials contaminated by EEG, muscle and movement artefacts. We used a Disa Neuromatic 2000 stimulator and averager unit made by Dantec Ltd to average 500 or 1000 tracings with a stimulus rate of 7 Hz intraoperatively and these were then transferred to an Apple micro computer for sequential storage.

A fairly standard anaesthetic technique was employed:

- Omnopon or temazepam premedication;
- halothane or isoflurane induction;
- maintenance halothane, nitrous oxide and alcuronium or atrocurium muscle relaxant;
- labetolol or nitroprusside used as appropriate to control the BP.

The routine use of spinal cord monitoring for spinal surgery remains somewhat controversial. All methods of spinal cord monitoring are highly sensitive to potentially damaging insults to cord function but are prone to different artefactual disturbance depending on the method employed[2].

PATIENTS

Of 308 procedures, 208 were anterior approaches, 77 posterior and 23 combined approaches. Some 209 patients had spinal deformity, 54 spinal trauma, 45 spinal tumour. There were 130 males and 178 females.

RESULTS

Results were analysed using conventional definitions (Table 1).

TP = 7 TF = 16 FP = 46 FN = 0. Sixty-eight patients had an SEP amplitude drop greater than 50% at some point intraoperatively (Table 2).

Table 1 Somatosensory evoked potentials:

Definitions
>50% drop with post-op neurological deficit = true positive
>50% drop no post-op neurological deficit = false positive
No drop more than 50% no neurology = true negative
No drop more than 50% with post-op neurology = false negative
No recognizable recording = technical failure

Table 2

Diagnosis: monitoring abnormal	
Fractures	4
Ankylosing spondylitis	4
Connective tissue	3
Congenital scoliosis	13
Thoracic discs	3
Idiopathic scoliosis	21
Neuromuscular disease with scoliosis	13
Tumours	4
Miscellaneous	3
TOTAL	68

True positives

		Recovery
HA	previous laminectomy – spondylolisthesis	Yes
JC	spondyloptosis	Yes
CJ	congenital scoliosis	Yes
EP	congenital scoliosis	Yes
SS	Reiters syndrome C1/C2 instability	No
CS	congenital scoliosis	Yes
MH	Duchenne's	Yes

Approach

3 posterior

3 combined event at posterior stage

1 anterior

Cause of drop	**Internal fixation**
Correction 2	2 Harrington rods
Positioning 1	1 Dick
Distraction 2	2 Luque
Unknown 1	1 AO plate
Wire tightening 1	1 Zielke

- Five had wake-up tests, two had no wake-up test.
- HA had no movement until release of instrumentation; postoperatively weakness and parasthesia in both legs.
- JC had no dorsiflexion of right foot on repeated wake-up test, and post-op hyparaesthesia both thighs and lateral aspect of both feet.
- CJ had no movement in right leg on wake-up, left instrumentation removed, postoperatively had grade 3 power right leg, hyperaesthesia below t2.
- EP had no movement, instrumentation removed, increasing paraplegia postoperatively.
- CS moved both feet, operation continued, has parasthesia right leg and minor weakness.
- SS had no wake-up test, had an unstable C1/C2 and was face-down with a tube and strapping covering his face. He did not understand English. Wake-up test impossible.
- MH had no wake-up test. Duchenne's non-walker. Leg hyperaesthesia post-op.

All had normal pre-op SSEPs. In six out of the seven cases the drop was 100% whereas in one case it was 50%. At end of procedure, three had no SSEP responses and three had reduced responses.

Clinical signs

Three had normal pre-op status. Three were abnormal. Two had minor weakness or increased reflexes in legs. One man had cervical myelopathy.

Two before the end of the operation had 2nd or 3rd normal wake-up tests but had postoperative neurological deficits: the SSEPs were more sensitive than the wake-up test.

False positives

25 could be related to physiological causes:

10 anaesthesia

13 hypotension

2 neuromuscular

15 attributed to surgical events, six no obvious cause

Only one in the physiological group had wake-up tests. This was in the first years of monitoring when the significance of correctable factors was not appreciated.

Six in the surgical group had wake-up tests. If the fall was for less than 30 minutes they did not have a wake-up. If it was prolonged or complete they had a wake-up test.

One in the unknown group had a wake-up test.

Table 3

Diagnosis in technical failures (i.e. no effective monitoring)

3 neuromuscular disease	1 ankylosing spondylitis
1 tumour	1 Marfans
2 idiopathic scoliosis	1 TB kyphosis
2 congenital scoliosis	

Of those who did not have wake-up test: two had tumours, two had only brief falls. In one case the responses disappeared *before* instrumentation, cause unknown.

Of the technical failure group (Table 3), seven had normal pre-ops, 1 had no response, 3 had no pre-ops.

Three had wake-up tests because they had good pre-op neurology and were undergoing high risk correction.

Thirteen did not have wake-up tests: five were having anterior discectomies/ anterior release. One had a tumour excision. Three occurred intermittently during operation. Two had severe pre-op neurology. One had interference from time of instrumentation. One unknown if had wake-up.

CONCLUSIONS

1. The method is highly versatile and applicable to a wide range of spinal operations and pathologies, including, in our series, many anterior and combined approaches.
2. With a technical failure rate of 5.5%, the method can be relied on for effective monitoring of most patients.
3. The interpretation of response changes by the Clinical Neurophysiologist, in consultation with the anaesthetist, allows us to discount those 'false positive' changes which are due to physiological or pharmacological factors[3]. Accurately identified, these factors are correctable without altering surgical proceedings in any way. Thus, no 'wake-up' test is required.
4. The sensitivity and specificity of our method of spinal monitoring using surface cortical and cervical electrodes, compares very favourably with larger series using invasive methods for a narrow range of procedures[4]. We had no false negative to date, thus the sensitivity is 100%. The specificity, calculated as $TN/(TN + FP) \times 100$, and discounting those changes in monitoring attributable to non-surgical, correctable causes, is 93%.
5. 'False positive' in monitoring as currently defined, is an unsatisfactory term in designating at least three separate sub-categories which it is vitally important to distinguish. These are: monitoring changes due to indentifiable and correctable non-surgical factors; changes associated with instrumentation or other critical surgical manipulation, which must

be reported to the surgeon, and may or may not modify his strategy; and finally, those potentially misleading changes in monitoring which cannot be related either to surgery or the physiological parameters. It is very possible that changes in monitoring associated with instrumentation or surgery, but not followed by postoperative neurological deficit, do reflect real but reversible transient cord dysfunction, and the term 'false positive' as applied to this group is thus highly misleading. If the term is kept for changes of category three (changes due to unidentified, irrelevant factors) then the specificity for our series using this redefined category of 'true' false positive would rise to 98%.

6. Only 13 patients in our series had 'wake-up' tests, previously the 'gold standard' by which cord integrity was assessed intraoperatively. Our monitoring method allowed the surgeon to avoid wake-up tests in 295 cases (95.8%). Three patients with postoperative neurological deficit had a first or subsequent 'normal' wake-up test during operation. This technique appears therefore to be more sensitive than the 'gold standard' wake-up test and may under most circumstances replace it.

References

1. Nuwer M. Monitoring spinal cord surgery with cortical somatosensory evoked potentials. In: Desmedt JE, ed. Neuromonitoring in Surgery. Amsterdam, The Netherlands: Elsevier Science Publishers BV; 1989:151–164.
2. Friedman WA, Grundy BL. Monitoring of sensory evoked potentials is highly reliable and helpful in the operating room. J. Clinical Monitoring. 1987;3:38–44.
3. Young W. Letter, J Neurosurg. 1986;64:987–988.
4. Forbes HJ, Allen PW, Waller CS et al. Spinal cord monitoring in scoliosis surgery. Experience with 1168 cases. J Bone Joint Surgery (Br). 1991:73–78, 487–491.

20
Spinal cord monitoring with a new device for automated signal analysis

J. HERDMANN, F. SONNENSCHEIN, W. KUNZE, M. KRZAN, M. KRÄMER, J. DVORÁK and D. GROB

INTRODUCTION

Assessment of the functional integrity of motor and sensory pathways using intraoperative evoked potential monitoring has been recognized as a means of minimizing the risk of damage to neural pathways during surgical and neuroradiological procedures. In the United States an increasing number of neurophysiologists and neurologists, but also neurosurgeons and ENT physicians offer their services for neurophysiological monitoring to hospitals, spine units, or surgeons in private practice. Most individuals engaged in monitoring are still looking for the 'unique, universal, and easy-to-handle' evoked potential (EP) and electromyography (EMG) machine designed for intraoperative monitoring. For this reason several centres throughout the world have developed their own monitoring machinery often based on commercial EMG/EP machines by writing special software programs.

Our aim was to develop an integrated neurophysiological monitoring system (hardware and software) for use in any standard monitoring situation. Requirements of first priority were:

1. multimodality,
2. automatic signal analysis,
3. trend analysis, and
4. easy handling.

IOM is based on a standard AT compatible computer and an analog-to-digital (AD) board that comes with a digital input/output (I/O) board and a programming langauge called *auto!Lab* (Soft- & Hardware Design, Arnoldstr. 19, Düsseldorf, Germany) designed for neurophysiological data acquisition, data processing, graphic display, and process control. In a first step the system was designed for spinal cord monitoring with the choice of somatosensory evoked potentials, motor evoked potentials, and continuous auditory feedback of electromyographic signals. Acoustic and visual stimulation together with the respective signal analysis will be implemented in later releases.

HARDWARE

IOM is based on a 50 MHz 486 AT computer with 4 MB RAM, fast hard disk drive, 5.25″ and 3.5″ floppy drives together with VGA graphics. The evoked potentials are fed into programmable AC amplifiers (Madaus Medizin Elektronik, Freiburg, Germany; gain: 100–500 000) which allow for conventional analog signal filtering (low cut-off 0.01–300 Hz; high cut-off: 30 Hz–15 kHz), before they are sent to a 12 bit AD converter (Soft- & Hardware Design, Düsseldorf, Germany) where further digital filtering can be performed. Four amplifiers are mounted in a headbox together with three programmable multiplexer boards (IOM TEAM, Düsseldorf, Germany), two of which allow for switching the signals from 22 electrode sockets onto the four amplifiers in various combinations, and one of which switches the electrical square wave stimuli of two constant current electrical stimulators (CI 15, AMPLAID, Milan, Italy) onto six pairs of stimulator electrodes. The headbox is programmed via an RS232 serial port. IOM has six TTL output channels to trigger any stimulating device (magnetic stimulator, electrical stimulators, acoustic stimulator etc.). In later releases acoustic and visual stimulators will be programmed via additional parallel and serial interfaces. The system is compatible with any Hewlett Packard laser printer; a HP LJ IIP is recommended. Colour printing is possible with a HP DJ 500 C.

All power supplies are fed by an earth free isolating transformer. Patient safety is fully ensured with isolating currents higher than 4 kV.

SOFTWARE AND FUNCTIONS

The IOM software is written in *auto!Lab* (Soft- & Hardware Design, Düsseldorf, Germany), a universal software toolbox designed for neurophysiological data acquisition, signal analysis and process control. Programming in *auto!Lab* is similar to PASCAL and readily learned. Commands in *auto!Lab* allow for 16 channel data acquisition, window orientated data display, signal transformation including fast Fourier transformation (16–8192 points), and digital filtering both finite and infinite impulse response. IOM allows for sampling rates up to 50 kHz.

The IOM software controls all programmable devices (stimulators, multiplexer boards, and amplifiers) via serial and parallel ports. Electrode impedance is checked via software and displayed on the screen.

IOM provides eight 'software channels' and four 'hardware channels' allowing for simultaneous recordings from up to four electrode positions and consecutive recordings with automatic signal analysis and trend functions from up to eight channels. Data is displayed on the screen in up to six windows showing the baseline and the last line recordings each. The user selects the channel or trend function to be displayed. Function keys allow for zooming, superimposition or subtraction of the baseline (from consecutive recordings), or waterfall display of the last fifty traces for each window. A playback function allows for subsequent judgement and possible adjustment of cursors.

Together with the initial baseline acquisition the signal analysis function allows for definition of the onset or peak latency as well as of the peak-to-peak or ground-to-peak amplitude to be detected and measured automatically in order to calculate trend curves. Warning criteria are defined for latency and amplitude trends separately for each software channel (percent amplitude decrease and milliseconds latency increase). An opto-acoustic signal is given if a warning criterion is exceeded. Automatic signal analysis is performed only within a time window, which is defined during the baseline set-up. Artefact rejection is based on selectable amplitude criteria and is only effective within this time window.

The settings for eight software channels each are saved in so-called 'statuses' or 'batteries'. Twenty such statuses can be quickly loaded from the hard disk in order to change or adapt the monitoring protocol. Each status contains all settings necessary to run the different tests defined for each of the eight software channels:

1. stimulating device (output port, position of the stimulating electrodes, stimulation parameters);
2. position of the recording electrodes (input socket and amplifier to be switched in the headbox);
3. amplifier settings (gain, low cut-off, high cut-off, notch filter);
4. number of averages;
5. digital signal transformation (invert, rectify) and digital filtering;
6. artifact rejection level;
7. automatic signal analysis for latencies (signal onset or predefined peaks) and for amplitudes (peak-to-peak, ground-to-peak);
8. alert ranges (warning criteria);
9. priority level for an automatic recording mode.

In the automatic recording mode the tests of the eight software channels are run through consecutively and repeated as often as selected with the priority level which can range from 1/10–10/10: for example control recordings can have a low priority of '1/10' so that they are run through only during every tenth trial.

The large number of available input sockets on the headbox, with the switching facilities of its multiplexer boards, allows for a unique versatility and complexity of monitoring with multiple recording sites. These functions are indispensable in the operating theatre, where the electrode set-up cannot be changed during the operation. All changes of status settings are immediately saved on hard disk so that in the case of a power interruption the monitoring can be picked up within seconds of the system being restarted.

The monitoring procedure is documented with printouts of original traces, trend curves or tables listing the results of automatic signal analysis of each trace. Listings and search functions allow for patient documentation and easy retrieval of information.

The system is primarily designed for routine spinal cord monitoring during spinal surgery, vascular surgery, or neuroradiological procedures on segmental vessels using standard neurophysiological techniques and montages. It is also helpful in the evaluation of comatose patients in the intensive care unit. After placing the appropriate stimulating and recording electrodes, a patient can be automatically monitored for as long as necessary to observe long term trends of the evoked potentials. Flexibility provided by software control allows for research applications and comprehensive developments.

Acknowledgement

This study was supported by a grant from the Swiss Orthopaedic Association (SGO Stiftung Stipendienfonds) and by Madaus Medizin Elektronik, Freiburg, Germany.

21
Evaluation of peroperative sensory evoked potentials recorded from nerve roots to the cervical epidural space during brachial plexus surgery

M. A. HETREED, L. A. HOWARD and R. BIRCH

INTRODUCTION

After trauma to the brachial plexus it is difficult to determine the exact location and severity of injury to the nerve roots and other proximal elements of the nerve plexus. Electrophysiological testing is an important supplement to clinical examination and sophisticated imaging techniques but it may not be possible to confirm the continuity of individual nerve roots with the spinal cord before they are exposed at operation. In this situation the macroscopic appearance of the nerve roots can be deceptive and electrophysiological monitoring again proves useful. The evoked muscle response (M-response) will fail in the presence of a peripheral lesion and Somatosensory Evoked Potentials (SEPs) in response to direct root stimulation have been used[1,2]. Recorded from the scalp overlying the contralateral cortex these potentials are impaired by anaesthetic agents and may require considerable averaging to be shown clearly. In these circumstances an apparently absent trace may be time-consuming and, if the M-response is present, confusing. SEPs recorded in the cervical epidural space for spinal cord monitoring during scoliosis surgery are minimally affected by inhalational anaesthetic drugs[3] and have been used during exploratory brachial plexus surgery. We have found them to be of greater amplitude than cortical responses simultaneously recorded by a factor of 4 to 10 and to be identifiable with very little averaging – typically less than 30 sweeps[4]. We have found epidurally recorded SEPs more consistent and reliable than cortical SEPs and we report their use in 51 patients undergoing surgical exploration of brachial plexus lesions.

METHODS

Patients studied had a median age of 23.2 years (range 14–51) and at operation the time elapsed from injury was (median) 4.8 months (range 2 days–30 months). 49 were men, two women. Patients known to have anatomical abnormalities or traumatic damage to their cervical spine were excluded.

Patients received premedication with papaveretum 15 or 20 mg and hyoscine 0.3 or 0.4 mg according to their weight one hour prior to anaesthesia. General anaesthesia was induced with thiopentone 3–5 $mg\,kg^{-1}$ and intubation facilitated with vercuronium (Norcuron) 0.1 $mg\,kg^{-1}$. Anaesthesia was maintained thereafter with enflurane 1–2% in nitrous oxide 66% in oxygen. No further muscle relaxant drugs were used but respiration was maintained by mechanical ventilation of the lungs. Before surgery commenced a 3FG bipolar electrode (S-Pace Ltd, temporary cardiac pacing electrode) was inserted percutaneously to the epidural space at C_{6-7} or C_7T_1 as described elsewhere[4]. The correct placement of the electrode was confirmed by recording SEPs in response to percutaneous stimulation of the median and/or ulnar nerves of the patient's uninjured arm.

After exposure of the injured brachial plexus to the required extent, elements of the plexus were stimulated using a bipolar silver wire electrode (Medelec). Stimuli of 0.2 ms duration at a fixed voltage of 3 to 20 volts were applied at a rate of 5 to 10 Hz until the continuously averaged response was clear. Typically 30 to 150 sweeps were required. The response was recorded using a Medelec MS91 Electromyograph; filter settings were 200 Hz to 2 kHz and a sweep time of 10 or 30 ms used according to the observed response.

RESULTS

In 9 of the 51 cases we encountered technical difficulties which resulted in no recordings in five cases. Our learning curve accounted for two of these and a lack of equipment for one. We had difficulty placing the epidural electrode in six cases including one dural tap but satisfactory recordings were obtained in four of these.

In 46 cases recordings were made from 141 nerve roots; no recordings were attempted where avulsion was obvious (except to validate the method) or where function was not compromised. In the nerve roots tested a reproducible evoked potential of any form was taken as evidence of continuity with the spinal cord. Such a response was recorded from 72 nerve roots. No response was obtained from 69. In 13 instances these results were contrary to surgical expectation – two were unexpectedly positive in one patient and 11 were unexpectedly negative in six patients. In all but one of the latter group subsequent histological examination of nerve root biopsies or the findings at follow-up confirmed the electrophysiological diagnosis. In one case after a myelogram which showed C_5 and C_6 intact these roots appeared intact at operation and M-responses were obtained on stimulation but no SEP could be elicited. The patient subsequently recovered muscle power but no sensation in the territory of the upper trunk sustaining the possibility that the sensory rootlets only had been avulsed. We present two illustrative case histories.

Case 1. A 23-year-old male suffered a right brachial plexus lesion among other injuries from a motorcycle accident. From history and examination he had a severe complete plexus lesion with weakness also of trapezius and the right hemidiaphragm. Preoperative myelography showed an intact C_7 nerve root and possible integrity of C_5, C_8 and T_1. At operation an electrode was inserted via the C_{6-7} interspace. After full exposure of the proximal plexus it was clear that C_6 and T_1 were avulsed from the spinal cord but the other roots appeared intact. SEP recordings at the epidural electrode however showed no response from stimulation of C_5 or C_8 but a clear response from C_7 (Figure 1). Nerve grafts were positioned from the C_7 root to the lateral and posterior cords and from C_8 (on the basis of its macroscopic appearance) to the posterior cord. Subsequent histological examination of the nerve root biopsies confirmed the SEP findings. At follow-up after 14 months he had some return of sensation but no power in his upper arm with an advancing Tinel sign. This recovery is not identifiably related to the 7th or 8th cervical nerve root.

Case 2. A 19-year-old man injured his right brachial plexus in a motorcycle accident 11 months prior to surgery. His presentation suggested a preganglionic lesion for C_5 and C_6 but the function of his forearm and hand indicated that the lower plexus was intact. At operation the upper trunk was seen to taper to fibrous cords at the C_5 and C_6 foramina suggesting preganglionic injuries and scar tissue was present throughout the posterior triangle. Stimulation of the lower plexus (C_8 root) gave the expected M-response and a clear short latency SEP at the epidural electrode. Stimulation of C_5 and C_6 gave no M-response but surprisingly a clear and repeatable epidural SEP (Figure 2). Upper trunk stimulation was then observed to produce an M-response in pectoralis major. This implied a lesion in continuity of the upper trunk but a severe one with no spontaneous recovery in 11 months. Neurotization of the suprascapular nerve using the spinal accessory nerve and of the musculocutaneous nerve using the third and fourth intercostal nerves was performed. After two years he had recovered weak supraspinatus function but not biceps activity – unsurprising after such a delayed referral – and he underwent triceps to biceps tendon transfer.

The form of the response has been variable. This is to be expected because of the uncertain position of the recording electrode in the epidural space and the variable exposure and condition of the nerve roots in these injured patients. Although any reproducible SEP was diagnostically acceptable two different characteristic forms of response were identified. A short latency response was identified in 49 of 72 positive recordings and was the only response in 38. This has a latency to the first peak of 1.5 to 2.5 ms and the first peak was often followed by a second flatter peak, the entire response being complete within 5 ms (Figure 3a, b.). Where such a second peak occurred the interpeak latency ranged from 1 to 2 ms with a mean of 1.4 ms. The condition of the nerve roots producing this response was usually judged to be good; of 38 nerve roots 24 appeared undamaged with no supraclavicular lesion distal to them while only seven had supraclavicular lesions in continuity in direct relation to them. A second characteristic response was identified in

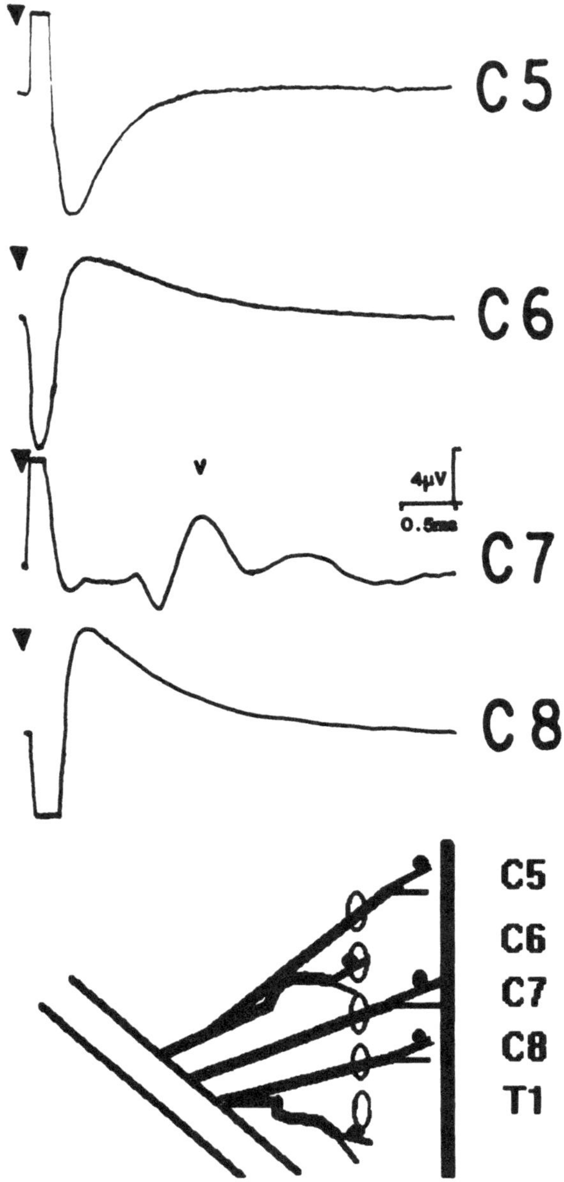

Figure 1 Case 1: Intraoperative SEP traces and diagrammatic representation of operative diagnosis. Latency of marked peak C_7 1.66 ms

21 recordings, being the only response in 10 of these. It has a latency to the first peak of 5 to 8 ms and a variable pattern comprising from one to four negative/positive waves with a duration of 3 to 13 ms (Figure 3c). These

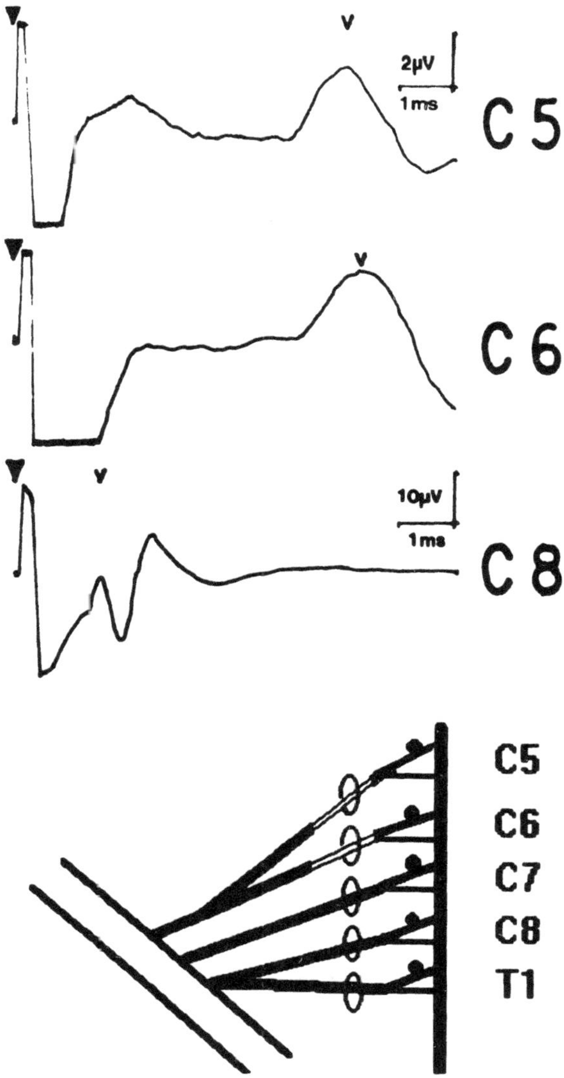

Figure 2 Case 2: Intraoperative SEP traces and diagrammatic representation of operative diagnosis. Latency of marked peaks C_5 6.04 ms, C_6 6.30 ms, C_8 1.52 ms

recordings were associated with supraclavicular lesions in continuity in 12 instances and with severe plexus injuries with avulsion of all roots but one in six cases. In only one case were two such recordings obtained from an apparently normal proximal plexus. In 11 of the 21 recordings this response was preceded by a short latency response as described above.

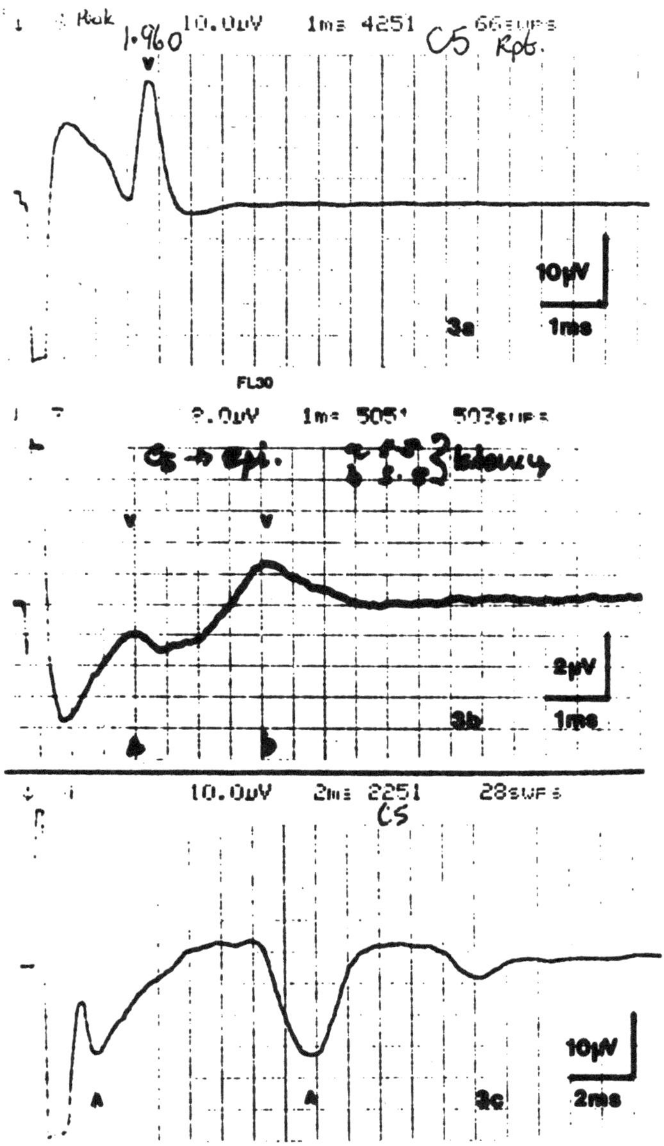

Figure 3 Typical traces: a. and b. are characteristic of short latency traces, c. is characteristic of longer latency traces (see also the top two traces of figure 2)

Thirteen recordings have not been included in this analysis because they were the result of stimulating components of the brachial plexus other than the nerve roots. The responses produced, while serving their diagnostic purpose, were variable in their form and latency.

DISCUSSION

On the basis of these results we believe that stimulation of a normal nerve root of the brachial plexus will give a characteristic short latency response at a recording electrode in the cervical epidural space (Figure 3a, 3b). We would expect this response to be similar to the segmental spinal cord potential described by Shimoji[5] in man following stimulation of peripheral nerves of the upper limb. He describes a triphasic signal $P_1N_1P_2$ (of which P_1 is not always seen) each peak representing the afferent root volley, synchronized activity in the dorsal horn and primary afferent depolarization respectively. N_1 has a duration of 5.5–8 ms and P_2 27 ms or more. If the responses we obtain equate to the N_1P_2 components of such a potential their brevity in our cases may be due to the short transmission distance we employed, probably about one-tenth of that used by Shimoji. This will diminish in proportion the spread of the signal arriving via the mixed nerve at the cord. The longer latency potentials that we have recorded (Figure 3c, Figure 2) were more often associated with apparent injury of the proximal nerve roots and may be indicative of injury at the nerve root entry zone to the spinal cord. These potentials may represent delayed transmission of the normal response or may be revealed elements of a complex waveform which is normally isoelectric due to destructive interference. Their form is very variable in contrast to the short latency potentials. Recording from the epidural space carries some risk related to electrode insertion compared to skin surface recording but it may be appropriate if better information is obtained. We have found the traces to be clearer, less sensitive to anaesthesia and to require less averaging than cortical SEPs but if they provide no more than the 'intact or avulsed' message that cortical SEPs can convey the risk may be judged too great. Proof of the value of the qualitative information we think may be provided by epidural recordings must come from the results of surgery in the few cases where grafts or repairs are made such that recovery of particular functions may be identified with specific nerve root origins.

The uncertainty resulting from electrophysiological findings at odds with preoperative and operative assessment is difficult to resolve because the instances are few and the outcome of nerve grafts and repairs is not wholly predictable. In our series 13 instances of disagreement between electrical and clinical findings occurred in seven patients but in only two cases was action taken whose outcome might be expected to decide the argument. In one patient the electrical findings seem vindicated after 14 months; the other patient has been lost to follow up. It is not possible from this to assess the clinical value of these recordings; more cases and more follow-up will be needed.

References

1. Landi A, Copeland SA, Wynn Parry CB, Jones SJ. The role of somatosensory evoked potentials and nerve conduction studies in the surgical management of brachial plexus injuries. Journal of Bone and Joint Surgery. 1980;62B:492–496.

2. Sugioka H. Evoked potentials in the investigation of traumatic lesions of the brachial plexus. Clinical Orthopaedics and Related Research. 1984;184:85–92.
3. Loughnan BA, King MJ, Grundy EM, Young DL, Hall GM. Effects of halothane on somatosensory evoked potentials recorded in the epidural space. British Journal of Anaesthesia. 1989;62:297–300.
4. Anderson SK, Loughnan BA, Hetreed MA. A technique for monitoring evoked potentials during scoliosis and brachial plexus surgery. Annals of the Royal College of Surgeons of England. 1990;72:321–323.
5. Shimoji K. Human spinal cord potentials (SCPs): Ascending recording variations – an update. In: Ducker TB, Brown RH, eds. Neurophysiology and Standards of Spinal Cord Monitoring. New York: Springer-Verlag; 1988.

22
Automated spinal cord monitoring system with multimodal evoked potentials

M. IMAI, Y. ATSUTA, Y. HARADA, T. IWAHARA and Y. TAKEMITSU

INTRODUCTION

Spinal Cord Evoked Potentials (SCEPs) induced by spinal cord stimulation are widely used in Japan to monitor spinal cord integrity[1,2]. When abnormal mechanical force is applied to the spinal cord, SCEP exhibits a decrease in amplitude and a delay of the peak latency. Evoked potentials to monitor spinal cord function may vary according to disease types and operative techniques. With manual monitoring, a specialist has to repeat meticulous operation of electrodiagnostic equipment including changes of electrodes for multimodal monitoring and analysis of recorded data many times to evaluate the changes of evoked potentials during surgery. The time required for one manual monitoring takes at least several minutes. Thus, for intraoperative monitoring to be used widely and easily, an automated system with multimodal SCEPs was developed.

SYSTEM CONFIGURATION

The equipment system consists of an electromyograph (DANTEC Cantata), a tape recorder, a channel selector, an electrical relay, a MODEM, a voice synthesizer, and a microcomputer which controls all elements (Figure 1). The channel selector is able automatically to turn on or off the connection between the patient's electrodes and the electro-stimulator or the amplifiers. The microcomputer is notebook size with 3.6 MB main memory, 80 MB hard disc drive, a 3.5″ floppy disc drive and colour liquid crystal display, and features such peripheral hardware control as the General Purpose Interface Bus (GPIB) and parallel I/O interface. Raw signals, ECGs, arterial pressure and the surgeon's voice are all recorded on FM tape. Foot switches are able to start or stop the monitoring and selection of electrode combinations for electrical stimulation and recording.

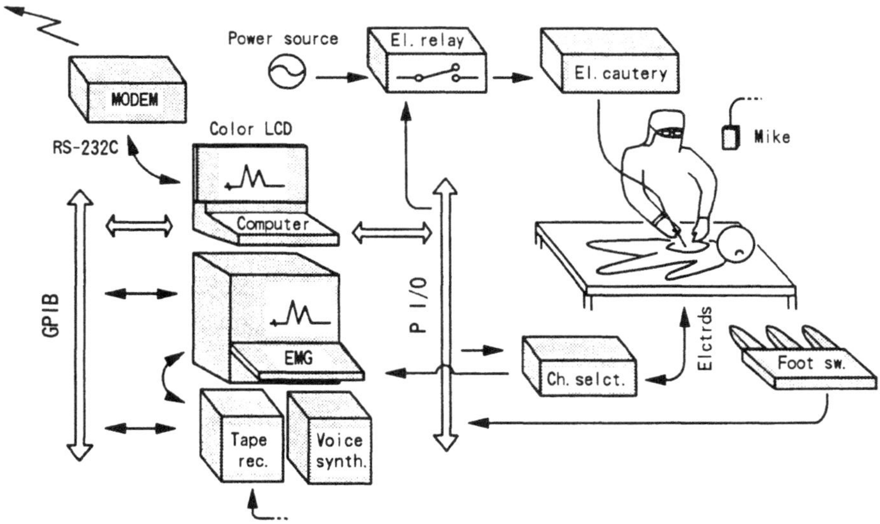

Figure 1 System configuration

MULTIMODAL MONITORING

In this system, SCEPs such as ascending, descending and segmental SCEPs can be previously programed into the computer. The program determines the stimulating electrode's location, frequency, current intensity, recording electrode's location, band pass filter, averaging time, input gain, and sweep speed of display. Any program can be selected by stepping on the foot switch when surgeons want to monitor SCEPs. Automatically repeated monitoring is also possible.

CRITERIA OF ALARM SIGNALS

Alarm signals are given by the microcomputer's flashing colour display and buzzer (Figure 2). There are three types of alarm signals according to the grade of change in amplitude and latency (ms or %):

a. safety range: a blue light which flashes slowly twice
b. caution range: a slightly faster yellow light flashing ten times
c. danger range requiring immediate action: a red light flashing quickly twenty times

Following alarms, announcements of the rate of amplitude changes, delays in latency and suggestions to surgeons are made by the voice synthesizer in English or Japanese.

Potential change				Alarm			State
Shape change	Ratio of amplitude (%)	Change in latency (ms)	(%)	Color	Times	Rate	
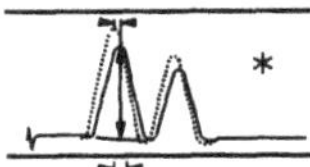	80 <	0.1 >	3 >	Blue	2	Slow	Safe
	50–80	.1–.3	3–5	Yellow	10	Moderate	Caution required
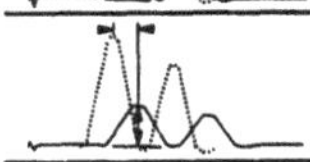	50 >	0.3 <	5 <	Red	20	Rapid	Immediate action required

Figure 2 Criteria of alarm signals. Solid line: latest potential, dotted line: control, negativity upwards. Each parameter is variable according to the disease type and operative techniques. After these alarms, announcements concerning the rate of amplitude change, delays in latency and suggestions to surgeons are made by the voice synthesizer

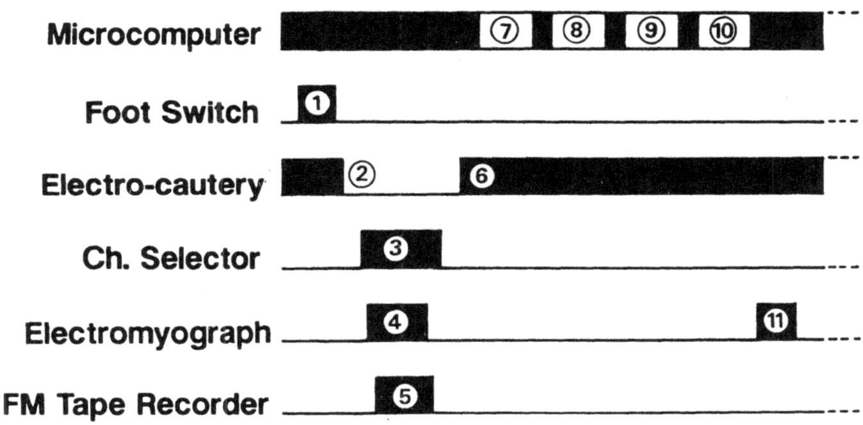

Figure 3 Operation sequence of the equipment

OPERATION SEQUENCE OF THE EQUIPMENT

The intraoperative automatic monitoring is carried out in the following order (Figure 3).

1. The surgeon steps on the foot switch to send the start signal to the computer. The computer is also able to generate starting signals automatically at pre-set intervals.
2. The electro-cautery is cut off by a relay switch.
3. The channel selector chooses the stimulating and recording electrode, and connects them to the stimulator and amplifier of the electrodiagnostic system.

4. The evoked potential is averaged and signal noise is eliminated.
5. The tape recorder records the raw signals, surgeon's voice, ECGs and the arterial pressure for later analysis.
6. Electrode lines to the patient are switched off and power to the electrocautery is restored.
7. The averaged potential is transferred to the microcomputer
8. Amplitudes and latencies of detected peaks are measured and compared with the control.
9. Alarm signals and announcements are given to the surgeon according to the pre-set criteria of conditions.
10. Data are stored on the floppy disc.
11. A hard copy of the SCEP is made which shows amplitudes and latencies.

RESULTS

Since 1986 automated spinal cord monitoring has been successfully performed in over 150 cases including cervical and thoracic spinal canal stenosis, spinal deformities, spinal tumours, traumatic spinal injuries, hip dislocations and aortic aneurysms. Using this automated monitoring system, analysed data information was available to the surgeon about 20 seconds after the starting signal.

In the following case, the system played a notably important role. A 45-year-old female patient with incomplete paralysis underwent removal of a spinal meningioma. The spinal cord evoked potential was used for the automated monitoring (electrical stimulation: C7, recording: T11/12). The yellow and red alarms were activated during removal of the tumour. After observing recovery of the potential, the surgery was completed without any neurological aggravation.

DISCUSSION

It has been reported that iatrogenic neurological complications may be caused by spinal surgery, total hip replacement, and aortic surgery. In Japan, multimodal SCEPs such as ascending, descending and segmental SCEPs have been used for the monitoring of neural integrity. The system described in this report includes fully automated operation of all equipment, automatic analysis of multimodal evoked potentials and automatic alarm generation.

There are three types of alarm signals according to the grade of change in amplitude and latency. The alarm with a flashing display and buzzer allows the surgeon to instantaneously recognize interpreted results. Following these alarms, a voice synthesizer provides the rate of amplitude change, delays in latency and suggestions to surgeons. Although the criteria for each alarm in relation to spike amplitude is still controversial, we chose a 50% decrease from the control as the dangerous limit (red alarm) according to Atsuta[3] and Iwahara[4]. However, even in the yellow alarm range, it would be advisable for the surgeon to re-check the procedure carefully.

Since amplitude decrease without any latency delay has been caused by averaging contaminated potentials in some cases, it was necessary to isolate recording conditions from electrical or mechanical noise (electro-cautery, surgical air tome and suction). Latency delays which were not accompanied by any amplitude decrease were probably due to dislocation of the recording electrode. Even in such cases, our system was capable of re-starting the monitoring immediately with the reset control. Furthermore, since this system is applicable, not only to the monitoring of surgery but also to later analysis of the data, it will contribute to studies of clinical outcomes.

CONCLUSION

1. A reliable automated spinal cord monitoring system based on a microcomputer was developed and over 150 spinal surgeries have been carried out with it since 1986.
2. Alarm signals and announcements were automatically given to the surgeon according to pre-set criteria of amplitude changes and latency delays of the evoked potentials, which allowed the surgeon to react quickly to possible compromise of the spinal cord.
3. With multimodal evoked potentials, this system is applicable to a variety of surgical procedures in which neurological complications may arise.

References

1. Imai M, Harada Y, Atsuta Y et al. Automated spinal cord monitoring for spinal surgery. Paraplegia. 1989;27:204–211.
2. Harada Y. Study on experimental spinal cord injuries by distraction using spinal cord evoked potentials. J Jpn Orthop Ass. 1983;57:685–701, (in Japanese).
3. Atsuta Y, Harada Y, Iwahara T et al. Relationship between locomotor function and spinal cord evoked potential during acute spinal cord compression. J. Electrodiagnosis of the Spinal Cord. 1987;10, 1:5–8 (in Japanese).
4. Iwahara T. The influence of spine distraction on cat spinal cord blood flow and evoked potentials. J Jpn Orthop Ass. 1991;65:44–55.

23
Pain-related somatosensory evoked potentials following CO_2 laser stimulation in myelopathy

R. KAKIGI, Y. KURODA and H. SHIBASAKI

INTRODUCTION

A challenge to workers studying somatosensory evoked potentials (SEPs) is to record pain-related SEPs, that is, responses resulting from impulses ascending through small myelinated fibres[1]. A low power and long wavelength CO_2 laser beam induces pain or heat sensation when applied to the skin, and several papers have reported SEPs induced by CO_2 laser stimuli (pain SEPs). In this study, we attempted to analyse results of pain SEPs in normal subjects and in patients with myelopathies to establish the usefulness of pain SEPs for clinical application

MATERIALS AND METHODS

Fifty normal volunteers ranging from 17 to 75 years of age (mean $\pm$ S.D. = 44.2 $\pm$ 16.6) served as normal controls. Thirty-nine patient were examined. They were eight patients with syringomyelia, 19 with HTLV-1-associated myelopathy (HAM) and 12 with multiple sclerosis (MS).

A special CO_2 laser stimulator for recording SEPs was designed by Nippon Infrared Industries Co. Ltd, Kawasaki, Japan. Its maximum power was 23 W, but an attenuator limited the output to approximately 12.5 W for safety reasons. Laser wavelength was 10.6 μm, and the diameter of the irradiation beam was approximately 2 mm. Stimulus duration was 10 ms, and stimulus interval was 3 s. Stimulus intensity was approximately 18 mJ/mm^2 which elicited sharp pain that normal subjects described as a tolerable 'painlike pin-prick'. Laser beam was routinely applied to the dorsum of hand and foot, but other various areas were also stimulated depending on clinical symptoms of each patient. In order to avoid habituation, irradiated points were moved slightly for each stimulus. The eyes were protected by swimming-goggles.

Silver disk electrodes (1 cm diameter) were attached to the skin with collodion and filled with electrode jelly; impedance was maintained at less than 5 kΩ. Exploring electrodes were placed at Cz (International 10–20 system), because pain SEPs are maximal around the vertex[1,2]. The linked earlobes (A1 + A2) were used as the reference. Amplifier frequency response was 0.2–30 Hz and analysis time was 1.5 or 2.5 s. The number of sampling points was 512 for each channel. Five to 20 responses were averaged in one session, and at least three sessions were done in each subject. After confirming a consistency of waveforms, all averages were added together and the group-averaged waveforms were analysed. Relative positivity at grid 1 resulted in a downward deflection in all records. All normal volunteers and patients gave informed consent. Conventional SEPs following electrical stimulation (electric SEPs) of the median and the posterior tibial nerve were also examined in all patients.

RESULTS

Normal subjects

As the positive components, P340 in hand-stimulated SEPs and P400 in foot-stimulated SEPs, were largest and more stable compared with the preceding negative peaks among normal individuals (Figure 1)[1–3], these positive components were mainly analysed in this study. Mean and SD of their peak latencies were 340 ± 20 and 399 ± 22 ms, respectively. As both P340 and P400 were maximal at Cz and distributed widely over the scalp with no significant asymmetry[1,2], those recorded at Cz electrode were mainly analysed in this study. Conduction velocity (CV) of the spinal cord could be indirectly calculated by measuring peak latencies of hand-stimulated SEPs and foot-stimulated SEPs. When CV of the peripheral nerve was estimated to be[3] 10 m/s, CV of the spinal cord[4] was 9.8 ± 2.4 m/s.

Patients:

(1) Syringomyelia

Both pain and electric SEPs were examined in eight patients with syringomyelia who showed various forms of dissociated sensory loss[5]. Pain SEPs were abnormal in all patients, and their results were generally compatible with the degree of the clinical impairment of pain-temperature sensation. Pain SEPs were useful to detect segmental lesions in the central grey matter and the spinothalamic tract (Figure 2). Subclinical abnormality could also be detected in three patients.

Conventional electric SEPs with the scalp reference were normal in six patients. However, anterior and posterior cervical responses using a non-cephalic reference were absent or small in seven patients. Electric SEPs following posterior tibial nerve stimulation were normal in seven patients.

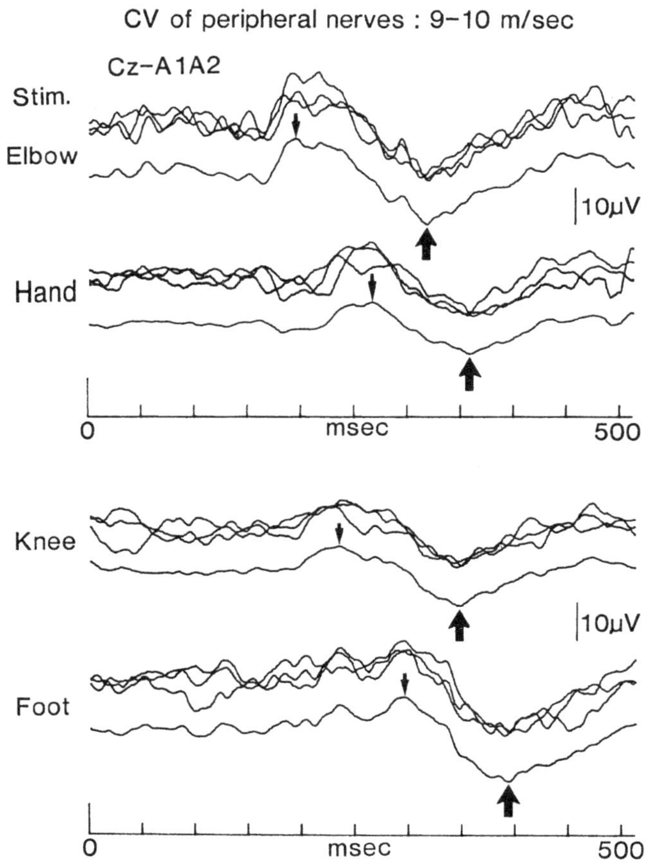

Figure 1 Pain SEPs following stimulation of the foot, knee, hand and elbow in a normal subject. The upper trace for each stimulus site is a superimposition of waveforms from three recordings, and the lower one is their group-averaged waveform. Relative positivity at grid 1 results in a downward deflection in this and all the following figures. Peaks of the negative and positive potentials are shown by small and large arrows, respectively. We termed the positive potential following the hand and foot stimulation, P340 and P400, respectively. Conduction velocity (CV) of Aδ fibres of the upper limb was calculated from the difference of the positive peak latencies following the hand and elbow stimulation and the distance between the two stimulus sites. Similarly, CV of the lower limb was calculated with respect to the positive potential following the foot and knee stimulation. CV of the upper and lower limb in this subject was 7.1 and 7.3 m/s, respectively[3]

These findings suggest that the function of the ascending fibres through the posterior column is intact in most patients, whereas the dorsal horn, where a fixed cervical potential is generated, is impaired. Pain SEPs combined with electric SEPs therefore appear to be extremely useful for investigating physiological function in the sensory pathways in patients who show 'dissociated sensory loss' such as in syringomyelia.

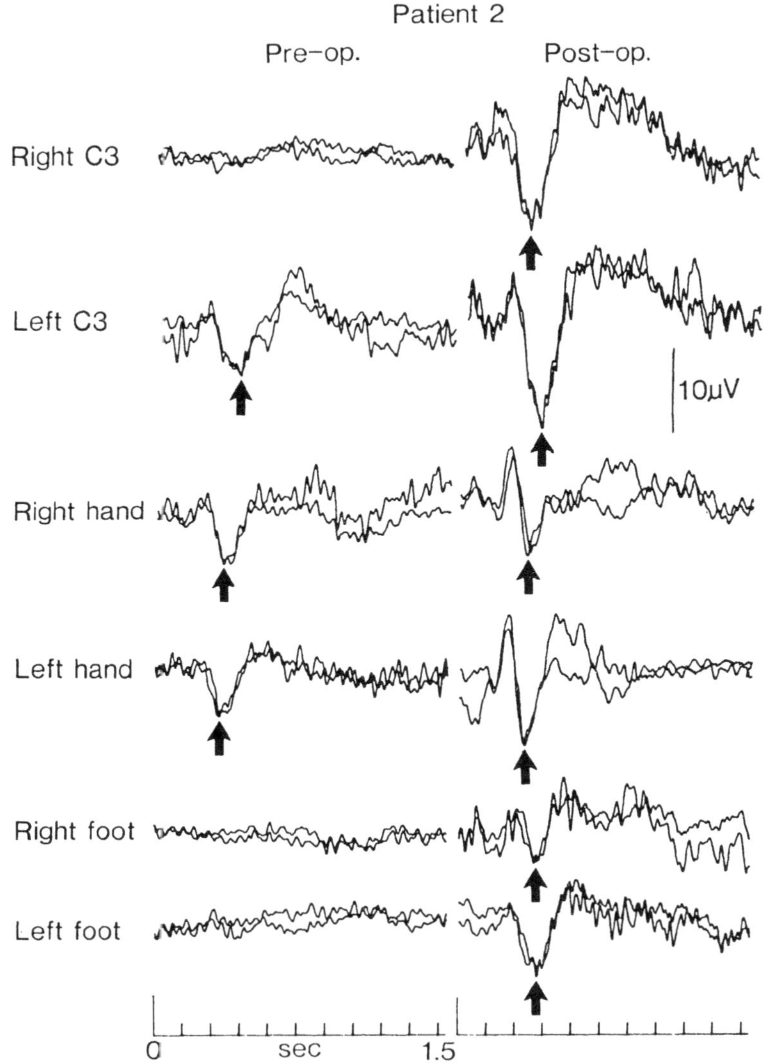

Figure 2 Pain SEPs before and after the operation in a patient with syringomyelia, whose syrinx was found from C2 to C4 by MRI. Before the operation, no significant response could be identified following stimulatin of the right C3 dermatome (over the clavicle), or right or left feet, and the response following left C3 stimulation was present with significantly prolonged latency. As pain-temperature sense of feet was judged to be normal by clinical examination, the finding indicated 'subclinical abnormality'. After the operation, the responses following stimulation of those areas were clearly identified[5]

(2) HTLV-I-associated myelopathy (HAM)

Pain and electric SEPs were examined in 19 patients with HAM[6]. The latency of P340 potential following hand stimulation was within the normal range in all patients. The P400 potential following stimulation of the foot was significantly prolonged in four patients. CV of STT was significantly decreased in nine patients, and eight among them showed no impairment of pain-temperature sensation (subclinical abnormality) (Figure 3). Mean CV of STT (6.4 ± 3.8 m/s) was significantly smaller than normal controls by Student's *t*-test ($p < 0.01$).

As for SEPs following median nerve stimulation, N9 recorded at Erb's point was of normal latency and amplitude in all patients. The interpeak latency between N9 and N13 was significantly increased in nine patients, but the interpeak latency between N13 and N20 (so-called central conduction time) was within the normal range in all patients.

As for SEPs following posterior tibial nerve stimulation, N19 recorded at 12th thoracic vertebra and N27 at 2nd cervical vertebra were recorded in 17 and 11 patients respectively. In those patients, peak latency of N19 was normal in all, but N27 was significantly prolonged in nine patients. The first cortical positive potential, P37, recorded at Cz' was significantly prolonged in 14 out of all 19 patients. Indirect measurement of CV of the posterior column (PC) was significantly decreased in 14 out of 19 patients, and vibration sensation was judged to be normal in seven out of them (subclinical abnormality). CV of PC (29.9 ± 8.7 m/s) was significantly smaller than normal controls by Student's *t*-test ($p < 0.01$).

(3) Multiple sclerosis (MS)

Pain and electric SEPs were examined in 12 patients with MS[7]. P340 in the hand SEPs was normal in 19 arms, significantly delayed in one arm, and could not be recorded in four arms. The presence of cervical cord lesions was estimated by clinical findings in all three patients whose P340 was abnormal. P400 could not be consistently detected in 11 feet (6 patients), and was significantly delayed in three feet (Figure 4). In five patients with a normal P340 but unidentified or delayed P400, the lesions were judged to be in the thoracic spinal cord by clinical findings. All of the patients in whom pain SEPs were abnormal showed impairment of pain-temperature sensation to some degree.

As for SEPs following median nerve stimulation, N9 recorded at Erb's point was normal in all of the patients. Neither N13 at the cervical vertebra nor N20 at the scalp was recorded in five arms of three patients, in whom vibration sense was impaired to some degree. In one patient only N20 was significantly delayed. Since vibration sense in her arm was judged to be normal by clinical examination, this finding indicated a 'subclinical abnormality'.

As for electric SEPs following posterior tibial nerve stimulation. N19 recorded at the 12th thoracic vertebra was normal in all of the patients. The first positive cortical potential, P37, was significantly delayed in ten feet (five patients) and could not be clearly identified in eight feet (four patients). Vibration sensation was impaired in the foot in four patients whose P37 was absent.

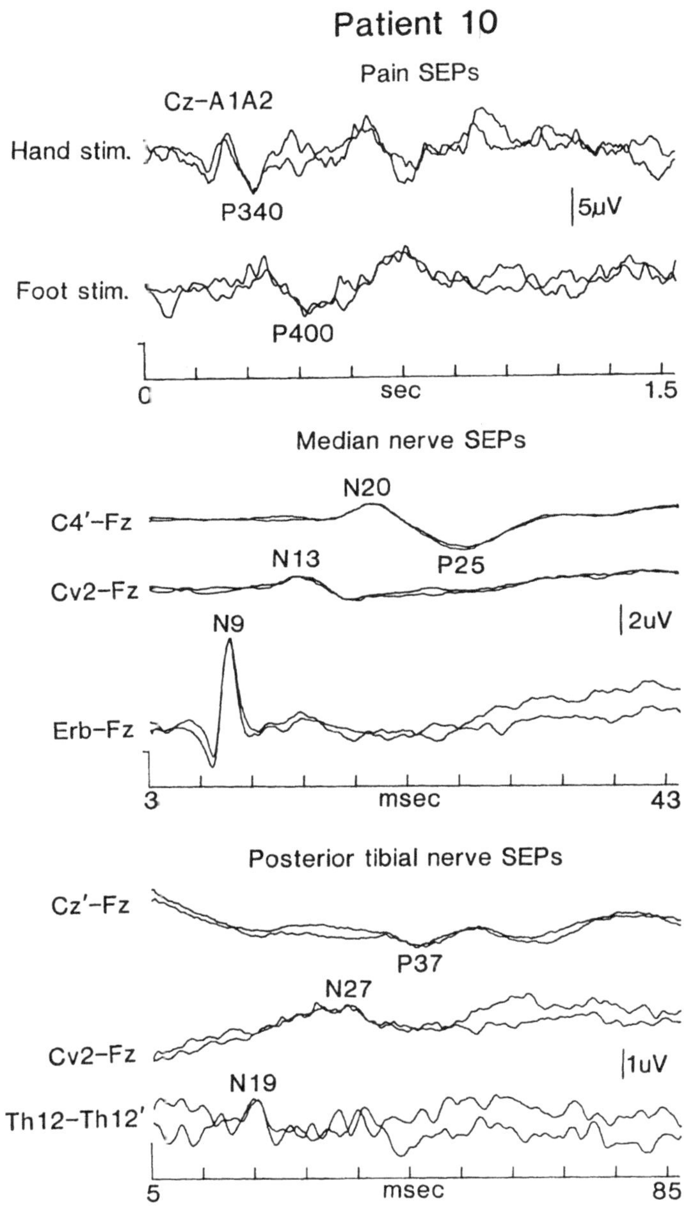

Figure 3 Pain SEPs, median nerve SEPs and posterior tibial nerve SEPs in a patient with HAM. Peak latency of both P340 and P400 is within the normal range, but CV of STT is significantly small (3.7 m/s) (subclinical abnormality). Peak latency of N9 is normal, but the interpeak latency between N9 and N13 is significantly prolonged (5.3 ms). Peak latency of N19 is within the normal range, but that of N27 as well as P37 is significantly prolonged. CV of PC, 30.6 m/s is significantly small[6]

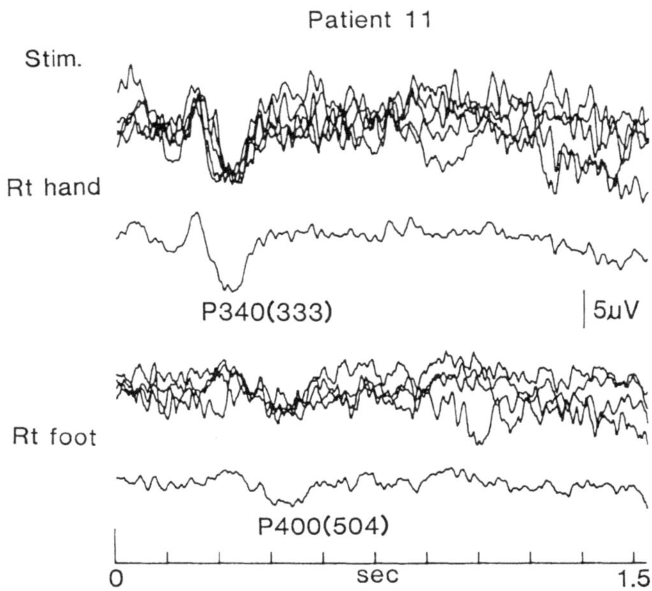

Figure 4 Pain SEPs following stimulation of the dorsum of the right hand and foot in a patient with multiple sclerosis. The upper trace is a superimposition of several averages and the lower one is their summed waveform. Exploring electrode is placed at Cz and the reference is the linked earlobes. P340 following the right hand stimulation was normal, but P400 following the right foot stimulation was significantly delayed[7]

DISCUSSION

As clinical evaluation of pain-temperature sensation by using pin-prick is neither objective nor quantitative, pain SEP following CO_2 laser stimulation is considered to be very useful to investigate physiological functions of the ascending sensory tracts which are responsible for pain-temperature sensation.

In normal volunteers, amplitude decreased and latency increased with reduction in the stimulus intensity and the subjective pain feeling. Lidocaine injection to produce anaesthetic nerve block resulted in loss of pain SEPs, but the potential was relatively preserved during ischaemic nerve block[1]. Approximate CV of the peripheral nerves obtained in the present investigation, 10 m/s, falls in the lower- to mid-range of CV of Aδ fibres[3]. Therefore, we consider pain SEPs to be generated by impulses arising from pain stimuli and ascend through Aδ fibres. This hypothesis was supported by results in patients with peripheral neuropathies. In that study, results of pain SEPs were consistent with an impairment of pain-temperature sensation and with a density of small myelinated fibres[3].

Indirect measurement of CV of the ascending spinal tract following CO_2 laser stimulation in normal subjects gave the value of approximately 8–10 m/s. This value is consistent with CV of the spinothalamic tract by which pain and temperature sensations are mainly mediated. Measurement of CV of the spinothalamic tract by using pain SEPs was extremely useful to detect subclinical abnormality in HAM. Pain SEPs were also useful to confirm lesions in the central grey matter of the spinal cord in patients with syringomyelia in whom conventional electric SEPs using a scalp reference were generally normal. Pain SEPs were also very useful to detect lesions in the spinothalamic tract in MS.

In conclusion, pain SEP following CO_2 laser stimulation is very useful to study the physiological functions of the spinothalamic tract, and its usefulness was much increased when conventional electrical SEPs were also examined to study the functions of the posterior column.

References

1. Kakigi R, Shibasaki H, Ikeda A. Pain-related somatosensory evoked potentials following CO_2 laser stimulation in man. Electroenceph clin Neurophysiol. 1989;74:139–146.
2. Kakigi R, Shibasaki H, Neshige R, Ikeda A, Kuroda Y. Pain-related somatosensory evoked potentials in cortical reflex myoclonus. J Neurol Neurosurg Psychiatry. 1990;53:44–48.
3. Kakigi R, Endo C, Neshige R, Kuroda Y, Shibasaki H. Estimation of conduction velocity of Aδ fibers in humans. Muscle Nerve. 1991c;14:1193–1196.
4. Kakigi R, Shibasaki H. Estimation of conduction velocity of the spino-thalamic tract in man. Electroenceph clin Neurophysiol. 1991;80:39–45.
5. Kakigi R, Shibasaki H, Kuroda Y, Neshige R, Endo C, Tabuchi K, Kishikawa T. Pain-related somatosensory evoked potentials in syringomyelia. Brain. 1991b;114:1871–1889.
6. Kakigi R, Kuroda Y, Takashima H, Endo C, Neshige R, Shibasaki H. Physiological functions of the ascending spinal tracts in HTLV-I-associated myelopathy (HAM). Electroenceph clin Neurophysiol. 1992a;84:110–114.
7. Kakigi R, Kuroda Y, Neshige R, Endo C, Shibasaki H. Physiological study of the spinothalamic tract conduction in multiple sclerosis. J Neurol Sci. 1992b;107:205–209.
8. Kakigi R, Shibasaki H, Tanaka K, Ikeda T, Oda KKI, Endo C, Ikeda A, Neshige R, Kuroda Y, Miyata K, Yi S, Ikegawa S, Araki S. CO_2 laser-induced pain-related somatosensory evoked potentials in peripheral neuropathies: correlation between electrophysiological and histopathological findings. Muscle Nerve. 1991a;14:441–450.

24
Level diagnosis of cervical myelopathy using somatosensory evoked potentials

N. KAWAHARA, H. BABA, S. NAGATA, Y. KIKUCHI, K. TOMITA and H. YUGAMI

INTRODUCTION

Cervical myelopathy is commonly diagnosed by neurologic examination and radiologic examination, such as myelography, computed tomography and MRI. However the specific level of the major lesion has sometimes been difficult to determine by these methods in the patients with multisegmental spondylosis, ossification of posterior longitudinal ligament (OPLL), and a narrow spinal canal. The purpose of this study is to improve identification of the level of major lesion in cervical myelopathy using somatosensory evoked potentials (SEP) techniques.

PATIENTS AND METHODS

Twenty-one patients with cervical myelopathy without radiculopathy due to unisegmental spinal lesion underwent SEPs test preoperatively. The causes of myelopathy were cervical disc herniation in nine cases, cervical spondylosis in seven cases, and localized OPLL in five cases. There were 17 males and four females aged 36 to 65 (mean 54 years). In all cases the specific level of the major lesion was known to be unisegmental by myelography followed by computed tomography and MRI. The level of lesion was C3–4 spinal level in four patients, C4–5 in six, C5–6 in nine and C6–7 in two. The patient's functional disability was graded as mild when daily activity only was affected, moderate when there was dependence on aids or help, and severe when the patient was bedridden, chair-bound or had useless upper extremities[1]. One patient with C3–4 lesion was graded mild and three patients were graded moderate. Two were mild, four were moderate with C4–5 lesion. Two were mild, six were moderate and one was severe with C5–6 lesion. One was mild and one was moderate with C6–7 lesion.

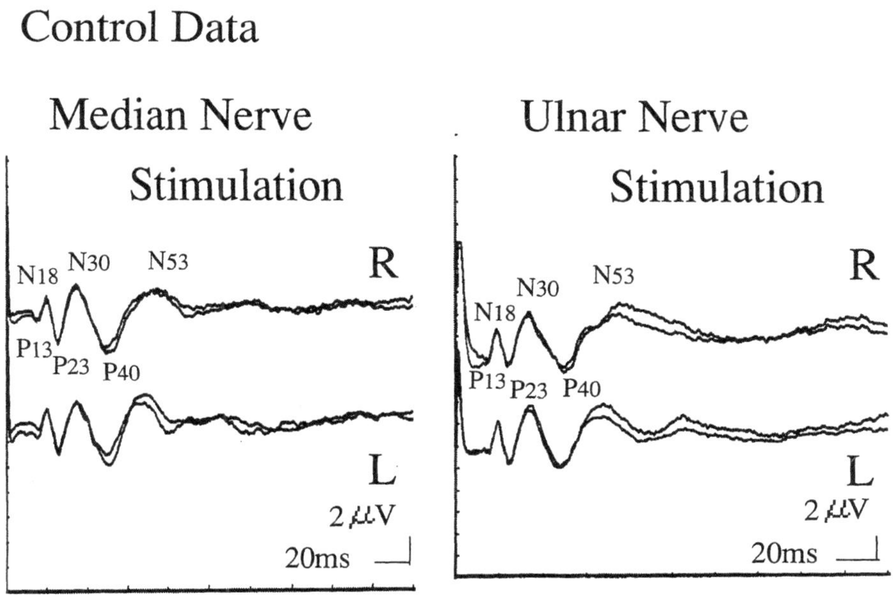

Figure 1 Representative examples of the SEP normal pattern. The initial deflection of each trace was an artefact following the electrical stimulation. In all traces an upward deflection is negative

ELECTROPHYSIOLOGICAL METHODS

Electrical stimuli (square-wave constant current pulses of duration 0.2 ms) were delivered at 3 Hz to the median and ulnar nerves at the wrist. The intensity was adjusted so as to produce a moderate twitch of the appropriate muscles. SEPs were recorded with a silver–silver chloride disc electrode attached over the hand area of the contralateral sensory cortex (7 cm lateral to a point 2 cm posterior of the vertex). A reference electrode was placed on the ear contralateral to the side of stimulation. Two hundred and fifty-six sweeps synchronized with the stimulus were amplified with the bandpass of 10 Hz to 10 kHz and averaged with a DISA 1500 system. Components within 200 ms from the stimulus artefact were averaged.

EVALUATION

Figure 1 shows representative SEPs after median nerve and ulnar stimulation. For the establishment of the normative values of the SEPs peak latencies (P13, N18, P23) and inter peak amplitudes (P13–N18, N18–P23, P23–N30), were studied in a control group of 100 healthy adults.

Limits of normal latency were defined by the mean ± 2SD of the control group compensated for height. Amplitude limits were also defined by the mean ± SD of the control[2].

RESULTS

Lesion at the C3–4 spinal level

Three had bilateral and one had unilateral abnormality of latency in median nerve SEPs. All four had bilateral abnormality of amplitude in median nerve SEPs. Three had bilateral and one had unilateral abnormality of latency in ulnar nerve SEPs. All four had bilateral abnormality of amplitude in ulnar nerve SEPs.

Lesion at the C4–5 spinal level

Four had bilateral and two had unilateral abnormality of latency in median nerve SEPs. Five had bilateral and one had unilateral abnormality in median nerve SEPs. Five had bilateral and one had unilateral abnormality of latency in ulnar nerve SEPs. Five had bilateral and one had unilateral abnormality of amplitude in ulnar nerve SEPs.

Lesion at the C5–6 spinal level

One had bilateral and one had unilateral abnormality of latency in median nerve SEPs. One had bilateral and two had unilateral abnormality of amplitude in median nerve SEPs. Eight had bilateral and one had unilateral abnormality of latency in ulnar nerve SEPs. Eight had bilateral and one had unilateral abnormality of amplitude in ulnar nerve SEPs.

Lesion at the C6–7 spinal level

Both patients had SEPs within normal limits following median nerve stimulation. One had bilateral abnormality of latency in ulnar nerve SEPs. One had bilateral abnormality of amplitude in median nerve SEPs.

Illustrative cases

1. A male office worker, aged 58 years. Examination revealed a mild spastic tetraparesis. He had paresthesiae and numbness in the fingers of both hands. Myelography demonstrated severe interior indentation of the cord at C4–5. Both median and ulnar nerve SEPs showed slightly prolonged latency and marked attenuation (Figure 2).
2. A male policeman, aged 42 years. He had a spastic gait, and developed paresthesiae and numbness in the ulnar site of both hands and forearm. Fine movements of both hands were impaired. Myelography demonstrated severe anterior indentation at C5–6. Median nerve SEPs were within normal limits but ulnar nerve responses were markedly reduced in amplitude and increased in latency (Figure 3).

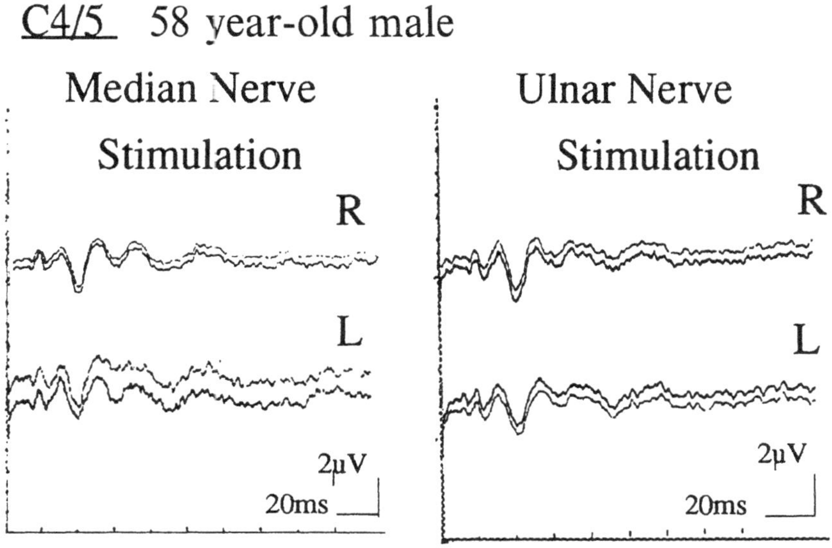

Figure 2 Case 1, a 58-year-old male with C4–5 cervical disc herniation. Both median and ulnar nerve SEPs showed slightly prolonged latency and very low amplitude

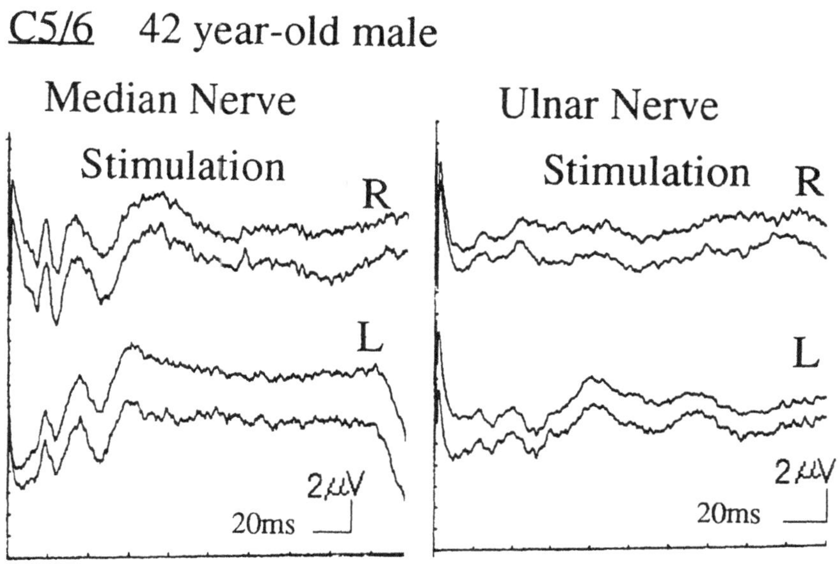

Figure 3 Case 2, a 42-year-old male with C5–6 cervical disc herniation. Median nerve SEPs were within normal limits but ulnar nerve responses were markedly reduced in amplitude and increased latency

DISCUSSION

Several papers[3,4] have reported segmental lesion identification in cervical myelopathy using spinal cord evoked potentials (SCEPs). However, SCEPs were recorded from the spinal epidural space using catheter electrodes with stimulation of peripheral nerves or the spinal cord. It is true that SCEPs are very sensitive methods for identifying segmental lesions in myelopathy but are invasive. Attempts have therefore been made to diagnose the responsible level of cervical myelopathy using SEP techniques after peripheral nerve stimulation which is non-invasive.

In this study all five patients with C4–5 spinal lesion had abnormal median and ulnar nerve SEPs. All nine patients with C5–6 spinal lesion had abnormal ulnar nerve SEPs, while only three of nine patients had abnormal median nerve SEPs.

The fibres of the median nerve mainly enter the C6 spinal cord segment, and the fibres of ulnar nerve mainly enter the C8 spinal cord segment. Tsuzuki et al.[5] reported the neuroanatomical relationships between the spinal column and entry zones of each spinal nerve. The spinal cord segment is situated about 1–1.5 segments above the corresponding vertebra. The C6 spinal cord segment which is the main entry zone of the median nerve is situated at the C4–5 vertebral level. So it may be difficult to detect a cord lesion at C5–6 vertebral level using median nerve SEP techniques.

CONCLUSION

1. Abnormal SEPs after median nerve stimulation suggest a lesion at or above C4–5 vertebral level in cervical myelopathy patients.
2. Abnormal SEPs after ulnar nerve stimulation with normal median nerve SEPs suggest a lesion at C5–6 or C6–7 vertebral level in cervical myelopathy patients.

References

1. Yu YL, Jones SJ. Somatosensory evoked potentials in cervical spondylosis. Brain. 1985;108:273–300.
2. Yugami H, Koshino Y, Yamaguchi N, Matsubara F. Somatosensory evoked potentials in normal young adults (in Japanese). Clin Electroencephalography. 1982;24:330–335.
3. Satomi K, Okuma T, Kenmotsu K, Nakamura Y, Hirabayashi K. Level diagnosis of cervical myelopathy using evoked spinal cord potentials. Spine. 1988;13:1217–1224.
4. Shinomiya K, Furuya K, Sato R, Okamoto A, Kurosa Y, Fuchioka M. Electrophysiologic diagnosis of cervical OPLL myelopathy using evoked spinal cord potentials. Spine. 1988;13:1225–1233.
5. Tsuzuki N, Honda H, Tanaka Y. Morphological variations of cervical spinal cord segments and their clinical significance (in Japanese). Seikeigeka. 1983;34:229–235.

25
Spinal cord monitoring during surgery for intramedullary spinal cord tumours

M. MURAKAMI, M. MOCHIZUKI, Y. OKAMOTO, A. MURATA and H. TORIKAI

INTRODUCTION

Spinal cord monitoring during spinal surgery using various methods has been performed widely to prevent iatrogenic spinal cord injury. At Chiba University Hospital, the spinal evoked potential elicited by direct spinal cord stimulation (conductive SpEP) was developed and has been applied clinically since 1972. During the past 21 years, 619 patients with various spinal disorders were monitored intraoperatively with this method. The purpose of this paper is to report the results of intraoperative spinal cord monitoring using this method for intramedullary spinal cord tumours.

MATERIALS AND METHODS

From May 1982 to December 1991, 25 patients with an intramedullary spinal cord tumour underwent spinal cord monitoring. They included 12 female and 13 male patients whose ages ranged from 7 to 56 years, with a mean age of 37 years. Tumours were located in the cervical or cervico-thoracic cord in 15 patients and in the thoracic cord in ten patients; of the latter, seven were midthoracic and three were at the level of the conus medullaris. The pathological diagnosis of the tumours included astrocytoma (11 patients), ependymoma (nine patients) and miscellaneous tumours (five patients).

After removing the laminae, both stimulating and recording electrodes were inserted under visual control in the posterior epidural space. The stimuli were delivered to the rostral part of the spinal cord and the evoked potential (conductive SpEP) was recorded at the caudal part of the spinal cord. Square-wave pulses of 0.2 ms duration and supramaximal intensity (5 to 15 mA) were applied at a rate of 10 to 30/s. The low and high frequency filter cutoffs were at 20 and 2000 Hz, respectively, and the evoked potentials were averaged 20 to 200 times to eliminate electric noise. A control SpEP was recorded before opening the dura mater.

RESULTS

Postoperative neurological complications

At operation, macroscopically complete tumour removal was believed to have been accomplished in 17 patients. However, 17 patients experienced neurological deterioration in the immediate postoperative period. Each postoperative neurological complication was recorded as one of two grades: mild, consisting only of deterioration of sensation, especially a dorsal column function, and severe, consisting of deterioration of both sensory and motor function. There were ten patients who experienced a mild complication and seven patients who experienced a severe complication. Two patients with postoperative Brown-Séquard syndrome were included in the latter group.

Basic SpEP pattern

It is impossible to specify, as in electrocardiography, a standard SpEP configuration. The wave configuration varied from patient to patient, being influenced mainly by the neurological status of the spinal cord. The basic SpEP pattern seen with an intramedullary tumour could be divided into two groups; type I, in which we could observe both an initial spike and a polyphasic component, and type II, in which only a polyphasic component could be seen. Sixteen cases were classified as type I and six as type II. Patients with type II SpEP had more severe symptoms preoperatively than did patients with type I. Three patients with paraplegia had no SpEP preoperatively.

Changes in spike component amplitude

The changes in amplitude of the spike component during surgery are illustrated in Figure 1. Just after opening the dura mater, the amplitude increased by 17% on average. Thereafter, as a result of myelotomy and preparation of the tumour, the amplitude decreased gradually. When the amplitude decreased to less than 50% of the control, we interrupted the operation. In such cases, we did not continue the operation if the amplitude did not recover.

The amplitude increased by 5%, on average, at the end of surgery. However, there were four cases in which the amplitude had decreased intraoperatively to less than 80% of the control. Three of these four patients experienced severe neurological complications, including the Brown-Séquard syndrome. Although the amplitude of spike component recovered and

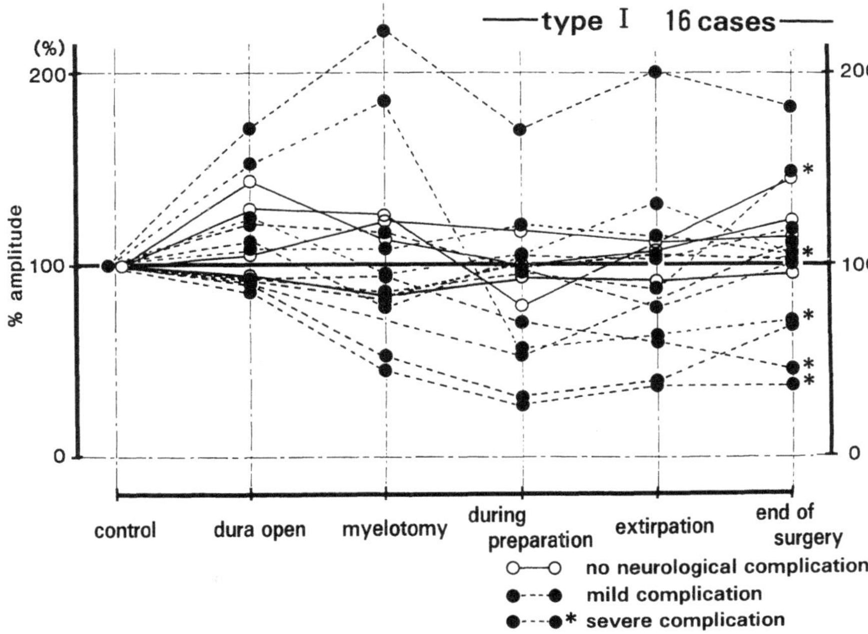

Figure 1 Changes in amplitude of the spike component which are related to the operative procedures. There were four cases in which the SpEP amplitude decreased to less than 80% of the control at the end of surgery. Three of these four patients experienced severe neurological complications, including the Brown-Séquard syndrome

increased above the control level by the end of surgery, two patients woke up with motor weakness in their upper extremities. In these cases, we could not detect any damage to the grey matter of the cervical spinal cord as reflected by the spike component of the SpEP. Operative procedures which induced a decrease in amplitude including myelotomy in three patients, manipulation and/or extirpation of the tumour in five patients and both procedures in two patients.

Changes in amplitude of the polyphasic component

The changes seen intraoperatively in the amplitude of the polyphasic component are illustrated in Figure 2. In 14 of 21 cases, the amplitude dropped below the control level at the end of surgery. The average percent decrease was 30%. Myelotomy was the major operative cause of a decrease in the amplitude of the polyphasic component. Seven of eight patients who developed an isolated sensory disturbance, especially of dorsal column function, experienced a decrease in the amplitude of the polyphasic wave at the end of surgery. In brief, deterioration of dorsal column function was related closely to a decrease in the polyphasic component amplitude.

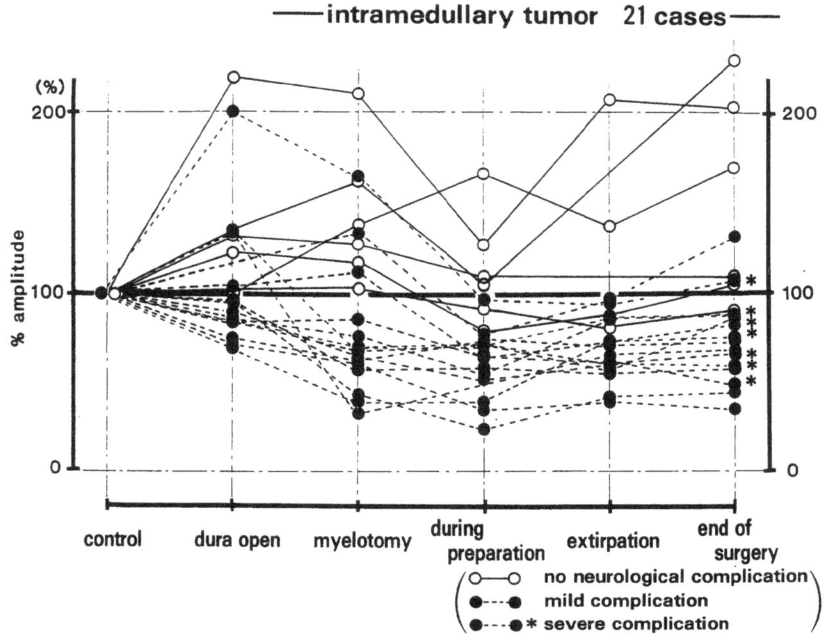

Figure 2 Changes in amplitude of the polyphasic component during surgery. Seven of eight patients who developed an isolated sensory disturbance, especially of dorsal function, experienced a decrease in the polyphasic wave amplitude at the end of surgery

Illustrative cases

Case 1. In a forty-eight-year-old man, spontaneous haematomyelia due to an intramedullary venous angioma at the lower cervical level was removed completely. Just after myelotomy, the amplitude of the polyphasic component decreased drastically and did not recover until the end of surgery. On the other hand, this patient exhibited only small, subcritical, changes in the amplitude of the spike component. After the operation, this patient lost dorsal column function in his right lower extremity without developing any gross muscle weakness in his lower extremities (Figure 3).

Case 2. A fifty-two-year-old man presented with an intramedullary ependymoma at the lower thoracic level. After opening the pia mater, the amplitude of both spike and polyphasic components increased. However, these potentials decreased drastically during extirpation of the tumour while the spinal cord was compressed with a retractor at the Th8 level. No recovery of the potential was observed. After the operation, this patient awoke from anaesthesia with a Brown-Séquard syndrome below the Th8 level (Figure 4).

DISCUSSION

Spinal cord monitoring during spinal surgery using various neurophysiologic methods has been performed widely to prevent iatragenic spinal cord damage.

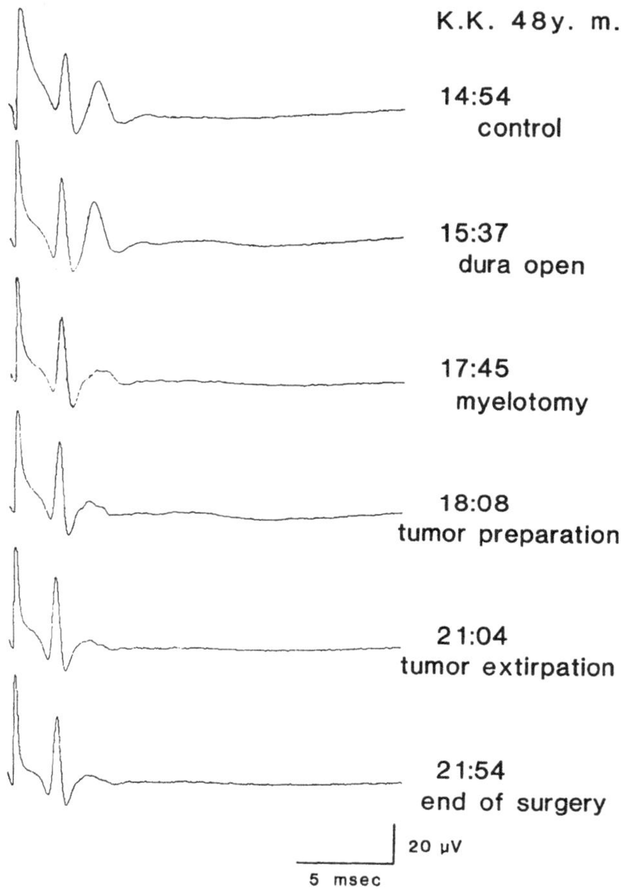

Figure 3 Intraoperative SpEPs in Case 1. Just after myelotomy, the amplitude of polyphasic component decreased drastically and did not recover until the end of surgery. After the operation, the patient lost dorsal column function in his right lower extremity without developing any gross muscle weakness in his lower extremities

Electrical potentials which can be measured and can be used clinically at the present time include the (1) somatosensory evoked potential (SEP) and (2) spinal cord evoked potentials (SpEPs), as described in the present study. Although SEP has the great merit of being non-invasive, it also has several disadvantages. The SEP appears to be too sensitive to changes of body temperature, haemodilution, blood pressure and anaesthetic drugs. Consequently the authors consider this technique too unstable for use in spinal cord function monitoring.

A method of spinal cord monitoring utilizing SpEPs elicited by direct stimulation of the spinal cord was established separately in1972 by Tamaki et al. and by Kurokawa[2]. Since then, this method has been widely accepted in Japan. During the past 20 years, 619 patients with spinal disease or injury

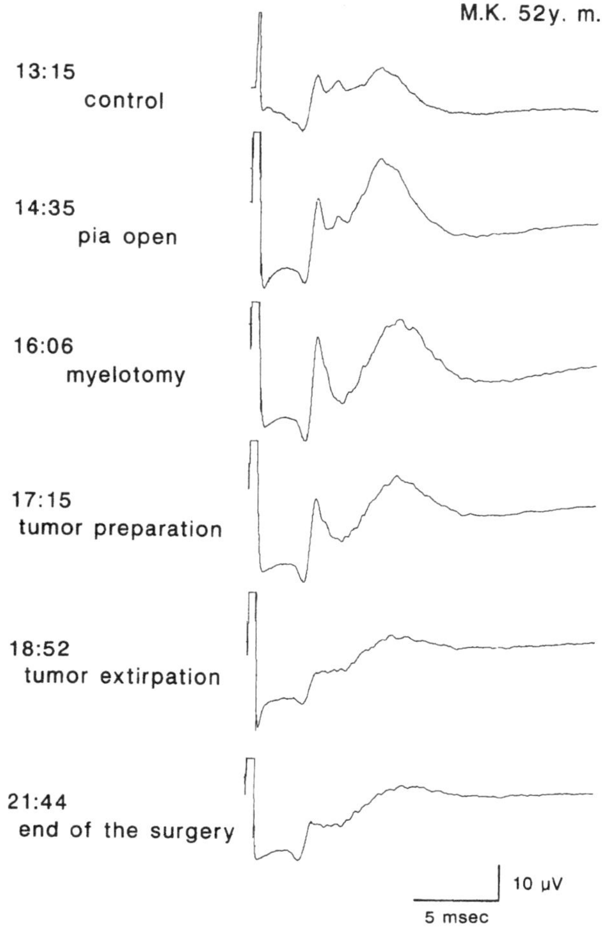

Figure 4 Intraoperative SpEPs in Case 2. The amplitude of both spike and polyphasic components decreased dramatically during extirpation of the tumour. At that time, the spinal cord was compressed with a retractor at the Th8 level. No recovery of the potential was observed. After this operation, the patient awoke with a Brown-Séquard syndrome below the Th8 level

were monitored at Chiba University Hospital. These included patients with spinal deformities (332 patients), spinal cord tumours (100 patients), ossification of spinal ligaments (43 patients), thoracic or thoraco-abdominal aortic aneurysms (24 patients) and miscellaneous pathological conditions of the spine and spinal cord. The effectiveness of SpEP monitoring during spinal operations has already been established[3,4].

It is not possible to specify a standard SpEP configuration. The SpEP configurations vary from patient to patient, being influenced by delicate differences in relations between the stimulating and recording electrodes and the neurological condition of the spinal cord. If the stimulating electrode is placed at the upper thoracic level and the recording electrode is placed in the subarachnoid space at the level of conus medullaris, a SpEP (descending conductive SpEP) from a neurologically intact patient consists of an initial spike component and the polyphasic component which follows. There have also been several studies to describe the pathways through which each component of the SpEP is transmitted[5,6]. The initial spike component of the SpEP has been recognized as the summation of action potentials of nerve fibres with a large diameter that are located not only in the postero-lateral region but in every quadrant of the spinal cord, including the anterior portion. The polyphasic component is considered to be conducted mainly through the dorsal column. However, postsynaptic potentials seem also to contribute to this component.

The parameters used to analyse the potential are the peak amplitude, wave configuration and latency. In practice, however, the peak amplitude of the spike component is the most reliable indicator, with enough sensitivity to reflect hazardous injury to the spinal cord. Recently the authors[3] analysed data obtained from 270 patients with scoliosis and concluded that a 50% reduction of spike component amplitude was the critical level associated with severe neurological injury in patients with normal spinal cord function. This opinion is similar to that of Yasukawa[7] and Jones et al.[8], but it may be necessary to modify this criteria for patients with neurologically deteriorated spinal cord, especially when intraspinal surgical manipulation is performed. In patients with a neurologically deteriorated spinal cord, conductive SpEP with atypical configurations are usually seen and every component should be closely scrutinized. Changes in amplitude of 20% or more are thought to signify neurological damage to the spinal cord in such patients.

In conclusion, intraoperative spinal cord monitoring using the conductive SpEP is a useful and reliable method to evaluate spinal cord function in order to prevent irreversible spinal cord damage. We cannot, however, relate every complex function of the spinal cord to a SpEP wave configuration. Thus, it is impossible to monitor completely spinal cord function by simply observing the evoked potential, and it may be necessary to monitor simultaneously other parameters in patients with more complicated risk factors.

References

1. Tamaki T, Yamashita T, Kobayashi H, Hirayama H. Spinal cord monitoring. Jpn J Electroencephalogr Electromyogr. 1972;1:196.
2. Kurokawa T. Spinal cord action potentials evoked by epidural stimulation of cord – A report of human and animal record. Jpn J Electroencephalogr Electromyogr. 1972;1:64–66.
3. Murakami M, Okamoto Y, Murata A, Minami S, Kitahara H. Spinal cord monitoring using spinal cord evoked potentials during surgery for scoliosis. J Jpn Scoliosis Soc. 1991;6:78–83.
4. Tamaki T, Noguchi T, Takano H, Tsuji H, Nakagawa T, Imai K, Inoue S. Spinal cord monitoring as a clinical utilization of the spinal evoked potential. Clin Orthop. 1984;184: 58–64.

5. Harada Y. Study on experimental spinal cord injuries by destruction by using spinal cord evoked potentials. J Jpn Orthop Assoc. 1983;57:685–701.
6. Toyoda A, Kanda K. Origins of spinal cord potentials evoked by stimulation of the cat spinal cord. In: Momma S and Tamaki T, eds. Fundamentals and Clinical Application of Spinal Cord Monitoring. Tokyo: Saikon Press; 1984:99–111.
7. Yasukawa K. Experimental and clinical studies on the evoked electrospinogram for monitoring spinal cord function. J Jpn Orthop Assoc. 1980;54:1661–1677.
8. Jones SJ, Edgar MA, Ransford AO. A system for the electrophysiological monitoring of the spinal cord during operations for scoliosis. J Bone Joint Surg. 1983;65B:134–139.

26
Graded risk levels of somatosensory evoked potentials during scoliosis surgery

Y. NODE, S. NAKAZAWA and S. UEMATSU

INTRODUCTION

It is essential to monitor spinal cord integrity throughout spinal surgery, particularly during spinal manipulation and instrumentation. However, because the procedures are carried out under general anaesthesia, the neurological status cannot be tested until the patient awakens after surgery. The application of intraoperative somatosensory evoked potential (SEP) monitoring makes it possible to assess the integrity of the peripheral nerve, spinal cord, and central neuronal pathway. Clearly, the total disappearance of the SEP is an unequivocal alarm signal, but it would be helpful to have a value that would constitute a warning before complete disappearance. Therefore, we followed a series of patients undergoing surgery with instrumentation for correction of scoliosis and developed risk levels by measuring SEP at different times during surgery. In this paper we provide information that will facilitate earlier awareness and timely detection of SEP deterioration to prevent catastrophic spinal cord injury.

MATERIALS AND METHODS

Eighty-six patients undergoing surgery with instrumentation for scoliosis were followed throughout surgery by SEP monitoring. SEP recordings meeting the following criteria were reviewed: 1) adult patients with idiopathic scoliosis; 2) no apparent neurological deficit before and after spinal instrumentation; 3) the main SEP components (P1 to N3) were reproducible; and 4) SEPs were recorded at least every 15 minutes throughout surgery.

Twenty-six patients were male and sixty were female. The mean patient age was 32 years (range 16–69 years).

Anaesthesia was induced with thiopental sodium and maintained by isoflurane or, in some cases, by addition of an intravenous injection of morphine sulphate or fentanyl. Systemic blood pressure, rectal temperature, and arterial blood gases were recorded.

Spinal instrumentation for reduction or stabilization of scoliosis was typically carried out with Harrington distraction and/or Luque Cotrel-Dubousset, or Wisconsin compression apparatus.

SEP recordings were made with a TECA-TE42 electromyograph equipped with a DE6 digital averager and a nerve stimulator. The posterior tibial nerves at the ankles were stimulated simultaneously. Technical details of the method are described elsewhere[1].

SEPs obtained after posterior tibial nerve stimulation could be divided into six major components for the first 100 ms. The components were identified by polarity and numbered by sequence of occurrence (P1, N1, P2, N2, P3, and N3). Downward polarity is designated as positive (P) and upward as negative (N).

SEP monitoring was divided into three surgical stages: Stage I, before instrumentation; Stage II, during instrumentation; and Stage III, after instrumentation. Stage I starts upon induction of general anaesthesia and the sustained stable anaesthetic state during exposure of the spine. The second stage begins upon application of spinal instrumentation. The third stage starts upon completion of the instrumentation, typically during wound closure. SEPs were recorded at least every 15 minutes, with more frequent recordings made during Stage II. Latencies and peak-to-peak amplitudes were recorded at each stage. The relative percentage change was calculated by comparing Stage II and III values with the value at Stage I (baseline value). Paired two-tailed t-tests were performed.

RESULTS

Latency

Table 1 summarizes relative percentage delay (RPD) of the latency at Stage II and III compared with that in Stage I. For example, the RPD of P1 during Stage II averaged 4.3% SD 1.7%. RPD values in Stage II and III were similar. Based on these data, acceptable latency deterioration, namely risk levels correlated with the latency delay, are determined as shown in Table 1. For P1, the RPD is 4.3% $\pm$ 1.7% in Stage II. Hence, to be acceptable, delay should not be in excess of 4.3% of the Stage I value. If it is above 4.3%, the value should be designated as requiring caution. If the value is at mean + 1 SD level, that is, 6.0% longer than the Stage I value, it is designated the warning level. When the delay is in excess of 7.7%, it is designated as the alarm value. Risk levels for Stage III can be determined in the same manner.

Amplitude

Relative decrease of amplitude at Stage II and III compared with the value of Stage I is listed in Table 2. Based on these values, the caution, warning,

Table 1 Relative prolongation (%) of latency at Stage II and III compared with the value of Stage I

SEP Component	*Stage*	*Mean ± SD*	*Caution* (Mean)	*Warning* (+1 SD)	*Alarm* (+2 SD)
P1	II	4.3 ± 1.7	4.3	6.0	7.7
	III	4.9 ± 2.6	4.9	7.5	10.1
N1	II	3.9 ± 1.9	3.9	5.8	7.7
	II	4.8 ± 3.3	4.8	8.1	11.4
P2	II	4.0 ± 2.5	4.0	6.5	9.0
	III	4.5 ± 2.9	4.5	7.4	10.3
N2	II	4.6 ± 3.3	4.6	7.9	11.2
	III	4.9 ± 3.8	4.9	8.7	12.5
P3	II	5.3 ± 4.6	5.3	9.9	14.5
	III	4.8 ± 2.8	4.8	7.6	10.4
N3	II	6.3 ± 6.3	6.3	12.6	18.9
	III	7.3 ± 6.5	7.3	13.8	20.3

Table 2 Relative decrease (%) of amplitude at Stage II and III compared with the value of Stage I

SEP Component	*Stage*	*Mean ± SD*	*Caution* (Mean)	*Warning* (+1 SD)	*Alarm* (+2 SD)
N1	II	29.6 ± 12.9	29.6	42.5	55.4
	III	29.6 ± 14.1	29.6	43.7	57.8
P2	II	33.3 ± 14.5	33.3	47.8	62.3
	III	34.9 ± 16.8	34.9	51.7	68.5
N2	II	36.6 ± 16.0	36.6	52.6	68.6
	III	33.6 ± 17.1	33.6	50.7	67.8
P3	II	26.3 ± 15.6	26.3	41.9	57.5
	III	26.5 ± 14.1	26.5	40.6	54.7
N3	II	19.0 ± 11.5	19.0	30.5	42.0
	III	21.1 ± 12.4	21.1	33.5	45.9

and alarm risk values for the amplitude deterioration can be determined. Caution is required if the amplitude of N1 is decreased more than 29.6% during Stage II; a warning level is reached if the amplitude decreases more than 42.5%; and the risk level should cause alarm if the value decreases by 55.4%. Risk levels during Stage III are also listed in Table 2.

Lastly, overall risk levels for combined Stage II and III values are summarized in Table 3.

DISCUSSION

How much change in the SEP waveform indicates significant disturbance of cord function? Although the answer is clear when the waveform disappears completely at the exact moment of spinal instrumentation, we need to have some guidelines before complete disappearance of the potentials. Most of the published data for the warning level appear to have been derived, or arbitrarily designated, based on laboratory normative values[2–8].

Table 3 Risk levels on SEP deterioration

	Caution	*Warning*	*Alarm*
Latency delay (%)*			
Stage II	4.3	6.0	7.7
Stage III	4.9	7.5	10.1
(Overall)	(5%)	(8%)	(10%)
Amplitude decrease (%)**			
Stage II	29.6	42.5	55.4
Stage III	29.6	43.7	57.8
(Overall)	(30%)	(45%)	(60%)

*P1 latency
**N1 latency

In our study, the values were calculated from those that had deteriorated in Stage II and III during and after spinal instrumentation. In the statistical calculations determining the latency delay that would require a warning, the values that did not change or were even faster than the baseline values were not included. Similarly the amplitude that did not change or actually increased during the surgical stages were excluded. Hence, our results should provide more practical risk levels than values derived from the normative data.

Furthermore, for practical application we have graded the deterioration to three risk levels, those requiring caution, warning, and alarm. We think that the percentage values for each component should be used separately rather than being assigned an overall value. However, for convenience, the overall value is presented here for an approximate estimate. We trust that these values will assist anaesthesiologists and surgeons to secure effective monitoring of the integrity of the spine during sophisticated spinal surgery. We conclude that latency delay of 5% and/or an amplitude decrease of 30% from the baseline value should alert the clinician to apply cautious observation (CAUTION), a latency delay of 8% and/or amplitude decrease of 45% is a warning about a potentially serious problem (WARNING), and a latency delay of 10% and/or amplitude decrease of 60% from the baseline necessitates immediate corrective action to prevent possibly irreversible spinal cord injury (ALARM).

References

1. Uematsu S, Tolo VT. Recording of the somatosensory evoked potential during surgery for scoliosis and midline myelotomy to monitor spinal cord function. Electromyogr Clin. 1981;21:253–266.
2. Brown RH, Nash CL Jr, Berilla JA, Amaddio MD. Cortical evoked potential monitoring: a system for intraoperative monitoring of spinal cord function. Spine. 1984;9:256–261.
3. Dinner DS, Luders H, Lesser RP, Morris HH, Barnet G, Klem G. Intraoperative spinal somatosensory evoked potential monitoring. J Neurosurg. 1986;65:807–814.
4. Gonzales EG, Hajdu M, Keim H, Brand L. Quantification of intraoperative somatosensory evoked potential. Arch Phys Med Rehabil. 1984;65:721–725.
5. Grundy BL, Nash CL, Brown RH. Deliberate hypotension for spinal fusion: Prospective randomized study with evoked potential monitoring. Can Anaeth Soc J. 1982;19:452–462.

6. Mostegl A, Bauer R, Eichenaur M. Intraoperative somatosensory evoked potential monitoring. A clinical analysis of 127 surgical procedures. Spine. 1988;13:396–400.
7. Schramm J. Spinal cord monitoring: Current status and new developments. Cent Nerv Syst Trauma. 1985;2:207–227.
8. York DH, Chabot RJ, Gaines RW. Response variability of somatosensory evoked potentials during scoliosis surgery. Spine. 1987:864–876.

27
Intraoperative spinal cord monitoring with spino-spinal evoked potentials

M. NOZAWA, Y. YAMAUCHI, T. TSUJI, Y. ARAI and H. KAJIHARA

INTRODUCTION

During the past eight years we have conducted intraoperative spinal monitoring in 56 cases of spinal or spinal cord surgery to prevent postoperative neurological complications using spino-spinal evoked potentials with electrical stimulation. This paper describes our experience with intraoperative changes of the evoked potentials. To detect changes of spinal cord function during spine or spinal cord surgery many investigators had reported different experimental or clinical methods. In 1972, Vauzelle described the useful intraoperative wake-up test for patients operated upon for severe spinal deformity[1]. Anaesthesia was adjusted for the wake-up test which allowed sufficient dissociation of pain and consciousness to permit spontaneous motion of the extremities in partial or temporary awakening during the course of anaesthesia. In 1984 Brown[2] reported the intraoperative use of somatosensory cortical evoked potentials to monitor spinal cord function during 300 orthopaedic surgical procedures. Brown described disadvantages of this method in which the wave-forms were frequently affected by medications and anaesthetic agents. Spino-spinal evoked potentials (S-SEPs) which are observed following direct stimulation of the spinal cord, from recording electrodes placed on the dura, has been developed mainly in Japan since 1972. These potentials are characterized by a high amplitude and sharply shaped wave-forms and changes in the potentials can be more easily evaluated. In 1983 Jones[3] reported 115 patients who underwent Harrington instrumentation and were monitored by spinal SEPS from the epidural space. Jones concluded that spinal somatosensory evoked potentials are sensitive to minor spinal cord impairment. Recently in order to monitor descending pathways during surgery, motor evoked potentials have been recorded from thenar muscles and from the epidural space. Recently, noninvasive motor cortex magnetic stimulation has become popular in Japan.

SUBJECTS AND METHODS

Spinal cord potentials were recorded in 56 patients. The diagnoses were as follows: scoliosis in 39 patients, spine and spinal cord tumour in 11, thoraco-abdominal aortic aneurysm in five, ossification of the posterior longitudinal ligament (OPLL) in one.

S-SEP monitoring by this method was the only spinal cord monitoring used over the entire period of the study. Usually we catheterize the epidural space with bipolar epidural tube type electrodes. Stimulating and recording electrodes are identical. According to the level of the operation we can decide the location for the electrodes. In most operations both electrodes are placed on the dura at rostral and caudal locations. Ascending or descending evoked potentials are selected according to the sharpness of the wave-form. We have used the Medelec MS6 electromyograph recorder for a long time in our operating room applied using a bipolar epidural tube type electrode. The constant current stimulator delivered square wave electrical impulses of 0.2 millisecond duration, and the intensity was between 15 and 100 volts. The stimulus rate was 20 Hz. The potentials recorded using this method consisted mainly of two components which had negative deflection peaks. In 1976, Imai[4] reported from his spinal transection experiments that the first component of the ascending potentials observed with this method reflected lateral column function. The second component mostly reflected dorsal column function.

Scoliosis cases

The operative methods for scoliosis were Harrington instrumentation in 29 patients and Cotrel-Dubousset instrumentation in 10 patients.

We compare the initial recordings with those obtained during the course of spinal instrumentation in scoliosis cases. A reduction of amplitude occurred in eight patients, and there was a more than 50% change in the amplitude of the response in two out of eight patients, but there were no postoperative neurological complications. In four patients there was an amplitude increase. Although we cannot explain this increase, it is possible to speculate that it was the result of displacement of the electrode when the spinal deformity was corrected or of changes in the blood circulation in the spinal cord after instrumentation. The latency of the response was increased in six patients out of 39 patients, but this was not very significant.

The spine and spinal cord tumour cases

We also compared the initial recordings with those obtained in the course of excision of spine and spinal cord tumours. The pathological diagnosis of the tumours were neurinoma in seven patients, ependymoma in one, neurenteric cyst in one, and metastatic thyroid cancer in one patient. A reduction of amplitude was observed in four out of 11 patients, but there was a more than 70% change in only one of these. This was one of two patients who developed neurological symptoms postoperatively. The latency

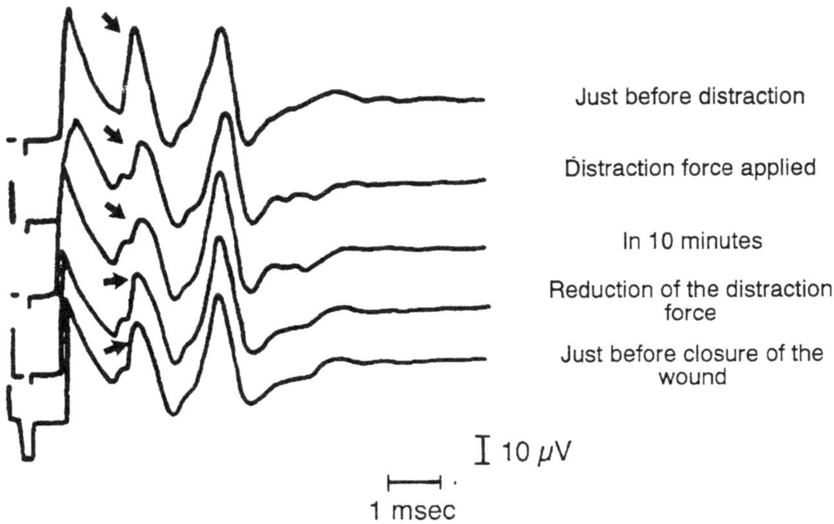

Figure 1 This patient was a 14-year-old boy with a hemivertebra. Harrington instrumentation was performed. A 25% reduction of the first component was observed within three minutes after initial Harrington rod distraction, and the latency of the first component was increased. Ten minutes after distraction the amplitude did not recover, therefore, the distraction force was reduced

of the response was increased in four out of 11 patients, the greatest change being 0.5 milliseconds.

The thoraco-abdominal aortic aneurysm cases

We had the chance to monitor the spinal cord function in five patients with thoraco-abdominal aortic aneurysm. In contrast with the spinal operations discussed above, a reduction of amplitude was rare. The latency of the response varied from 1.0 milliseconds to 1.5 milliseconds. In one case, the potentials gradually diminished and completely disappeared during aortic clamping. This patient developed complete paraplegia after the operation.

Case presentations

Case 1. The patient was a 14-year-old boy with a hemivertebra. During a one-year period there had been significant progression of the curve. Harrington instrumentation was performed. He was fused from T7 through L3. A 25% reduction of the first component was observed within three minutes after initial Harrington rod distraction, and the latency of the first component was increased. Ten minutes after distraction the amplitude did not recover, therefore the distraction force was reduced. Before wound closure, 85% of the amplitude had recovered to 85% of the original wave-form. There were no complications postoperatively (Figure 1).

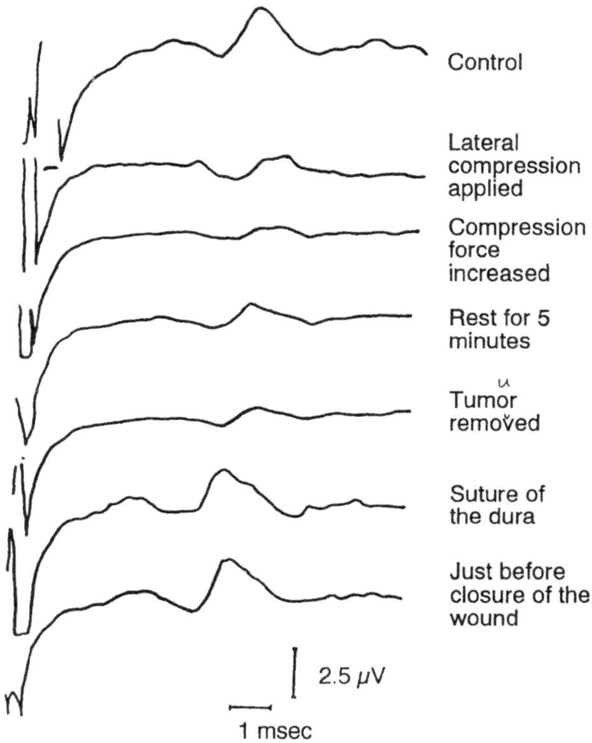

Figure 2 This was a 57-year-old woman with an intradural extramedullary neurinoma located at T8. The amplitude of the response decreased suddenly after lateral compression was applied to the spinal cord for excision of the tumour. At the completion of the manoeuvre within five minutes, the wave-form had partially recovered

Case 2. This was a 57-year-old woman with an intradural extramedullary neurinoma located at T8. MRI showed that the tumour was located at the anterior portion of the spinal canal. The amplitude of the response decreased suddenly after lateral compression was applied to the spinal cord for excision of the tumour. The reduction of the amplitude was more than 70% when compared to the initial recording. At the completion of the manoeuvre within five minutes, the wave-form had partially recovered. Just before wound closure the amplitude had returned to the initial value and the latency was decreased about 0.5 milliseconds (Figure 2).

Case 3. This patient was a 79-year-old man with a thoraco-abdominal aortic aneurysm. This patient had replacement of the aneurysm with a Gore-Tex artificial vessel. During aortic clamping the potentials gradually diminished and completely disappeared. Even after release of the clamp electrical silence continued. The patient developed complete paraplegia and died 14 days after surgery from other complications (Figure 3).

Case 4. This case was not included in this series but deserves description. This was a case of spinal cord monitoring during spinal cord angiography

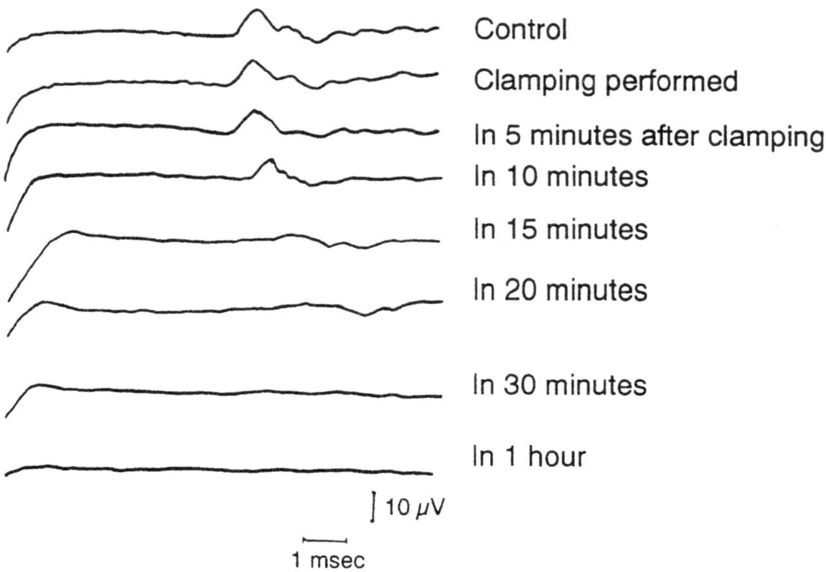

Figure 3 This patient was a 70-year-old man with thoraco-abdominal aortic aneurysm. This patient had replacement of the aneurysm with a Gore-Tex artificial vessel. During aortic clamping the potentials gradually diminished and completely disappeared

with embolization. This patient was a 52-year-old man with A-V malformation of the spinal cord vessels. He had bilateral dysesthesia in his lower extremities since he was 48-years-old. Myelogram and spinal cord angiography showed abnormal findings of the spinal cord vessels from C6 to L1. The main feeding artery was supplied from the right 7th intercostal artery. Blood flow through this artery was stopped with a balloon tipped catheter. During this temporary disruption of blood flow the wave-form did not change. Two weeks thereafter, the patient had embolization of this artery. He had a good recovery from his symptoms (Figure 4).

DISCUSSION

We have conducted intraoperative spinal cord monitoring in 56 cases by spino-spinal potentials with electrical stimulation. There were three cases with postoperative neurological complications. These complications occurred in two cases of spinal cord tumour and one case of OPLL. The amplitude of the potentials had decreased markedly after the manoeuvre compressing the spinal cord laterally. We have thought that the red signal representing spinal cord impairment was decrease of the amplitude more than 50%. However, in spine or spinal cord tumour cases spinal cord dysfunction was already present preoperatively, so that whenever we monitor these cases, we should pay attention to any slight decrease in amplitude during the operation.

Although S-SEP is supposed to represent function mainly in the dorsal columns and in the posterior half of the lateral column, our experience

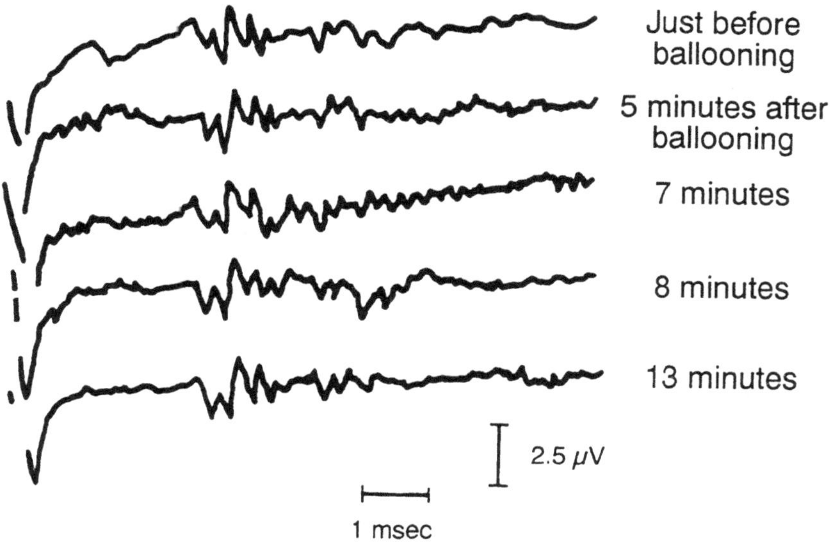

Figure 4 This was a case of spinal cord monitoring during spinal cord angiography with embolization. The patient was a 52-year-old man with A-V malformation of the spinal cord vessels. The main feeding artery was supplied from right 7th intercostal artery. Blood flow through this artery was stopped with a balloon tipped catheter. During this temporary disruption of blood flow the wave-form did not change

showed that it is a dependable procedure in view of the good correlation with clinical results and the absence of false negative cases. It is important to identify which monitoring method is ideally appropriate for monitoring spinal cord function in different patients undergoing different procedures.

References

1. Vauzelle C et al. Functional monitoring of spinal cord activity during spinal surgery. Clin Orthop. 1973;93:173–178.
2. Brown RH et al. Cortical evoked potential monitoring. A system for intraoperative monitoring of spinal cord function. Spine. 1984;9:256–261.
3. Jones SJ et al. A system for the electrophysiological monitoring of the spinal cord during operations for scoliosis. J Bone Joint Surg. 1983;65-B:134–139.
4. Imai T. Human electrospinogram evoked by direct stimulation on the spinal cord through epidural space. J Jap Orthop Ass. 1976;50:1037–1056.

28
Monitoring cortical evoked potentials (EPs) in operations on the cervical spine

N. J. SMITH, D. BEER, S. A. CLARKE, L. M. HENDERSON, A. JARDINE and A. MATHEW

INTRODUCTION

Cortical evoked potential recording was the first method of spinal cord monitoring to be introduced[1,2]. Cortical responses are affected by halogenated anaesthetic agents and by hypotension, which may be responsible for variability of amplitude and latency. To avoid this disadvantage of cortical recordings, invasive methods of monitoring have been devised, involving electrodes in the epidural space or interspinous ligament. Although these methods are widely used for operations on the thoracic and lumbar spine, they are not usually applicable in cervical surgery because electrodes cannot be placed in the epidural space or interspinous ligament above the level of operation. We have therefore used cortical responses to monitor operations on the cervical spine.

PATIENTS AND METHODS

Between January 1983 and December 1991 38 operations on the cervical spine of 33 patients were monitored at University Hospital, Nottingham. Preoperative and intraoperative records were reviewed, together with clinical notes and anaesthetic records where possible. Patients' ages ranged from eight years to 82 years (mean 45.3 years). The indications for operation were traumatic fractures (13 patients), rheumatoid subluxation (eight patients), secondary carcinoma (three patients), chordoma (one patient), cervical spondylosis (three patients), congenital anomalies (three patients) and subluxation due to torticollis (two patients).

All operations involved fixation of the spine, either with sublaminar wires or pedicle screws, and in most cases a bone graft was also used. No laminectomy operations were included in the present series.

Patients were given fentanyl 100 μg, induced 1 min later with thiopentone 4 mg/kg, and intubated following atracurium 0.6 mg/kg, ventilation until intubation being maintained on a mask and airway using 1% isoflurane with nitrous oxide and oxygen. Increments of fentanyl and atracurium were administered as required. Anaesthesia was maintained with either halothane 0.5% or isoflurane 1% as the volatile agent. In early cases hypotension was induced with a beta-blocker followed by glyceryl trinitrate (GTN) infusion, but more recently labetolol has been used preferentially with addition of a GTN infusion if necessary.

Stimulation was applied bilaterally to the tibial nerves at the ankles, and to the median nerves at the wrists. Nerve action potentials were recorded at the knee (tibial nerve) and in the supraclavicular fossa (median nerve) to ensure that stimulation had been applied correctly. The stimulus was a square wave pulse from a constant current stimulator (Dantec Neuromatic), pulse duration 0.2 ms, amplitude 30–40 mA. The stimulation rate was usually set at 7 Hz, but if cortical response amplitudes were small the rate was reduced to 3 Hz.

Responses were recorded from silver/silver chloride disc electrodes applied with collodion at the vertex (Cz, International 10–20 System) for the legs, and from the contralateral central region (C3 or C4) for the arms. A frontal reference electrode was applied, midway between Fz and Fpz. Electrodes were connected to a locally produced electrode selector box, having switches so that any electrode could be chosen as either active (lead 1) or reference (lead 2) without the need to insert or remove plugs during the operation. Responses were recorded on an electrically isolated recording system (Dantec Neuromatic). The input sensitivity was 0.5 μV per division, low pass filter 1000 Hz, high pass 50 Hz. This limited bandpass was chosen to give increased stability of responses, as recommended by Nuwer and Dawson[3]. These authors used a high pass filter at 30 Hz, but our equipment did not include a high pass filter at this frequency, so 50 Hz was chosen. The raw trace was displayed continuously so that obvious artefacts could be identified. Usually 500 responses were averaged, but occasionally 1000 responses were needed to give adequate noise elimination. Each average was transferred through the IEE488 link to an Apple IIc computer which was programmed to display a reference trace from the beginning of the operation together with the five averages acquired most recently. All traces were stored on disc and printed out later.

A significant change in response was identified according to the criteria of Veilleux et al.[4]. If initial responses were stable a 20% reduction in amplitude was regarded as significant, or a 1 ms increase in latency. If early responses were unstable a 50% drop in amplitude was used as the limit. In practice responses were usually so unstable that the 50% level was chosen. The surgeon was only informed that a change had occurred if it could be confirmed in repeated recordings, and if all extraneous factors could be excluded as the cause.

The spontaneous EEG as monitored continuously with a Cerebral Function Analysing Monitor (CFAM; Health Care Developments)[5]. This provided a plot of the amplitude and frequency content of a single channel

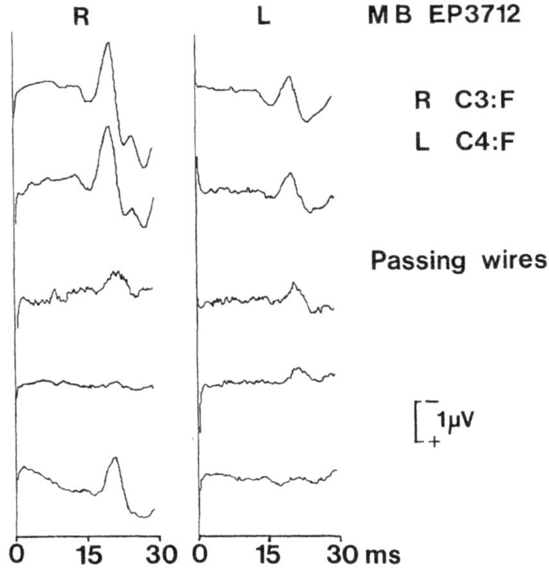

Figure 1 Cortical potentials recorded from the arms during an operation for rheumatoid atlanto-axial subluxation. The amplitude was initially smaller from the left arm, and responses from both arms showed a significant reduction in amplitude when sub-laminar wires were passed. There was some recovery in the response from the right arm before the end of the operation

of EEG on a chart moving at 302 cm/h and permitted detection of periods of gross amplitude suppression, which would account for disappearance of cortical evoked potentials.

RESULTS

No significant change in evoked potentials was detected in 36 of the 38 operations, and these patients awoke without new neurological deficit. One man aged 38 years with traumatic dislocation at C4/5, whose monitoring had been uneventful, awoke without weakness or sensory symptoms in any limb. The next morning, about 12 hours after the completion of the operation, he was found to have a profound quadriparesis, and evoked potentials from all limbs were absent. Some improvement occurred over the next few months, but one year after the operation he was still unable to lead an independent existence.

In two other cases significant changes were identified. The first was a woman of 56 with rheumatoid atlanto-axial subluxation, in whom there was a marked difference in amplitude from the two arms initially (smaller from the left arm) (Figure 1). When sublaminar wires were passed the evoked potential amplitude from the left arm fell by 50%, and the response from the right arm disappeared completely. The response from the right arm re-

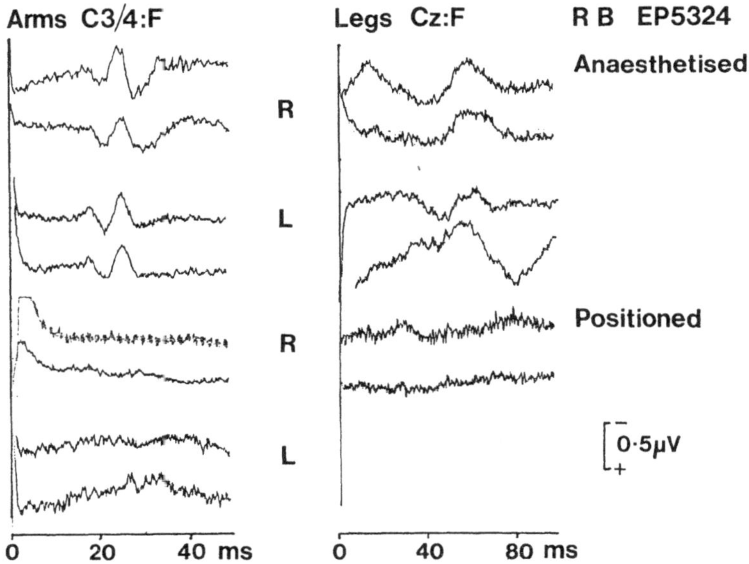

Figure 2 After induction of anaesthesia clear cortical responses were recorded from arms and legs in this man with bi-facetal fracture of C5. Responses disappeared after reduction of the fracture and positioning on the table

appeared before the end of the operation, but that from the left arm showed no recovery. When the patient regained consciousness post-operatively no new neurological signs could be detected.

The second patient was an alcoholic man of 47 with a bi-facetal fracture of C5, who had marked instability on neck flexion. There was evidence of a C4 root lesion, but no evidence of a cord lesion. Skull traction was applied, but he developed delirium tremens and removed the tongs, violently flexing his neck at the same time. No new neurological signs were found after this incident. After induction of anaesthesia symmetrical responses were recorded from arms and legs (Figure 2). The fracture was then reduced and he was positioned on the table, after which no response could be recorded from the right leg or either arm. A wake-up test confirmed severe quadriparesis. A CT scan showed a large disc fragment protruding into the spinal canal at C4/5.

One further patient showed a marked change in evoked potentials, but in this case the change was apparent only by comparison with the pre-operative recordings. This was a woman of 32 with a traumatic fracture of C2. Pre-operative recordings from the legs were symmetrical (Figure 3a), but immediately after positioning on the operating table no reproducible response could be recorded from the right leg (Figure 3b). This response remained unrecordable until the neck was straightened, after which stimulation of the right leg produced a response similar to that from the left leg. This patient awoke without deficit.

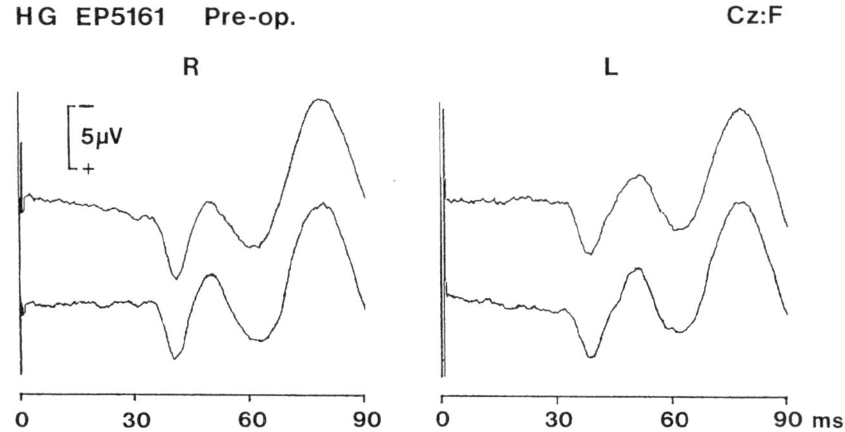

Figure 3a Symmetrical preoperative responses in a patient with a fracture of C2

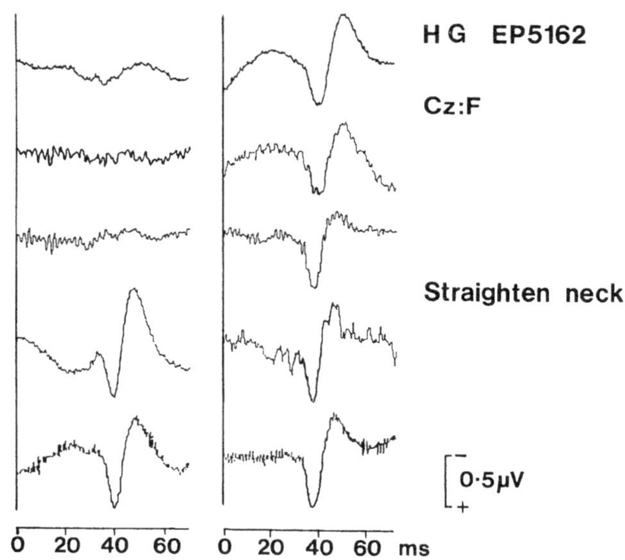

Figure 3b After positioning on the table no response could be recorded from the right leg (upper left traces), but the response returned when the neck was straightened (lower left traces). There was no change in potentials recorded from the left leg (traces on the right)

DISCUSSION

The method described here has proved reliable in operations on the cervical spine in which invasive methods of monitoring could not be used. Under anaesthesia with nitrous oxide and either halothane or isoflurane (up to 1%) cortical responses could be recorded from arms and legs, although the amplitude was considerably smaller than that recorded preoperatively. The narrow bandpass undoubtedly contributed to this reduction in amplitude, but its use was justified to obtain greater noise reduction in recordings made in the hostile environment of an operating theatre. A further factor tending to reduce the amplitude of responses was the high rate of stimulation (7 Hz), but this rate permitted 500 responses to be averaged in just over 1 minute, and this advantage outweighed the amplitude reduction.

The patient who became quadriparetic 12 hours after the operation should not be regarded as a false negative, since he had no new deficit at the end of the operation, and the neurological signs developed long after monitoring had been discontinued. Monitoring could not therefore be expected to detect the onset of the deficit, and certainly could not have prevented it[6]. This patient is therefore regarded as a true negative, and the series thus includes a total of 35 true negative cases. There were three positives, of which one was a true positive, and a second was clearly a false positive. The third awoke without deficit in spite of a prolonged period of absence of response from one arm. This would normally be regarded as a false positive, but may in fact have been a true positive, since she was not awakened while the response was absent, and so we do not know whether she had a deficit at that time. This case emphasizes the importance of undertaking preoperative recordings, particularly in patients with instability of the cervical spine, so that abnormalities developing during positioning on the table can be detected more reliably.

References

1. Nash CL Jr, Lorig RA, Schatzinger LA, Brown RH. Spinal cord monitoring during operative treatment of the spine. Clin Orthop Rel Res. 1977;126:100–106.
2. Engler GL, Spielholz NI, Bernhard WN, Danziger F, Merkin H, Wolff T. Somatosensory evoked potentials during Harrington instrumentation for scoliosis. J Bone Joint Surg. 1978;60-A:528–532.
3. Nuwer MR, Dawson EC. Intraoperative evoked potential monitoring of the spinal cord. Enhanced stability of cortical recordings. Electroenceph clin Neurophysiol. 1984;59:318–327.
4. Veilleux M, Daube JR, Cucchiara RF. Monitoring of cortical evoked potentials during surgical procedures on the cervical spine. Mayo Clin Proc. 1987;62:256–264.
5. Maynard DE. Development of the CFM: the cerebral function analysing monitor (CFAM). Ann Anesth Franc. 1979;20:253–255.
6. Friedman WA, Grundy BL. Monitoring of sensory evoked potentials is highly reliable and helpful in the operating room. J Clin Monit. 1987;3:38–45.

29
A computer simulation of conduction block – conduction block induced augmentation and killed-end potential

T. TANI, T. USHIDA and H. YAMAMOTO

INTRODUCTION

In electrodiagnosis of cervical compression myelopathy during posterior surgery, we usually recorded the evoked spinal cord potentials with monopolar needle electrodes inserted into the ligamentum flavum at serial intervertebral spaces after epidural stimulation of the thoracic or lumbar spinal cord or cauda equina. A large monophasic positive waveform known as killed-end potential often discloses the level responsible for a myelopathic condition. Interestingly enough, in about 80% of such cases, augmentation of the negative phase of the response can also be detected with the recording just below the responsible level (Figure 1). In order to elucidate the mechanism underlying such alterations of waveform, a computer simulation of conduction block was attempted.

THEORY

According to physics, the potential at a point a distance R from a point charge Q is given by:

$$V = K \cdot Q/R$$

where K is a constant. When a dipole which is a pair of equal and opposite point charges, +q and −q, separated by a distance δ is considered, its contribution to the potential at the point P is given by:

$$Vp = K(q/r_{+} - q/r_{-}) \tag{1}$$

because the potential of a number of charges is equal to the algebraic sum of the potentials due to each charge. If the distance r between P and the

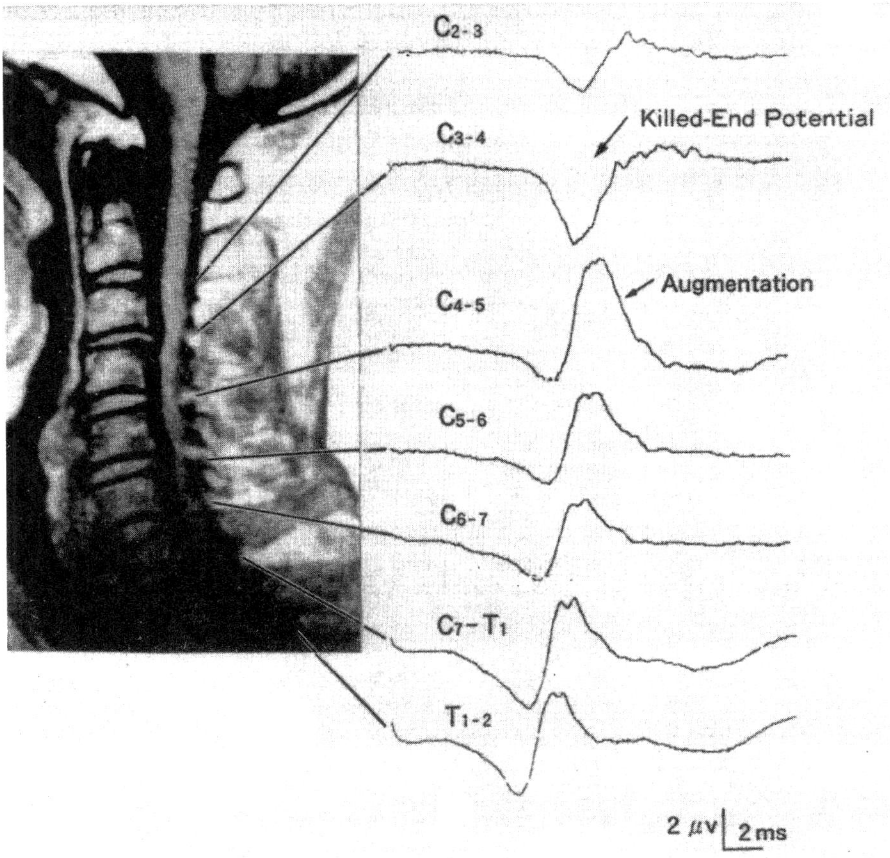

Figure 1 Monopolar recording of averaged evoked spinal cord potentials (ESCPs) elicited from the ligamentum flavum at serial intervertebral spaces after epidural stimulation of the thoracic spinal cord in a 57-year-old man with cervical myelopathy. Despite extensive spinal cord compression shown in T_1 weighted MRI, an abrupt change in waveform of the potential at C_{3-4} indicates the level responsible for myelopathy. The distinct augmentation of the response at C_{4-5} is also worthy of notice

centre of the dipole O is much greater than δ, we can rewrite (1) as:

$$Vp = Km \cdot \cos\theta / r^2 \tag{2}$$

where m is a dipole moment $q\delta$ and the angle space θ is the angle between the dipole vector and the line OP. We now consider a dipole layer of thickness δ which carries the constant surface charge density $\pm\sigma$. The contribution of a sufficiently small area dS of dipole layer to the potential at point P can be obtained by referring to the formula equation (2) for the potential of a dipole.

$$\mathrm{d}Vp = K(ma \cdot \mathrm{d}S) \cdot \cos\theta / r^2 \tag{3}$$

where *ma* is the dipole moment per unit area $\sigma\delta$. the potential Vp due to the entire dipole layer is therefore given by integration of equation (3),

$$Vp = \int Kma \cdot \mathrm{d}S \cos\theta / r^2 \quad (4)$$

Since

$$\mathrm{d}S \cos\theta / r^2 = \mathrm{d}\Omega \quad (5)$$

which is a solid angle subtended by area dS at distance r, we can rewrite equation (4) as:

$$Vp = \int Kma\mathrm{d}\Omega.$$

If ma is constant over the dipole layer, the result of integration is

$$Vp = Kma\Omega \quad (6)$$

which shows that the potential Vp is proportional to the solid angle Ω subtended by a dipole layer[1].

A nerve fibre carrying an impulse can be divided into two regions, active and inactive. In the inactive region, the axon membrane can be viewed as dipole layers with positive charges on the outer surface and negative charges on the inner surface, whereas dipole layers of the active region have reversed polarity, with negative charges on the outside surface. Approximating the action potential with the square wave[1], the transition between the two regions is assumed to occur instantaneously, and each region is assumed to have a different but constant potential. Assuming the idealized impulse travelling from left to right, potential at a point 'P' depends only on the solid angle Ω_1 and Ω_2 subtended by the boundaries between the active and inactive regions[1], because the nearer and farther membranes subtending solid angle α, β or γ contribute equal magnitude but opposite polarity, resulting in cancellation (Figure 2A). Therefore, we have only to calculate the solid angles subtended by a pair of wave-fronts of opposite polarity representing depolarization and repolarization. The potential is consequently obtained by adding the contribution of each wave-front.

$$Vp = Kma(\Omega_1 - \Omega_2). \quad (7)$$

In accordance with equation (5), equation (7) is expressed as

$$Vp = AKma\{X/(x^2 + y^2)^{2/3} - (x+s)/[(x + s)^2 + y^2]^{3/2}\} \quad (8)$$

(modified from Brown [1968][2]), where A is wave-front area, x is the perpendicular distance from P to the depolarization wave-front, y is the depth of the wave-front and S indicates a distance between each wave-front. Finally, assuming the situation in which a depolarization wave-front originates at a distance D from p and then approaches, reaches and leaves the point at a constant velocity V, we can rewrite (8) on replacing x by $D - vt$ as:

$$Vp = AKma\{(D - vt)/[(D - vt)^2 + y^2]^{3/2} - (D - vt + S)/[(D - vt + S)^2 + y^2]^{3/2}\} \quad (9)$$

(Figure 2B). Therefore, the potential Vp produced at a point p by a propagating

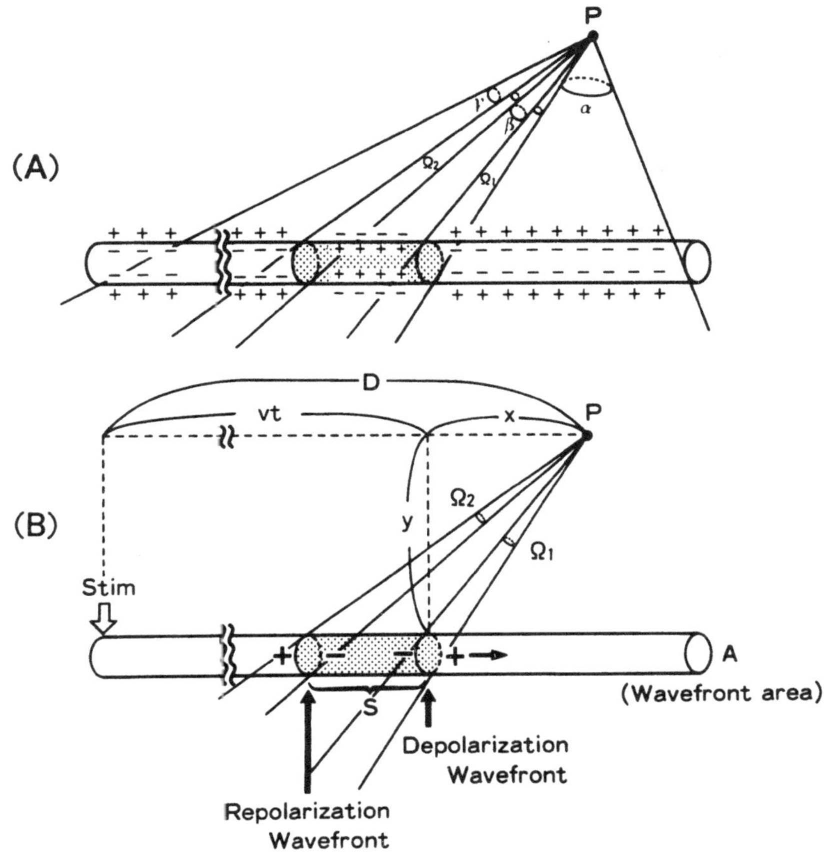

Figure 2 Simple dipole membrane model of a nerve fibre, square-wave approximation of an action potential, and solid angle analysis. A. Potential at P depends only on the solid angle Ω_1 and Ω_2 subtended by the boundaries between the active (shaded area) and inactive regions. The nearer and farther membranes subtending solid angle α, β or γ contribute equal magnitude but opposite polarity, resulting in cancellation. (Adapted from Woodbury[1]). B. Geometrical arrangement of a detecting point P and an idealized impulse (shaded area) travelling along a nerve fibre from left to right (Modified from Brown[2])

nerve impulse is given as a function of the time t taken for the impulse to propagate. It is equation (9) that we entered into the program for computer simulation.

METHOD

The program for computer simulation was written in Quick Basic V. 4.5 (Microsoft) running under MS-DOS V. 3.3C on an NEC PC9801. In equation (9) the location of point P determines D and y, whereas nerve fibre diameter defines A, V and S. In this study we have chosen 5 mm for y and 4.6 as the scaling factor which is the ratio of conduction velocity in m/s to total fibre diameter μm. S was calculated by multiplying the conduction velocity V of a nerve fibre by 0.5 ms which is a duration of a nerve action potential[3]. Equation (9) can therefore be an expression for computing a sequence of potential change due to a travelling impulse of a single nerve fibre as a function of time with the detecting point fixed, just like an actual situation of an action potential recording. In the process of digital-to-analog conversion for displaying calculated values on a CRT screen, the sample rate was set to 20 kHz. For simplicity, a simulation of the evoked compound spinal cord potentials (ESCPs) was carried out on the assumption that they were derived only from dorsal spinocerebellar tract. The anatomical data of fibre-diameter distribution of the tract in a human[4] were therefore entered into the program (Figure 3). In order to yield the ESCP, we computed the algebraic sum of each calculated single fibre potential which was weighted in proportion to its frequency distribution. A computer simulation of conduction block was attained in such a way that the conduction velocity V of each impulse was abruptly turned to zero exactly when the depolarization wave-front reached the point where D was 120 mm. The conduction block was produced progressively from larger to smaller axons, because mechanical compression is known to involve the largest myelinated fibres first.

RESULT

A triphasic waveform of each single fibre action potential was produced with an arbitrary but standardized amplitude scaling. The larger the fibre diameter, the greater its action potential which obviously contributes greatly to the amplitude of the compound nerve action potential. The ESCP was produced as an algebraic sum of simulated individual nerve action potential components weighted with frequency distribution. Basically the configuration of its major component inherited a positive-negative-positive triphasic waveform from constituent single fibre potentials, although the peaks were of serrated appearance and were followed by some smaller late components (Figure 3).

A simulation of a serial recording of ESCPs in increments of 10 mm along the length of the spinal cord was carried out by varying the value of D in equation (9) between 70 and 160 mm. The result of the simulation clearly demonstrates physiologic temporal dispersion, which means with increasing distance between stimulating and pick up electrodes, the recorded potentials become smaller in amplitude and longer in duration (Figure 4A). The effect of conduction block, progressively from larger to smaller axons, at the point of propagated distance 120 mm, on each simulated ESCP waveform is shown in Figure 4. The ESCPs were altered not only at the points after the block, but also at the points before the block except for the bottom traces. When conduction block was simulated in a group of fibres, 12 to 18 μm in diameter, which corresponds to about 15% of the population, the amplitude of the ESCP did not decrease, but increased at every point after the block. In addition, such a conduction block also increased the amplitude of the ESCP

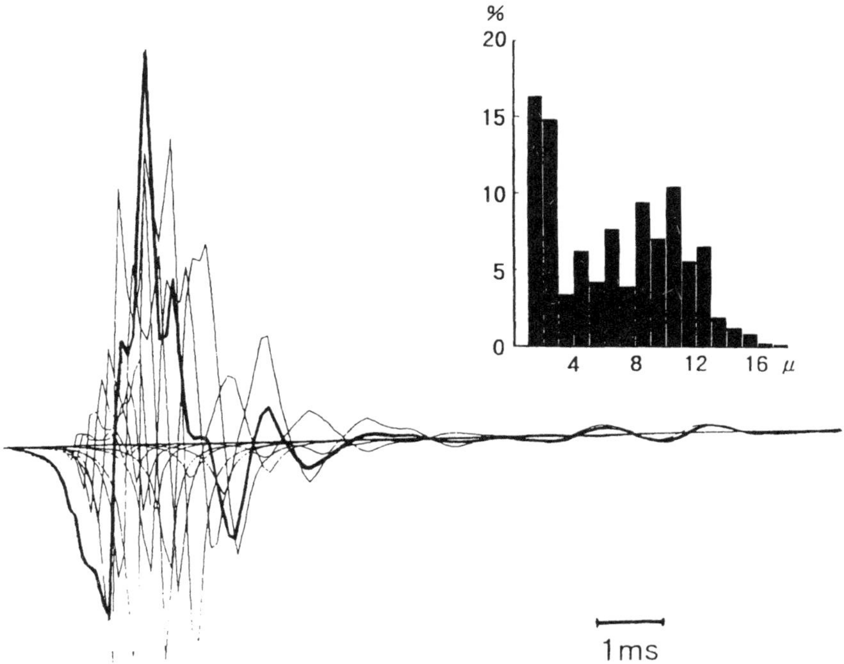

Figure 3 Fibre-diameter histogram of dorsal spinocerebellar tract in a human (top). (From Häggqvist[4]). Simulated nerve fibre action potentials weighted in proportion to their frequency distributions (thin lines) and their algebraic sum (thick line) produced at the point where 'D' and 'y' are 100 mm and 5 mm, respectively (bottom)

at the points 30 and 40 mm before the block, where D is 90 and 80 mm, respectively (Figure 4B). When the conduction in about 50% of axons was blocked, involving the fibres with diameter ranging from 7 to 18 μm, the ESCP produced at exactly the same point as the block, altered the monophasic positive waveform which is known as the killed-end potential. In such an extensive block, the waveform produced at the points 30 and 40 mm before the block still remained augmented, despite marked reduction of the negative phase in amplitude just before and after the block (Figure 4C). The effect of a conduction block in 100% of axons on simulated ESCPs was about the same as the 50% block except that the negative potentials were completely abolished at every point after the block. Again, the waveform produced at the points 30 and 40 mm before the block still remained augmented (Figure 4D).

DISCUSSION

It is commonly believed that conduction block results in a reduction in size of the compound nerve action potential. However, this study has shown that this is not always the case, and that conduction block can paradoxically

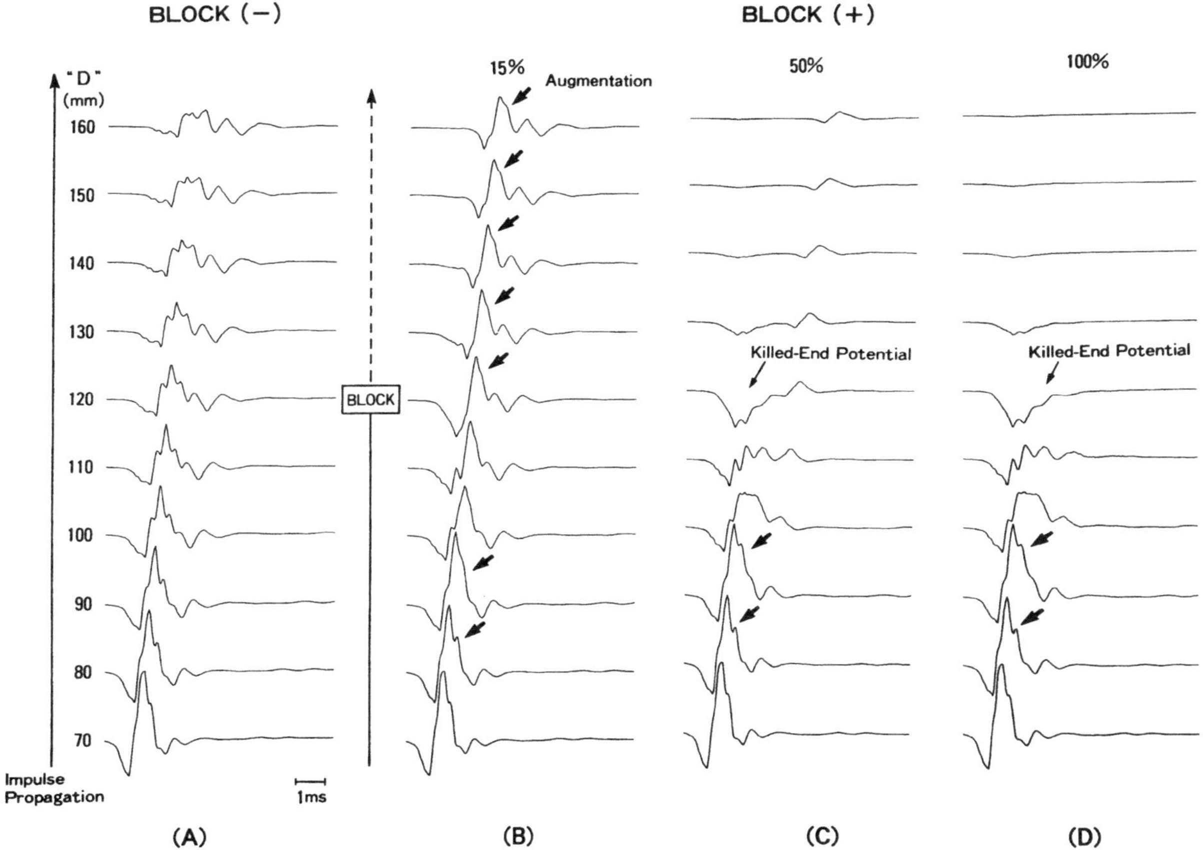

Figure 4 Simulation of a serial recording of ESCPs at the points of 5 mm for 'y' in increments of 10 mm along the length of the spinal cord. A. Without conduction block, physiologic temporal dispersion alone affects the waveforms considerably. B. C. D. With progressive percentage of conduction block from larger to smaller axons at the point of 120 mm for 'D', 15, 50 and 100% block, as indicated above, alter the waveforms not only at the points after the block, but also at the points before the block except for the bottom traces. Note distinct augmentation (thick arrows) instead of reduction in amplitude and killed end potentials (thin arrows)

increase the amplitude (Figure 4). In previous studies little attention has been paid to this phenomenon except by Lee and Kurokawa[5], who noticed clinically the various kinds of augmentation of the ESCP during spinal cord monitoring, including that related to conduction block. Kurokawa[6] also demonstrated the augmentation of the response experimentally, during the graded compression of peripheral nerve as well as spinal cord. More importantly, he found that the compound nerve action potential of the peripheral nerve, when it was covered with a layer of paraffin oil, did not show such an augmentation with nerve compression, which implies that an effect of volume conduction will be needed for the phenomenon to occur. This fact taken together with the result of the present study has excluded the possibility, regarding the aetiology of the augmentation, that individual nerve fibre action potentials themselves become larger, being associated with factors such as hyperpolarization. For the same reason, either slowed Na^+ inactivation which has been proposed as the cause of cold-induced augmentation of sensory compound action potential[7] or greater synchronization of the constituent nerve fibre potentials is equally unlikely. Then what is behind the phenomenon?

Considering that phase cancellation normally dictates waveforms of the compound action potential[8], an explanation can be offered. We should now focus our attention on the waveform of the potential due to a single nerve fibre carrying a propagating impulse when registered at a point through a volume conductor. Without conduction block, the simulated single fibre action potential was always triphasic, positive–negative–positive, regardless of the location of the recording point along the nerve fibre. With conduction block, however, its waveform altered to a positive–negative diphasic sequence, as was expected, at points just before the block, because at those points the wave-front passes but does not leave sufficiently. If the terminal positive phases of fast conducting fibres are normally out of phase with the main negative phase of the compound action potential, their absence could contribute to its augmentation, because cancellation which normally occurs fails to occur. If this is the case, it is understandable that the simulated ESCPs produced at the points 30 and 40 mm before the block in the present study remained augmented even when every axon was blocked (Figure 4D). There may be a similar mechanism underlying the equivalent phenomenon which we frequently encountered clinically in cervical myelopathy (Figure 1). This study also demonstrated ESCP enhancement at the points after conduction block only in the early stages of progressive percentage of conduction block from larger to smaller axons (Figure 4B). The explanation of this phenomenon can be made in the same way. If the entire triphasic sequences derived from some faster conducting fibres are not registered due to conduction block, and their terminal positive phases are normally out of phase with the main units, the conduction block could increase the amplitude of compound action potentials even at the points after the block. This effect, however, is inevitably more than offset with further involvement of fast conducting fibres which contribute to the main negative phase of the compound action potential, resulting in reduction in amplitude of the potentials, and finally they are abolished (Figure 4C, D). Furthermore, the killed-end potential can also be interpreted based on the concept of phase cancellation. At exactly the same point as conduction block, a blocked axon produced only an initial positive phase, the so-called approaching wave. If a majority of fast conducting axons are blocked, their approaching waves are now set free from the phase cancellation with negative phases, resulting in a large monophasic positive potential. It is clear that the killed-end potential in the clinical domain does not necessarily imply that every axon tested is affected, because conduction block in about 50% of constituent axons gave rise to a killed-end potential in the present study (Figure 4C).

Although we employed Woodbury's square wave approximation of a nerve fibre action potential, actual repolarization process is slower than depolarization, indicating that the terminal positive phase is smaller and longer than the initial positive phase. In addition, contrary to the assumption on which the computer simulation was made, neither the scaling factor nor spike duration is constant for all sizes of fibres. These factors will affect the above discussion partly, but not substantially. As Kimura[8] has already predicted in his paper on the fascinating concept of phase cancellation, the present study has demonstrated that a conduction block could paradoxically increase the amplitude of a compound action potential. An awareness of this possibility will help to interpret the electrophysiological data properly in both the experimental and the clinical domain.

References

1. Woodbury JW. Potentials in a volume conductor. In: Ruch TC, Patton HD, Woodbury JW, Towe AL, eds. Neurophysiology, 2nd ed. WB Saunders: Philadelphia; 1965:85–91.
2. Brown BH. Theoretical and experimental waveform analysis of human compound nerve action potentials using surface electrodes. Med Biol Eng. 1968;6:375–386.
3. Dumitru D, Delisa JA. AAEM Minimonograph #10: volume conduction. Muscle Nerve. 1991;14:605–624.
4. Haggqvist G. Analyse der Faserverteilung in einem Ruckenmarkquerschnitt (Th3). Z Mikrosk Anat Forsch. 1936;39:1–34.
5. Lee CS, Kurokawa T. Augmentation of spinal evoked potentials during surgery. Electroenceph Clin Neurophysiol. 1990;75:S80.
6. Kurokawa T. Augmentation of spinal evoked potentials (in Japanese). 14th Congress of Japan Society of Electroenceophalography and Electromyography. 1984.
7. Louis AA, Hotson JR. Regional cooling of human nerve and slowed Na^+ inactivation. Electroenceph Clin Neurophysiol. 1986;63:371–375.
8. Kimura J, Sakimura Y, Machida M, Fuchigami Y, Ishida T, Claus D, Kameyama S, Nakazumi Y, Wang J, Yamada T. Effect of desynchronized inputs on compound sensory and muscle action potentials. Muscle Nerve. 1988;11:694–702.

30
Quantifying the SSEP – the area under the curve

B. A. TAYLOR, K. GORDIEV and R. O'BRIEN

INTRODUCTION

The use of neurophysiological evoked potentials for intraoperative monitoring of the spinal cord is rapidly increasing. There are a multitude of different techniques in use, each with claimed or proven advantages, and it is likely that a combination of complementary methods will eventually provide the most accurate and reliable surveillance of the cord at risk. A problem common to all techniques, whether sensory or motor, is the way in which the recorded response should be quantified. Response quantification is becoming increasingly important with the impending use of computerized or automated monitoring machines, a development which is being driven by the need to provide greater monitoring services as efficiently as possible. Computerized automation can certainly ease the burden on the technician or clinician of repetitive response acquisitions and probably improve the consistency of record keeping, but to provide any automation of the interpretation stage, including the plotting of trend lines, it is necessary to quantify the responses. The reliability and accuracy of this quantification is of medico-legal as well as clinical significance.

At present the most commonly measured parameter is amplitude, and the usual quoted criterion for concern is a loss of greater than 50% of the initial, or control value[1,2]. This apparently straightforward criterion is not so easy or reliable in practice, because the evoked potential is a complex waveform generated by conduction of activity in several adjacent neural pathways and summated in a way dictated by the position and characteristics of the recording electrode. This means that when quantifying the recordings from an individual patient the relevant peaks must first be identified in the control waveform and then reliably re-identified in each successive recording. This highlights two problems:

1. Which peaks are to be chosen so that the 50% criterion is comparable to that identified as important by previous investigations?

2. What significance should be placed on a change in the whole waveform that is not reflected in the amplitude of the particular peak that has been chosen?

The problem of which peak to choose was highlighted in a previous study by one of the authors in which four electrophysiological technologists regularly involved in spinal cord monitoring were asked to individually mark the peaks that they would use in 100 SSEP recordings. Despite the fact that they all worked in two closely linked institutions the inter-observer disagreement rate ranged from 20 to 42%[3].

Intraoperative spinal cord monitoring is subtly different from most forms of clinical neurophysiological investigation. The evoked potentials are recorded repeatedly with exactly the same montage over the course of several hours. The initial form of the response is not particularly important as there is no need to make a diagnosis from it, but what does matter is that this response should not then change during the subsequent course of the procedure. The requirement therefore is for a measuring technique that can reliably detect a systematic change anywhere in the response and thus be a general surveillance parameter to direct attention to a more detailed analysis only when a change has occurred. Such a parameter could make initial computerized interpretation a possibility, and then only require highly skilled neurophysiological interpretation when a change is detected. At this point it does not seem justifiable to delegate the decision about spinal cord integrity to the computer and a final clinical decision must be retained.

MATERIALS AND METHODS

The measurement method investigated in this paper is the integral or area under the curve of the segment of the recording sweep that contained the evoked response, and it is compared to several different ways of measuring the amplitude. The method used to monitor the spinal cord was that developed at Stanmore[4] and involved recording the Spinal Somatosensory Evoked Potential (SSEP) with a bipolar epidural electrode in the lower cervical or upper thoracic area. Supramaximal stimulation of the posterior tibial nerve at the popliteal fossa with surface electrodes was applied separately to left and right sides and responses were recorded with a Medelec MS91 by averaging several hundred sweeps at 20 per second. Filter settings were LF 200 Hz, HF 2 kHz.

Analysis was carried out retrospectively on hard copy recordings selected from a database of over 1000 cases acquired over nearly ten years at The Royal National Orthopaedic Hospital, Stanmore. From these, 32 cases were selected with the knowledge only of the patient's name, hospital number, date of surgery and whether or not there was judged to have been a significant change in SSEP at the time of surgery. None of the patients were known to the selecting author (K.G). Sixteen cases were chosen from the subset who had significant changes at surgery and 16 from the remaining subset. Eight traces were chosen for each case – four for right leg stimulation and four for left. The four pairs were chosen as those marked at the time of recording as

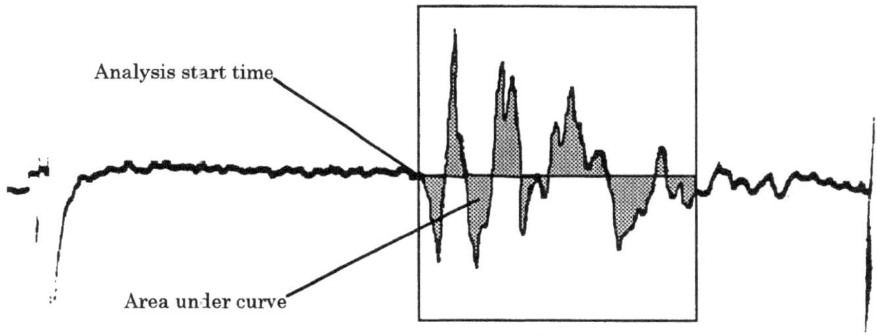

Figure 1 Illustration of the area measurement method. The integral is calculated about the trace mean, ignoring sign. Units are microvolt.milliseconds. Baseline calculation performed with the 10 ms analysis window start time at response latency. Subsequent calculations made after shifting the window by 2 and 4 ms before and after latency

'control', 'initial distraction', 'final distraction' or 'wire tightening', and 'end of procedure'. This yielded 256 traces for analysis.

Each trace was digitized at a resolution of 200 dpi and the resultant image file was scanned and converted into a co-ordinate file of 256 × 256 points[3]. This file was then presented to a previously written program on an Acorn Master computer that calculated the integral (ignoring sign) of the waveform in a 10 ms window onto the 30 ms sweep (Figure 1). This window was set to start at the latency of the response, judged as the start of the first recognizable deviation from the baseline part of the response. It was then shifted 2 and 4 ms before and 2 and 4 ms after this point and the integral calculated again for each setting. This was done to assess the need to accurately place the analysis window in relation to the response and thus give some indication of the robustness of the technique.

The maximum peak-to-peak, N1–P1, N2–P2, and N3–P3 amplitudes were measured directly from the original paper records. All data were stored in a spreadsheet (Microsoft Excel 4.0) and analysed using the same package.

RESULTS

Data from all cases were pooled to plot scatter diagrams of area against each of the amplitude measurements, giving 256 data points for each graph (Figure 2). In each case there was a recognizable relationship between the area of the whole response and the amplitude of each SSEP peak, or indeed the maximum peak-to-peak amplitude. The correlation coefficients were highest for peak-to-peak and P2–N2 (0.83 and 0.84 respectively). The area function was minimally affected by the position of the analysis window in relation to the response (Figure 3). This was especially true when the window was started earlier than the response latency.

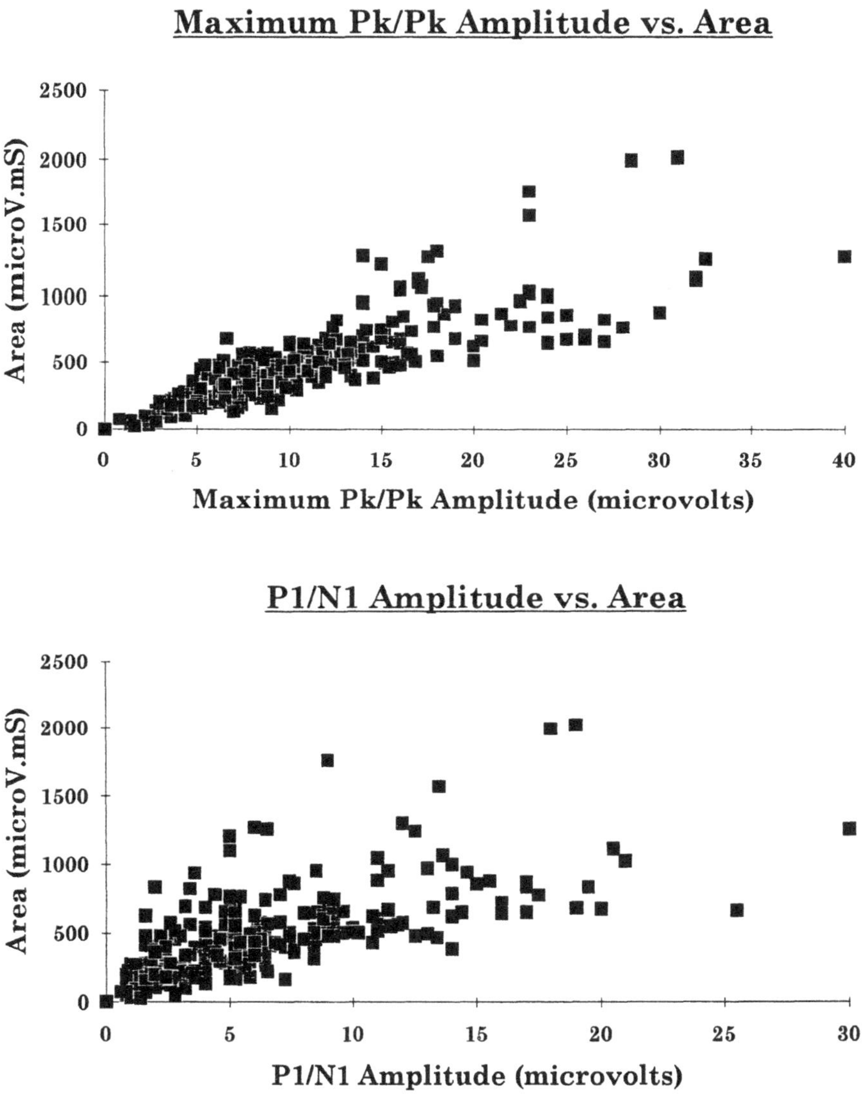

Figure 2 Scatter diagrams of area against individual peak amplitudes (n = 256). Cases include those with normal and abnormal postoperative neurology as described in the text

DISCUSSION

The area under the curve correlates closely with conventionally measured amplitudes and might thus be reasonably substituted when quantifying evoked potentials during spinal cord monitoring. The criteria for significant change should also be similar to those for amplitude. It is a function that is difficult to obtain manually but easily calculated by computer, and has the

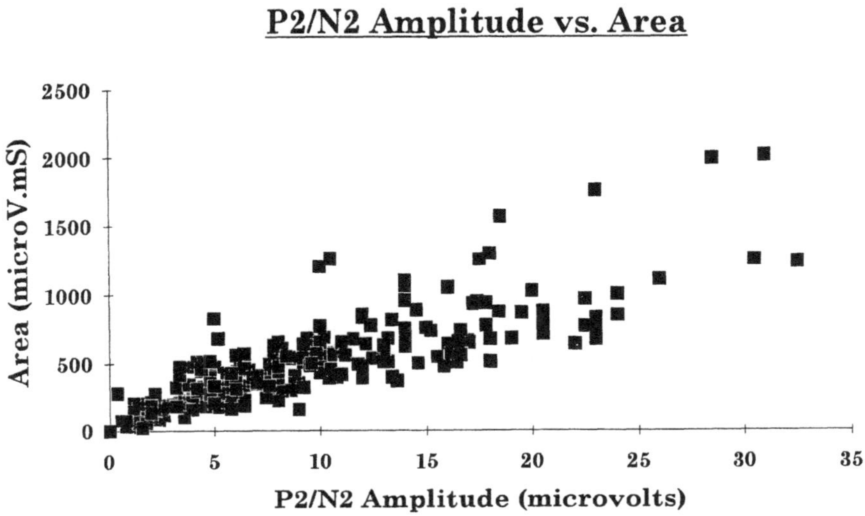

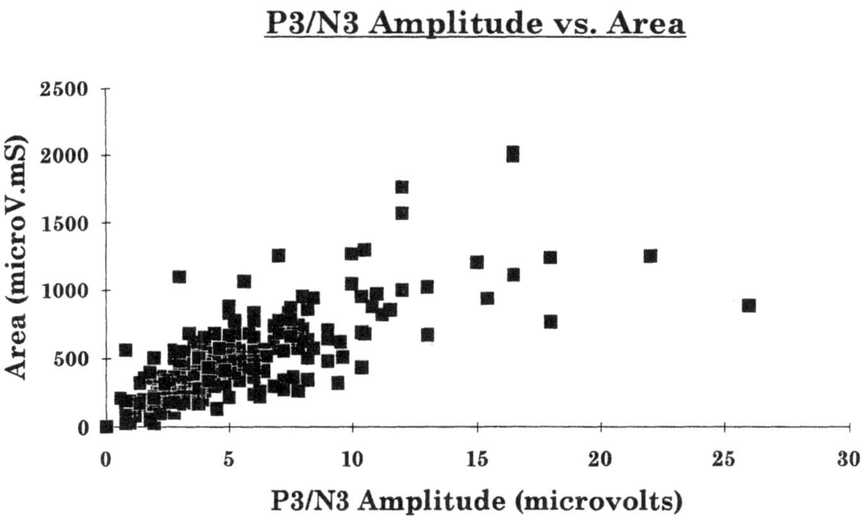

Figure 2 *Continued*

advantage that individual peaks do not need to be tracked reliably, which is difficult in computerized analysis.

The area could be regarded as a measure of neurological power or multipathway conduction and is thus appropriate when seeking the generalized neurological injury that may occur during correction of spinal deformities. Its use should not be limited to the SSEP as investigated here and the method could equally be used for MEP spinal cord monitoring.

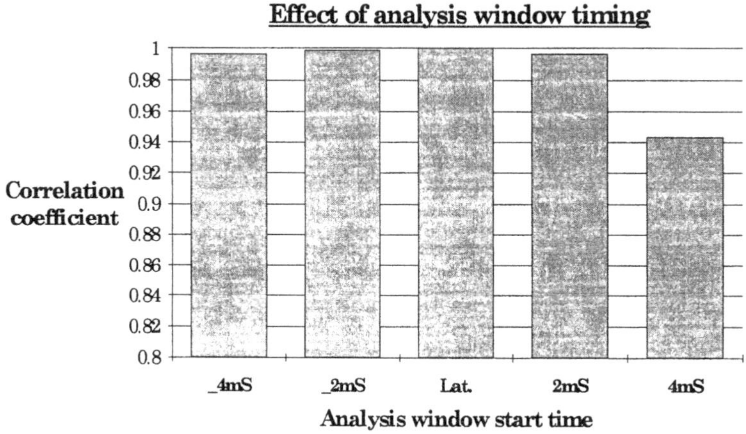

Figure 3 Correlations of calculated area under the response curves when analysis window start time was at latency with those at 2 and 3 ms earlier or later. There is very little effect on the area function when the analysis window start time preceeds the response latency by up to 4 ms. This should make the area function resistant to the effects of latency jitter

References

1. Jones SJ, Carter L, Edgar MA, Morley TR, Ransford AO et al. Experience of epidural spinal cord monitoring in 410 cases. Spinal Cord Monitoring. Springer-Verlag; Berlin Heidleberg New York Tokyo: 1985:215–220.
2. Forbes HJ, Allen PW, Waller CS, Jones SJ, Edgar MA et al. Spinal cord monitoring in scoliosis surgery. Experience with 1168 cases. J Bone and Joint Surg. 1991;73-B:487–491.
3. Taylor BA. Spinal cord monitoring. An investigation to identify the techniques and limitation of computer assisted interpretation. Thesis; Liverpool; 1990.
4. Jones SJ, Edgar MA, Ransford AO. Sensory nerve conduction in the human spinal cord: epidural recordings made during scoliosis surgery. Journal of Neurology Neurosurgery and Psychiatry. 1982;45:446–451.

31
The effects of propofol and nitrous oxide on median nerve somatosensory evoked potentials during fentanyl anaesthesia

M. ZILELI, T. BALCIOĞLU, M. ÇALIŞKAN and F.ÖVÜL

INTRODUCTION

Intraoperative somatosensory evoked potential (SEP) monitoring is a useful adjunct to spinal cord and vascular neurosurgery. Alteration of somatosensory central conduction time (CCT = for median SEP, N20 latency minus N13 latency) and amplitude of cortical responses (for median SEP, N20 − P25 difference) may indicate impaired conduction in pathways from the cervical spinal cord to the cortical somatosensory strip. The anaesthetic techniques used during surgery requiring SEP monitoring should have the least possible effect on amplitudes and latencies. Nitrous oxide and propofol are anaesthetics proposed to have a moderate depressant effect on SEPs. Propofol has been reported to be less depressant on SEPs than nitrous oxide[1,2].

We have compared the effects of nitrous oxide and propofol in combination with a bolus injection of fentanyl on median SEPs in each group.

MATERIAL AND METHOD

Anaesthetic procedure

Twenty consenting healthy patients with no pulmonary, cardiac and neurological problems (ASA class 1) scheduled for elective lumbar disc surgery were selected for the study (ages 21–60; 11 males, 9 females). Patients were randomly divided into two groups of ten each.

None of the patients were premedicated. None were having any drug known to effect SEPs. After baseline mSEP recordings at the operation table, anaesthesia was induced with atropine 0.5 mg, etomidate 0.25 $mg.kg^{-1}$ and a bolus dose of fentanyl 8 μg.kg^{-1} given in two minutes. Pancuronium 0.1 mg.kg^{-1} was given to provide neuromuscular blockade. In both groups the lungs were initially ventilated with 100% O_2 automatically by an anaesthetic machine (Engström ER 311). After prone positioning and at least ten minutes after etomidate doses, 60% nitrous oxide in oxygen inhalation was started for ten patients (Group I). The other ten patients received 0.5 mg.kg^{-1} bolus propofol (Group II). Following the loading dose, propofol was given at a rate of 10 mg.kg^{-1}.h^{-1} for the first ten minutes, 8 mg.kg^{-1}.h^{-1} for the next 10 minutes and 6 mg.kg^{-1}.h^{-1} thereafter. No additional doses of fentanyl or pancuronium were given to either group within the 60 minutes' recording time. Patients in Group II (propofol) received 40% O_2 in air.

All patients were monitored with ECG, rectal temperature and pulse oxymetry. Patients were kept normothermic (36–37°C), normocapnic ($PaCO_2$ 40 $\pm$ 2 mmHg) and normotensive (systolic blood pressure within 20% of the baseline). The lungs were ventilated using tidal volumes of 10 ml.kg^{-1} at a rate of 14–16 breaths per minute.

SEP procedure

Right or left median nerves were stimulated at the wrist with bipolar surface electrodes. Square waves of 0.1 ms, at twice motor threshold with 3.2 Hz repetition rates were used. Two channel recordings were made from C2 spinous process (negative), and from C3′ or C4′ scalp electrodes (negative). Fpz was used as reference (positive) for both electrode sets with a forearm ground electrode. Platinum subdermal needles were used for recordings with measured impedances less than 3 kΩ. An ESAOTA (Florence, Italy) EP unit with hard disc control and automatic artefact rejection was used. 200–400 traces with 50 ms analysis time, 20 μV/division amplification, 20–5000 Hz band pass filters were averaged.

Ten measurements were made at the following time points: Before induction (1), 2 minutes after fentanyl (2), 2 minutes after etomidate (3), before N_2O or propofol (4), 5 minutes (5), 15 minutes (6), 20 minutes (7), 30 minutes (8), 45 minutes (9), and 60 minutes (10) after N_2O or propofol (Figure 1).

An off-line analysis was done from two parameters: First, somatosensory central conduction time (cortical N20 latency minus cervical N12 latency) and second, cortical amplitude (N20 amplitude minus P25 amplitude).

Statistical analysis

CCT and N20 – P25 amplitudes of all patients at each measurement point were plotted on graphs and analysed with two-sample t tests for independent variables. The significance of differences was established between two sets of data: 1) Comparison of Group I with Group II. 2) Comparison of the 4th measurement values in each group (before N_2O or propofol) with other values.

Table 1 CCT and N20 − P25 amplitude value (mean ± standard deviation) of both groups of patients at 10 measurement points

Time scale	CCT N_2O $\bar{X} \pm SD$	P^1	CCT Propofol $\bar{X} \pm SD$	P^1	N20 − P25 amplitude N_2O $\bar{X} \pm SD$	P^1	N20 − P25 amplitude Propofol $\bar{X} \pm SD$	P^1
1	5.59 ± 0.62	ns	5.63 ± 0.66	ns	5.95 ± 2.42	*	3.98 ± 1.93	ns
2	5.82 ± 0.70	ns	5.84 ± 0.89	ns	6.9 ± 3.19	ns	3.93 ± 2.24	ns
3	6.56 ± 0.60	**	6.45 ± 0.72	**	16 ± 5.43	**	9.5 ± 4.3	**
4	5.74 ± 0.58		5.76 ± 0.59		7.49 ± 3.32		5.28 ± 3.23	
5	5.66 ± 0.57	ns	5.87 ± 0.67	ns	3.84 ± 1.63	**	3.16 ± 1.46	*
6	5.59 ± 0.65	ns	6.14 ± 0.61	*	3.14 ± 1.86	**	3.19 ± 1.31	*
7	5.64 ± 0.73	ns	6.12 ± 0.60	*	3.2 ± 1.70	**	3.14 ± 1.07	*
8	5.68 ± 0.51	ns	6.27 ± 0.69	ns	3.33 ± 1.28	**	3.08 ± 1.23	*
9	5.70 ± 0.61	ns	6.19 ± 0.80	ns	3.43 ± 1.35	**	3.12 ± 1.28	*
10	5.89 ± 0.59	ns	6.14 ± 0.60	ns	3.37 ± 1.39	**	3.39 ± 1.65	ns

p values (p^1) are a comparison of the 4th measurements with others in each group. ns = non significant, $p > 0.05$, * = $p < 0.05$, ** = $p < 0.001$. The meanings of time scales are in the text

RESULTS

Twenty patients entered this study. Their average age was 45 ± 11 (21–60) years and average height was 167 ± 7.3cm. Reproducible median SEPs could be recorded in all 20 patients. CCT and cortical N20 − P25 amplitudes are shown in Table 1. This table also includes statistical significance levels of values from each point compared to the 4th time point (before propofol or N_2O administration).

Statistical analysis shows that neither nitrous oxide nor propofol altered the CCT significantly. The amplitudes, however, were significantly depressed both with nitrous oxide and propofol. The decrease in amplitudes was more pronounced with nitrous oxide (approximately 60%) than with propofol (ca 20%), although it was not significant ($p > 0.05$). The depressive effect of nitrous oxide on cortical mSEP amplitudes was, however more pronounced ($p < 0.001$) compared to the propofol effect under the same conditions ($p < 0.05$).

When we plotted the time sequences of latency and amplitude changes, we saw that etomidate has a significant effect on both latency and amplitudes that lasts approximately 4–5 minutes (Figures 2, 3).

DISCUSSION

Other studies comparing volatile with IV anaesthetics are not designed solely to compare the anaesthetic techniques[1-3]. For that reason we wanted to design a comparative study with a stricter experimental protocol without using other volatile and intravenous agents known to effect SEPs.

As in other studies, we have observed better preservation of amplitudes with propofol than with nitrous oxide. Although this difference was statistically not significant, a slightly better response with propofol could clearly be seen.

Kalkman et al. have reported that propofol-alfentanil combination pre-

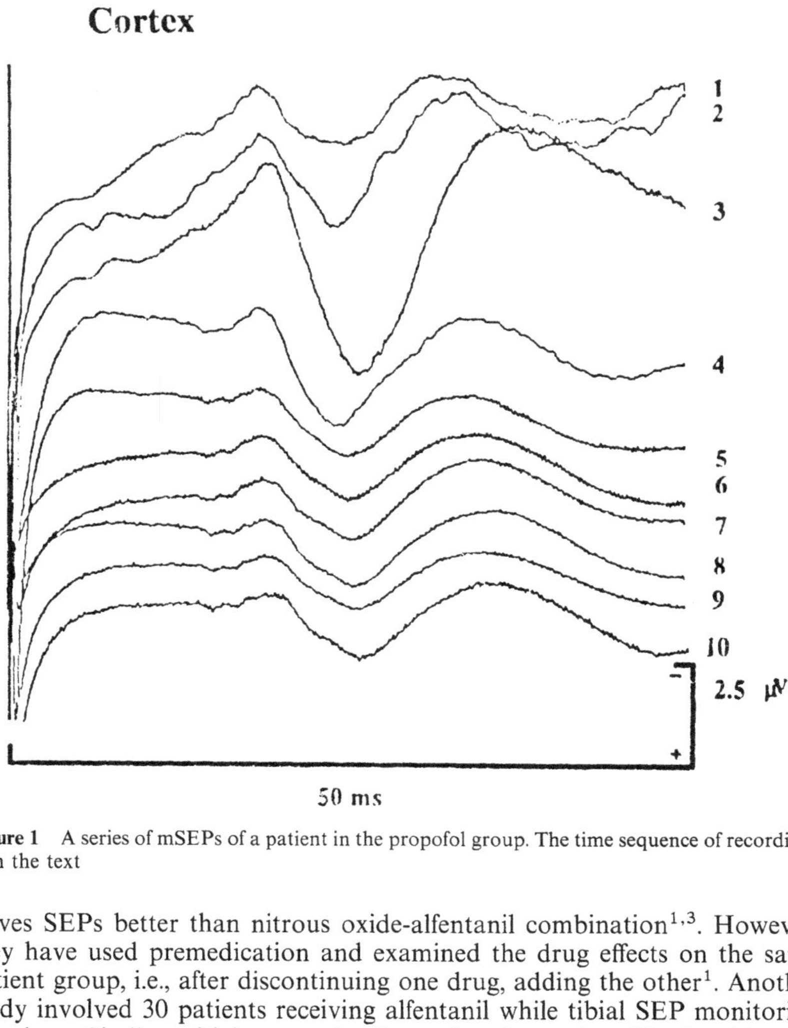

Figure 1 A series of mSEPs of a patient in the propofol group. The time sequence of recordings is in the text

serves SEPs better than nitrous oxide-alfentanil combination[1,3]. However, they have used premedication and examined the drug effects on the same patient group, i.e., after discontinuing one drug, adding the other[1]. Another study involved 30 patients receiving alfentanil while tibial SEP monitoring was done. Similar criticism may be directed at the study of Taniguchi et al., who describe a good effect with propofol[2]. They have compared intravenous and inhalation anaesthetics in a high risk patient group during a routine monitoring procedure. They used additional volatile agents (enflurane or isoflurane) other than nitrous oxide in so-called 'balanced anaesthesia' that are known to attenuate SEPs[4,5]. They also used a variety of narcotic analgesics. In contrast to these two studies, we used a bolus injection of fentanyl only during a controlled study monitoring mSEPs.

It has been reported that propofol causes a marked increase in latency of mSEP peaks with latencies greater than 25 ms, whereas those with latencies less than 20 ms were unaffected[6]. This finding is in accordance with our study, although we did not examine peaks with latencies greater than 20 ms

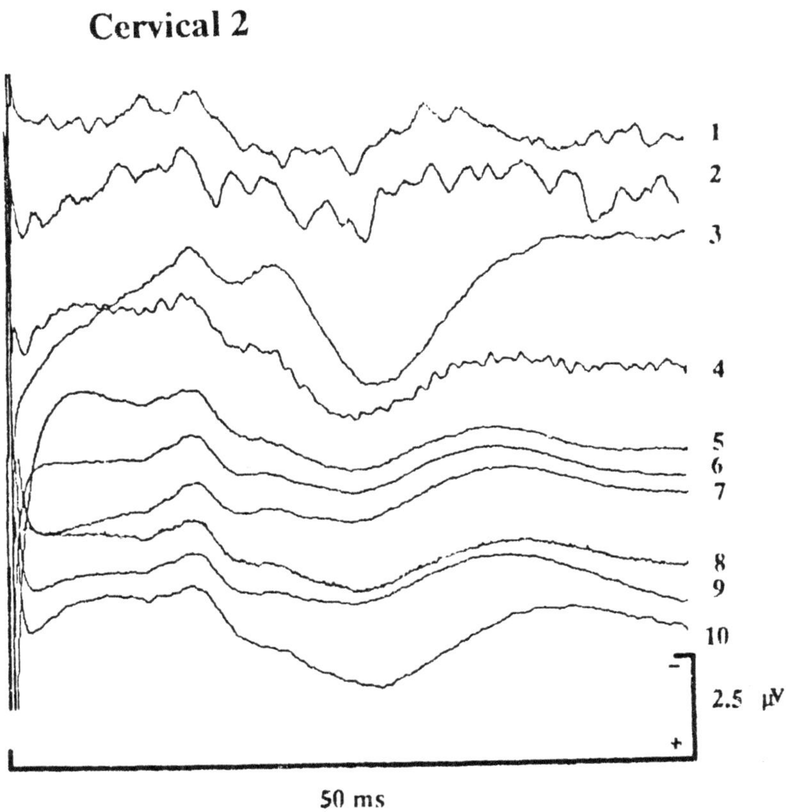

Figure 1 *Continued*

since they are not very useful for monitoring purposes.

The facilitative effect of etomidate on EEG and evoked potentials is well known[4,7]. We observed facilitation to about 250% of pre-induction SEP amplitudes, in accordance with other studies. Its site of action is thought to be cephalic to the spinal cord and continues for 4–5 minutes[4]. When we plotted the amplitude values, we saw that the amplitude of the propofol group had recovered after 60 minutes of the monitoring period. This might be due to disappearance of the effects of fentanyl after 30–45 minutes.

CONCLUSION

In contrast to earlier studies reporting better latency and amplitude preservation with propofol anaesthesia, this study concludes that fentanyl–propofol or fentanyl–nitrous oxide combinations are equally suitable anaesthetic protocols, when monitoring with mSEP is indicated.

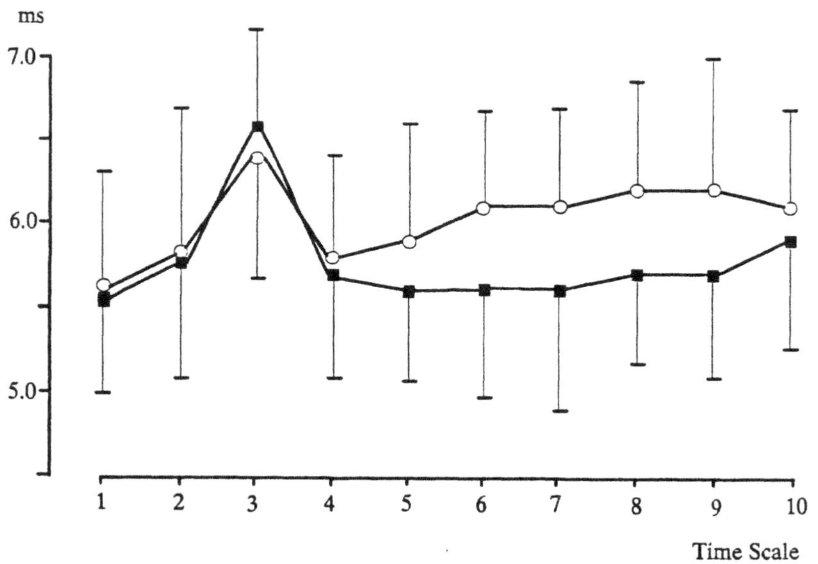

Figure 2 Graph showing the means ± standard deviations of CCT values of both groups. Squares = nitrous oxide; circles = propofol

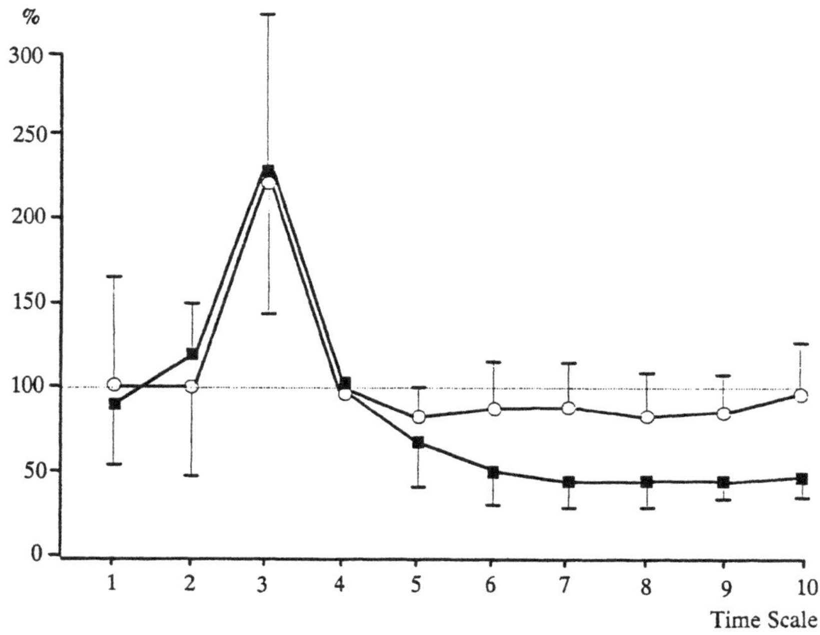

Figure 3 Graph showing the means ± standard deviations of cortical amplitudes of both groups. The amplitudes are normalized taking the before medication values (4th point) as 100%. Squares = nitrous oxide; circles = propofol

Acknowledgement

This study was partly supported by Alexander von Humboldt Foundation (Bonn, Germany).

References

1. Kalkman CJ, Traast H, Zuurmond WA, Bovill JG. Differential effects of propofol and nitrous oxide on posterior tibial nerve somatosensory cortical evoked potentials during alfentanil anaesthesia. Brit J Anaest. 1991;66:483–489.
2. Taniguchi M, Schramm J, Nadstawek J, Fromm G. Inhalation anesthetics in intraoperative SEP monitoring during aneurysm surgery. (In press).
3. Kalkman CR, Brink SA, Been HD, Bonill JG. Variability of somatosensory cortical evoked potentials during spinal surgery. Effects of anaesthetic technique and high-pass digital filtering. Spine. 1991;16:925–929.
4. McPherson RW, Sell B, Traytsman RJ. Effects of thiopental, fentanyl, and etiomidate on upper extremity somatosensory evoked potentials in humans. Anesthesiology. 1986;65: 584–589.
5. Peterson DO, Drummond JC, Todd MM. Effects of halothane, enflurane, isoflurane, and nitrous oxide on somatosensory evoked potentials in humans. Anesthesiology. 1986;65: 35–40.
6. Sheepstra GL, de Lange JJ, Booij LHDJ, Ros HH. Median nerve evoked potentials during propofol anaesthesia. Brit J Anaesth. 1989;62:92–94.
7. Samra SK, Sorkin LS. Enhancement of somatosensory evoked potentials by Etomidate in cats: An investigation of its site of action. Anesthesiology. 1991;74:499–503.

Studies of the human motor system

32
Effects of anaesthesia on the EEG – bispectral analysis correlates with movement

S. M. BOWLES, P. S. SEBEL, V. SAINI and N. CHAMOUN

INTRODUCTION

There is considerable interest in the effects of anaesthetic drugs on the computer-processed EEG or quantitative EEG (QEEG). Sidi[1] used the 95% spectral edge frequency (95% SEF) to administer fentanyl during open heart anaesthesia, thereby reducing the amount of anaesthesia given, and providing a more stable haemodynamic state. Although Rampil[2] predicted the haemodynamic response to tracheal intubation using the 95% SEF, Dwyer and Rampil[3] were unable to predict movement in response to a surgical incision under isoflurane anaesthesia with the 95% SEF.

The lack of ability of the 95% SEF to predict movement to surgical incisions may in part be related to the fact that most studies have considered the power spectrum only. The power spectrum does not adequately quantitate the EEG. The power spectrum expresses amplitude as a function of frequency, and is obtained from the Fourier transform. In addition to the power spectrum, the Fourier transform also contains absolute phase information expressed as a function of frequency[4]. Absolute phase often carries little information and is usually ignored. Recently a new technique has become available to evaluate and quantitate inter frequency phase relationships (relative phase). This technique is called bispectral analysis. We have utilized bispectral analysis of the EEG to evaluate whether or not this technique could predict movement in response to a surgical incision in patients anaesthetized with isoflurane and oxygen.

METHOD

Forty-two ASA class I–III consenting patients between the ages of 17 to 70 years were studied. Patients were randomly assigned to three isoflurane dose groups of 0.75, 1.0 and 1.25 MAC (age adjusted).

After preoxygenation, anaesthesia was induced with thiopental 5 mg/kg. Endotracheal intubation was performed under neuromuscular blockade produced by 100 mg of succinylcholine. The return of neuromuscular function was documented by return of train of four and absence of fade to 100 Hz stimulation of the ulnar nerve. End-tidal isoflurane was measured with a Datex gas analyser or mass spectrometer and maintained at a predetermined end-expired concentration for 5 min prior to skin incision. End-tital oxygen and carbon dioxide were also measured.

Gold cup electrodes were attached to the scalp in a standard manner, positioned at P3, P4, F3, F4, (GND) and CZ (REF) and secured with collodion (impedance $\leq 6\,k\Omega$). Two to five minutes of awake baseline recording of EEG was obtained. The raw EEG signal was obtained from a device provided by Aspect Medical, digitized, and stored on a Toshiba 5200 computer along with oxygen saturation, heart rate, and Dinamap systolic, diastolic, and mean blood pressures. The stored data was later transferred to a streaming tape and analysed off-line by Aspect Medical. The EKG and body temperature were also monitored.

Patients were classified as movers if there was any movement within 60 seconds of the first incision.

BISPECTRAL ANALYSIS

Bispectral analysis determines phase coupling between individual frequencies and quantifies this coupling with the bispectral coherence. Bispectral analysis uses a quadratic or second order function of two frequencies, theta 1 and theta 2, to determine if a third frequency (harmonic) exists which is the sum of the two component frequencies, theta 1 and theta 2[5,6]. Figure 1 shows two complex waveforms representing the EEG. Each waveform appears different from the other, but each is composed of identical components (frequency and amplitude). In the top waveform, the component frequencies are independent, and in the lower waveform, the two and three Hz frequencies are phase locked; thus,the five Hz frequency is in part a harmonic of the two and three Hz frequencies. Bispectral analysis of the lower waveform reveals a spike in bispectral coherence at the intersection of the two and three Hz frequencies. The power spectrum is identical in both waveforms. Bispectral analysis is performed on the output of the Fourier transform and retains the power information present in the Fourier transform. Unlike the power spectrum, the bispectrum contains both power and phase information. Therefore, bispectral analysis quantitates the EEG better than the power spectrum. The bispectral index (BIS), as it was developed in this study, is a probability index with values between zero and one. Thirty-three bispectral variables plus a burst suppression ratio are combined by stepwise discriminant analysis into a single value (BIS).

Analysis

Bispectral variables and traditional power spectral variables were calculated off-line and evaluated as predictors of movement. A bispectral index (BIS) was constructed retrospectively from a database of 160 patients and consisted of 33 bispectral variables plus a burst suppression ratio. Similarly, a power spectral index (PSI) was created from eight power spectral variables plus the burst suppression ratio. The database included our 42 patients plus other patients who had been studied using bispectral analysis. These other patients had been anaesthetized with different anaesthetic drugs.

The indices were created using logistical stepwise discriminate analysis. The statistical evaluation was performed using T test and ANOVA. Sensitivity, specificity and accuracy were calculated.

Sensitivity = Number of correctly predicted movers/Total number of movers

Specificity = Number of correctly predicted non-movers/Total number of non-movers

Accuracy = Total number of correctly predicted patients/Total number of patients

RESULTS

There was no incident of recall of operative events. During the surgical prep, three patients in the 0.75 MAC group moved so violently that they required neuromuscular paralysis prior to incision. Steady state was maintained and these patients were classified as movers.

Traditional power spectral and bispectral variables were evaluated as predictors of movement. Except for relative delta power, no single power or bispectral variable nor ratio of variables perform as well as the corresponding index. The remaining results are described in Tables 1 and 2. Relative delta is expressed as a logarithmic function.

DISCUSSION

Dumermuth[5] demonstrated phase coupling of component frequencies of the EEG. Barnett[7] demonstrated that phase coupling of EEG frequencies decreased as subjects fell asleep and increased as they awakened. A mean BIS of 0.65 in the group of patients who moved is higher than the BIS of 0.46 in the group of patients that did not move with incision. Thus, as the patients were more deeply anaesthetized the BIS was lower, suggesting less phase coupling. In this study, no singe parameter was adequate to predict movement in response to an incision. Although relative delta power is significant between movers and non-movers, its sensitivity, specificity, and accuracy are no better than for median frequency which was not significant between the two groups. A combination of power spectral or bispectral variables always performed better than a single variable alone.

The bispectral analysis uses the phase information of the Fourier transform

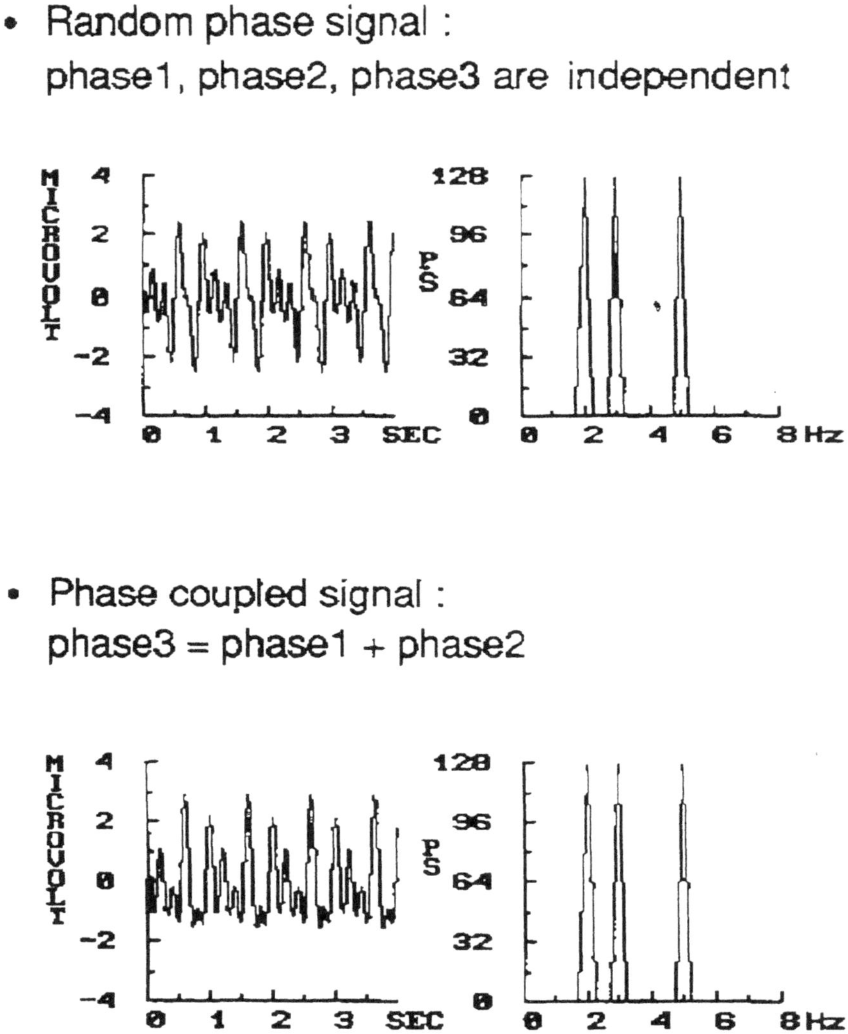

Figure 1 Bispectrum vs. power spectrum

Table 1 Comparison of move and no move for EEG variables–isoflurane anaesthesia (Data are mean ± SE)

EEG	*Move*	*No move*	$p<$
BIS	0.65 ± 0.03	0.46 ± 0.04	0.001
PSI	0.56 ± 0.01	0.52 ± 0.02	0.162
SEF 95%	15.42 ± 0.72 Hz	14.07 ± 0.73 Hz	0.220
Median Freq.	4.50 ± 0.24 Hz	3.87 ± 0.39 Hz	0.148
Rel. Delta	−3.47 ± 0.23 db	−2.64 ± 0.20 db	0.016

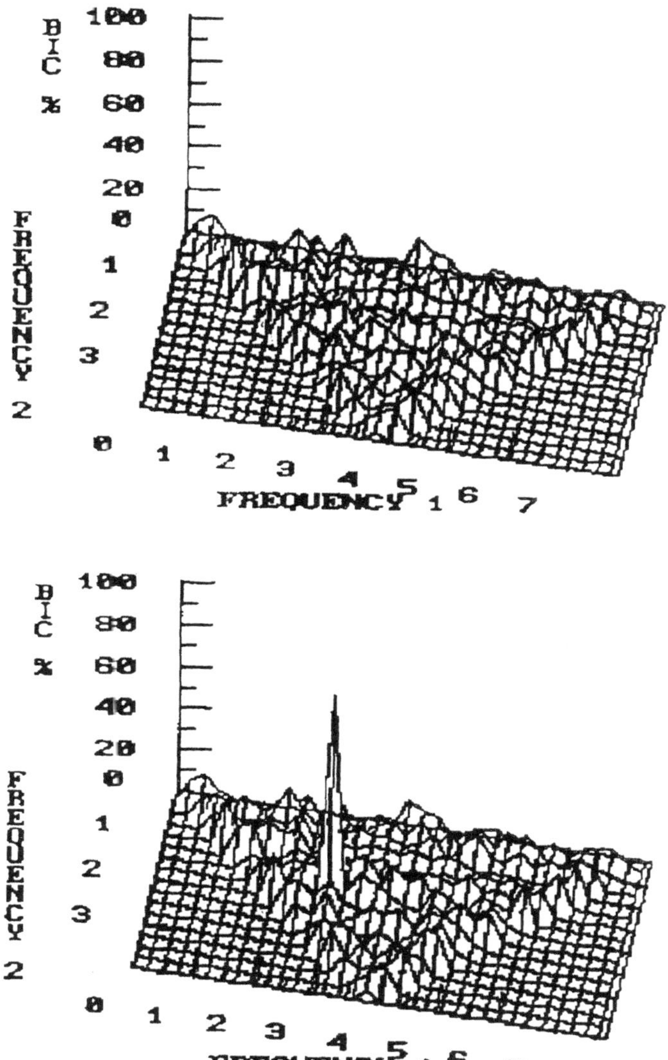

Figure 1 *Continued*

Table 2 Diagnostic accuracy of EEG variables

	Sensitivity	*Specificity*	*Accuracy*
EEG			
BIS	96%	63%	83%
PSI	96%	25%	69%
SEF 95%	62%	81%	69%
Median Freq.	73%	75%	74%
Rel. Delta	73%	75%	74%

that has been ignored in the clinical application of EEG signal processing until now. The representation of phase as a function of frequency is conceptually difficult to grasp. However, by evaluating the EEG for phase coupling of the component frequencies, the phase relationships are more easily represented. Bispectral analysis determines and quantifies phase coupling in the EEG. Bispectral analysis preserves both the power and phase information present in EEG[5,6]. Therefore, bispectral analysis is a powerful tool with which to evaluate the EEG effects of drugs.

References

1. Sidi A, Halimi P, Cotev S. Estimating anaesthetic depth by EEG during anesthetic induction and intubation in patients undergoing cardiac surgery. Journal of Clinical Anesthesia. 1990;2:101–107.
2. Rampil IJ, Matteo RS. Changes in EEG spectral frequency correlate with the hemodynamic response to laryngoscopy and intubation. Anesthesiology. 1987;67:139–142.
3. Dwyer R, Rampil I, Eger EI, Bennett HL. The EEG does not predict movement to surgical incision at 1.0 MAC isoflurane. ASA Abstracts. 1991;75:No 3a 1025.
4. Aron Cohen Biomedical Signal Processing. CRC Press. 1986;vol 1:4–5.
5. Dumermuth G, Huber PJ, Kleiner B, Gasser TH. Analysis of the interrelations between frequency bands of the EEG by means of the bispectrum. A preliminary study. Electroencephalography and Clinical Neurophysiology. 1971;31:137–148.
6. Huber PJ, Kleiner B, Gasser T. Statistical methods for investigating phase relations in stationary stochastic processes. IEEE Transactions on Audio and Electroacoustics. 1971;19:78–86.
7. Barnett TP, Johnson LC, Hicks N, Nute C. Bispectral analysis of electroencephalogram signals during waking and sleeping. Science. 1971;172:401–402.
8. Edmonds HL, Wauquier A. Computerized EEG monitoring in anesthesia and critical care. Helsinki: Instrumentation Science Foundation; 1986.

33
H-reflex spinal cord monitoring during vertebral column stabilization surgery

F. BRACCHI, P. A. GROSSI, L. TROVATI and P. VIGANO

INTRODUCTION

A serious complication that can follow vertebral column stabilization surgery is the appearance of spinal cord lesions[1,2]. Without considering surgical errors, these lesions can be due to an impairment of the spinal cord blood supply caused by mechanical stress, e.g. vertebral distraction by the Harrington or Cotrel-Debousset (C.D.) rods, or a combination of this with anaesthetics. At an early stage this kind of impairment can be reversible if prompt action is taken.

Controls can be made by briefly interrupting surgery and anaesthesia, soon after vertebral arthrodesis, to check voluntary leg movements. Even this procedure however is not a guarantee since a vascular impairment can happen later during the final stages of the operation. The most serious consequence is complete paraplegia.

For this reason it was thought desirable to monitor continuously spinal cord function, during surgery, by checking the Hoffmann reflex (H-reflex). The assumption was that its variability in amplitude would be a good indication of spinal cord function, because the excitability of the extensor muscles' alpha motoneurons is constantly under the facilitatory action of supra-spinal centres. In addition, the technique to elicit and record the H-reflex is simple, known and standardized (for a review see Schieppati[3]).

MATERIALS AND METHODS

Since 1971 monitoring of the H-reflex has been a part of our clinical procedures during surgery for vertebral column stabilization. In 679 patients (498 females and 181 males with ages ranging from 10 to 60) both the Harrington and C.D. techniques have been employed to treat cases of

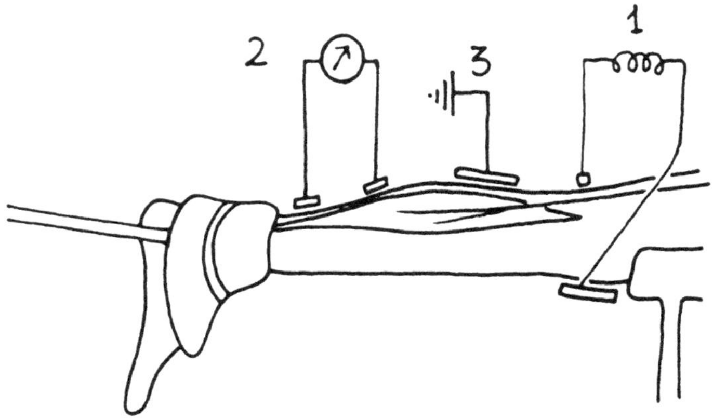

Figure 1 Arrangement of stimulating and recording electrodes on the leg: 1) stimulating electrodes (cathode at the popliteal fossa over the posterior tibial nerve; anode over patella); 2) recording electrodes (reference over the Achilles tendon and the second over the soleus muscle); 3) grounding. The ankle restraint is used to exert skeletal traction

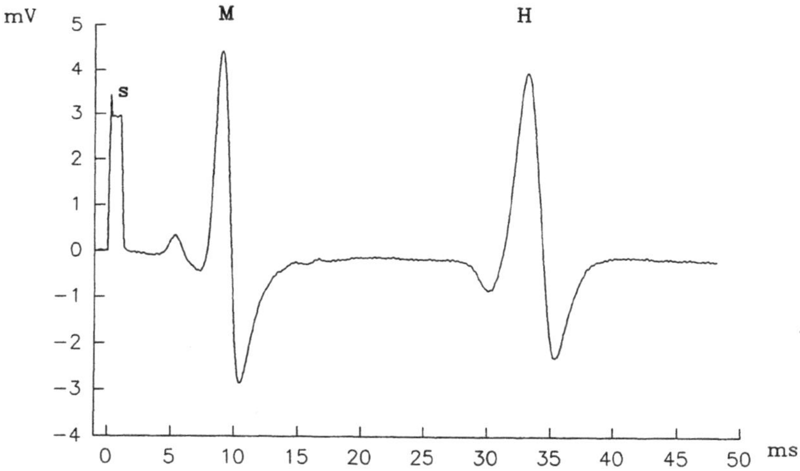

Figure 2 M wave (direct stimulation of soleus motor nerve fibres) and H wave (H-reflex) recorded from a subject before anaesthesia; s is the stimulus artefact. The trace is triggered 1 ms before onset of the stimulus

scoliosis, vertebral fractures and vertebral neoplasm. None of the patients showed signs of sensory or motor deficit before surgery.

The stimulus is a constant voltage square wave of 1 ms duration delivered every 10 s at the popliteal fossa (Figure 1). Its intensity is selected to obtain the highest H-reflex amplitude with a well-defined M wave (Figure 2).

The H-reflex excitability curve is monitored by delivering every 10 s two stimuli of the same characteristics; a conditioning stimulus followed (after a variable delay of 10 to 500 ms) by the test one. The amplitude of the H-reflex wave elicited by the test stimulus is measured as a percentage of the one elicited by the preceding control stimulus. Stimulation parameters were established on all patients, after premedication, before the induction of anaesthesia. H-reflex parameters were monitored the day before surgery in 20 patients only.

The H-reflex monitoring is carried out continuously from induction of the anaesthesia until complete return of consciousness.

The patient is kept in a prone position on the surgical table. For the treatment of scoliosis a collar and ankle restraints are used. In this way a certain amount of traction (10 to 25 kg) can be applied if necessary to help the setting of the Harrington or C.D. rods. Vertebral distraction is exerted by an outrigger device before insertion of the Harrington rod.

The first hundred patients treated were awakened as a routine after the Harrington instrumentation and tested for voluntary foot movements. On the patients that followed, the above procedure has been carried out only if called for, as in the case of a complete disappearance of the H-reflex (see results).

The e.m.g. recording apparatus is a one-channel unit (Medelec MS-90). In the averaging mode the real time analog trace, and the normalized averaging of responses to the last few stimuli, can be simultaneously displayed.

Anaesthesia (for 50 kg body weight)

Premedication: atropine 0.5 mg; droperidol 2.5 mg; fentanyl 0.05 mg.

Induction: droperidol 5 mg; fentanyl 0.1 mg; thiopental 200 mg; vecuronium bromide or pancuronium bromide 4 mg.

Maintenance of anaesthesia: Inhalation anaesthesia with nitrous oxide 65% and oxygen 35% and occasionally enflurane 0.2 to 1% max. to lower blood pressure; vecuronium or pancuronium 0.5 to 1 mg doses when necessary to lower respiratory resistance; fentanyl 0.01 to 0.02 mg when necessary.

Ventilation is regulated to maintain alveolar pCO_2 between 35 and 40 mmHg.

Alveolar pCO_2, inspired oxygen, pulse oximetry, blood pressure, EKG and heart rate are continuously monitored.

RESULTS

1. Controls before anaesthesia

Twenty patients were tested the day before surgery for control purposes. Stimulus intensity for the H-reflex and M wave threshold, and the H-reflex excitability curve, were within the values described in the literature for normal subjects.

2. After premedication

In these 20 patients, threshold values were not significantly different from the control ones. In all patients an increase of the complete H-reflex inhibition period (50 to 100 ms) was observed in the excitability curve.

3. After induction of anaesthesia

The behaviour of the H-reflex depends upon the drugs used. After a full curare dose the H-reflex and the M wave are usually completely abolished for up to 40–45 minutes with pancuronium bromide and up to 15–20 minutes with vecuronium bromide. During this period, depending on the subject and dosage used, the H wave can still be observed with the averaging technique. Its amplitude is only about 20–30 μV compared with a few millivolts for the control value.

During surgery

After recovery from curare the H-reflex is usually deeply depressed (down to 50–20% of the control value), but it never disappears completely with the drugs and dosages used. An excitability spectrum inversion is usual (i.e. the M wave appears at a lower stimulus intensity than the H wave). The excitability curve shows a full inhibition period of more than 100 ms. A single fentanyl maintenance dose lowers the H-reflex amplitude to one-half its previous value for 10–15 minutes. Enflurane deeply depresses the H wave. Its use is limited to the period before vertebral distraction. Curare in 0.5 mg doses does not significantly lower the reflex.

Most of the surgical procedures, traction (up to 25 kg) applied between the collar and the ankle restraints, vertebral distraction by an outrigger or by the Harrington or C.D. rods, have no significant influence on the H-reflex amplitude.

The H-reflex is always facilitated (up to 3–4 times its previous value) during the surgical manoeuvres to reach the ileum bone for retrieving additional bone graft material.

Hammer strokes on the osteotome chisel, when decortication of the posterior elements of the vertebrae is carried out, have a deep depressing action on the reflex. Following a single hammer stroke, over the lower dorsal vertebrae, the H-reflex is abolished for 2–3 minutes. No depression is detected when a bone rongeur or a Cobb spinal gouge is used.

Recovery from anaesthesia

The H-reflex progressively increases in amplitude in parallel with metabolism of the anaesthetic drugs showing low sensitivity to nitrous oxide by itself. It sharply returns to pre-anaesthesia control values with return of consciousness and appearance of voluntary movements.

Particular cases

In three cases, during surgical treatment for scoliosis, the H-reflex completely disappeared. This occurred about half an hour after the Harrington rod was set in place and while the wound was being sutured. The reflex could not be elicited even with modification of the stimulation parameters. Two of these patients had been previously routinely awakened for a brief period after rod insertion to test for voluntary foot movements. The anaesthesia was then suspended and, when awake, all showed signs of paralysis of the lower extremities with the H-reflex still completely abolished. Blood pressure was within normal range. In two of these cases the H-reflex reappeared, 3 minutes after the return to full consciousness, followed soon after by movements of the legs and feet. In the remaining case no H-reflex or leg movement reappeared even five minutes after return of consciousness. After re-induction of anaesthesia the wound was reopened, the Harrington rod cut and removed. Within three minutes the H-reflex reappeared and a new Harrington rod was set in place exerting less vertebral distraction. After recovery from anaesthesia, movements and sensitivity were normal.

CONCLUSIONS

The H-reflex seems to be a sensitive means of detecting at an early stage a spinal cord impairment when it is still in a reversible stage. Waking patients after vertebral distraction and fusion does not provide any guarantee against a spinal cord impairment soon afterwards. In the three cases described, the deep H-reflex depression could probably be attributed to a reversible vascular impairment directly involving the spinal cord reflex centre, or a higher cord segment. In humans it has been shown that the H-reflex is abolished, for a few hours, as a consequence of the reflex centre functional isolation from supra-spinal centres following spinal cord trauma or brain damage[4,5].

Hammering on the osteotome chisel when decorticating the vertebrae should be limited and if possible avoided, since signs of spinal cord concussion are always evident.

Bone graft retrieval from the ileum should be carried out after vertebral distraction because of the facilitatory effect on the H-reflex caused by surgical hip stimulation. This facilitation seems to be due to a long-loop circuit involving a supra-spinal centre[6]. Therefore this provides a useful test for the functionality of the spinal cord throughout its length.

References

1. MacEwen GD, Bunnell WP, Sriram K. Acute neurological complications in the treatment of scoliosis. A report of the Scoliosis Research Society. J Bone Joint Surg (Am). 1975;57: 404–408.
2. Michel F, Rubini J, Grand C, Bérard J, Koler R, Michel CR. Les complications neurologique de la chirurgie des déformations du rachis. Rev de Chir Orthopédique. 1992;78:90–100.
3. Schieppati M. The Hoffmann reflex: a means of assessing spinal reflex excitability and its descending control in man. Progress in Neurobiology. 1987;28:345–376.

4. Cadilhac J, Georgesco M, Benezech J, Duday H, Dapres G. Potential évoqué cérébral somesthésique et réflexe d'Hoffmann dans les lésions médullaires aiguës. Intérêt physiopathologique et pronostic. Electroenceph clin Neurophysiol. 1977;43:160–167.
5. Crenna P, Conci F, Boselli L. Changes in spinal reflex excitability in brain-dead humans. Electroenceph clin Neurophysiol. 1989;73:206–214.
6. Delwaide PJ, Crenna P. Cutaneous nerve stimulation and motoneural excitability. II: evidence for non-segmental influences. J Neurol Neurosurg Psychiatry. 1984;47:190–196.

34
Central motor conduction time to upper versus lower extremities

D. CLAUS, C. B. BRUNHÖLZL, M. WEIS, A. JASPERT and E. BIANCHI

INTRODUCTION

Transcranial magnetic stimulation of the human motor cortex allows investigation of central motor pathways. Latencies of muscle action potentials evoked by cortical stimulation represent total motor conduction time from the cortex to the target muscle. The central motor conduction time (CMCT) can be calculated by subtracting the peripheral motor latency after stimulation of spinal motor roots from the total motor conduction time. CMCT has proved to be a valuable diagnostic aid in various diseases. Prolonged CMCT has been observed in multiple sclerosis, amyotrophic lateral sclerosis and in hereditary ataxias[1]. As far as spinal cord disorders are concerned, there exist reports about traumatic lesions as well as syringomyelia. Cervical spondylosis has been studied by several authors [2,3]. In general, however, motor conduction was measured to upper limbs only. In the present study CMCT was investigated to upper as well as lower extremities.

PATIENTS AND METHODS

Transcranial magnetic stimulation was carried out with a Magstim 200 stimulator (The Magstim Company, Whitland, Dyfeld SA34 0HR, England) discharging into a circular high power coil (mean diameter 95 mm, maximum magnetic field 2.5 Tesla). The coil was positioned flat on the scalp with its centre over the vertex for recordings from the upper limbs and more rostrally for stimulation of motor areas representing the lower extremities[4]. Stimulus strength was 20% above motor threshold[4]. The ulnar nerve was stimulated electrically and supramaximally at the wrist and the peroneal nerve at the head of the fibula (Digitimer D180 stimulator, Welwyn Garden City). Cervical and lumbar motor roots were stimulated at C7 and L1 respectively with the electrical method described elsewhere[4]. Recordings were done from the voluntarily contracted abductor digiti minimi muscle (ADM) in the hand and anterior tibial muscle (TA) in the leg. Four compound muscle action potentials (CMAPs) were recorded. The largest amplitude of the first negative potential's component and the shortest onset latency were measured. CMCT was calculated by subtracting latencies after spinal stimulation from that after cortical stimulation (Figure 1).

Thirty-three patients had cervical myelopathy (31–77 yrs.). According to MRI (magnet resonance imaging) or myelography stenosis of the spinal canal was located at or above C7. In ten cases only signs of radiculopathy were present (depressed tendon reflexes, peripheral motor and sensory disturbance), while in 15 cases, upper motor neuron signs (brisk tendon jerks, spasticity, extensor plantar response) were present in the legs and in six cases in upper extremities. Cord compression was diagnosed by MRI scan in 25 cases.

In the syringomyelia group (11 cases, 24–79 yrs.) one patient had central signs in the arms and nine in the legs (seven cases extensor plantars). The diagnosis had been proved by MRI investigation in all cases. The diagnosis of HSP (hereditary spastic paraparesis) covers different disorders with varying age of onset and inheritance[5]. Clinical features are spastic gait with scissoring, exaggerated patellar tendon reflexes as well as achilles tendon jerks. There is ankle clonus and plantar responses are extensor. There are no other additional signs in pure HSP. We investigated 14 cases with the pure form of HSP (18–67 yrs.). Other disorders were excluded by normal neuroradiology and CSF as well as by follow-up investigation. Fourteen cases had central signs in lower extremities and four in the upper limbs. Statistical analysis was carried out by Fisher's exact test.

RESULTS

In the 33 patients with cervical myelopathy CMCT was prolonged to arms and/or legs in 24 cases (72%). CMCT was more often abnormal to legs (20/32 cases, 33/64 sites) than to arms (17/33 cases, 27/66 sites) (Figure 2). There is a correlation between abnormal CMCT to ADM and central signs in hands ($p < 0.05$) as well as between CMCT to TA and central signs in legs or extensor plantars respectively ($p < 0.001$). Furthermore, CMCT is more often abnormal to ADM ($p < 0.01$) as well as to TA ($p < 0.01$) in cases with cervical cord compression in MRI scan. In the group with syringomyelia eight patients (72%) presented with abnormal CMCT in one or more extremities. The frequency of abnormal CMCT was higher in legs (6/11 cases, 10/22 sites) compared with upper extremities (3/11 cases, 5/22 sites). In HSP CMCT was abnormal to the hand in only one case (Figure 3). To leg muscles CMCT was impaired in eight of the nine investigated cases (14/18 sites). CMCT was normal to three hands with central signs.

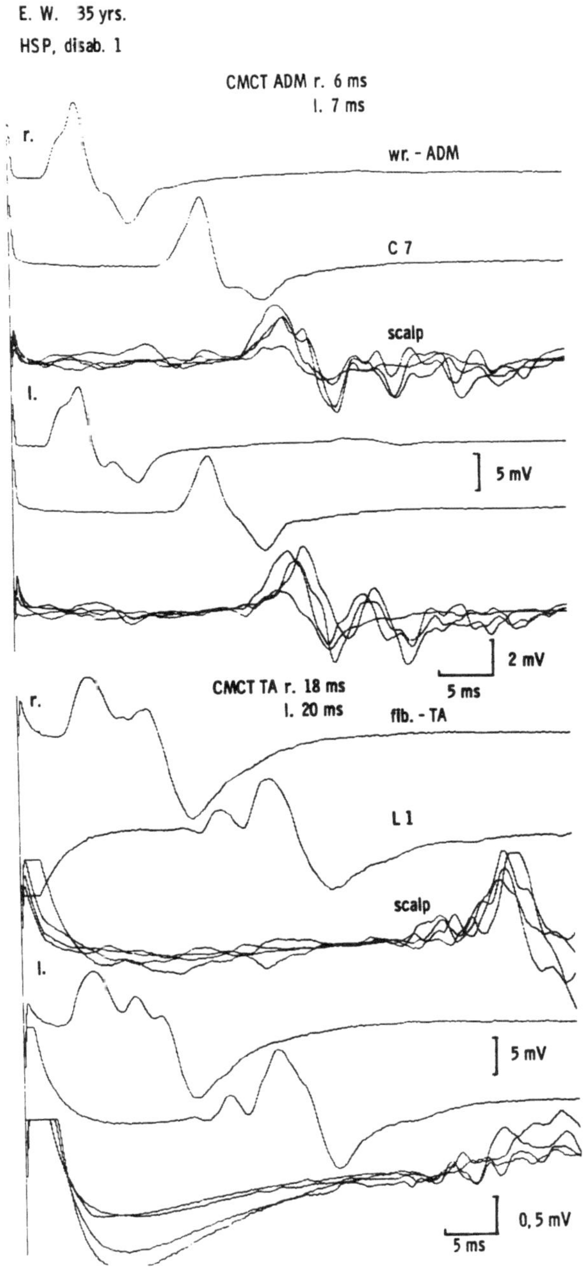

Figure 1 CMCT recording in a patient with HSP. CMAPs are recorded from right and left ADM as well as TA after peripheral (wrist, fibula head), spinal electrical stimulation (C7, L1) and transcranial magnetic stimulation (scalp). CMCT (latency difference C7-scalp, L1-scalp) is normal to ADM and prolonged by TA

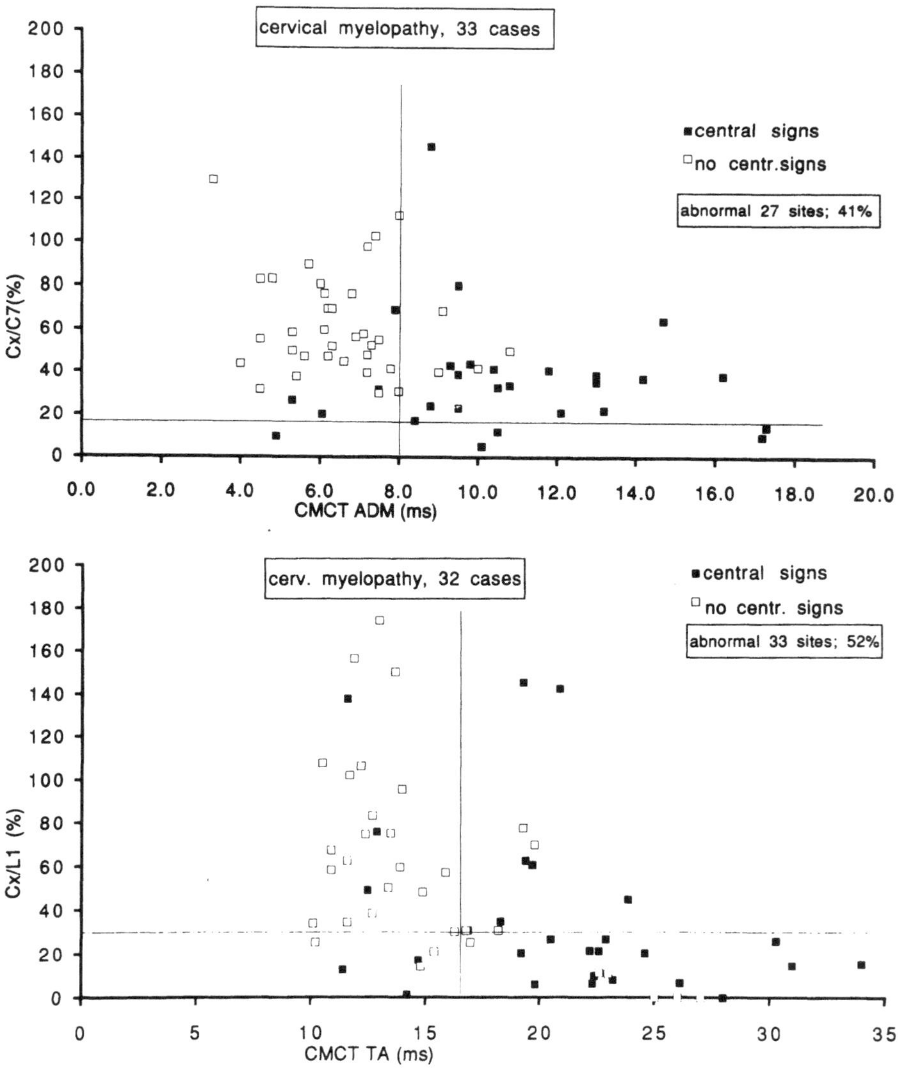

Figure 2 CMCT results in 33 cases with cervical myelopathy (in 4 TA no CMAP could be evoked by cortical stimulation). Limits of normal are marked by lines (influence of height on CMCT not taken into consideration). Abnormal results are clearly correlated with central signs in the legs

DISCUSSION

Chang and Masur[3,7] examined central conduction time in cord injury as well as cervical myelopathy. In cervical myelopathy CMCT results are sensitive. This is especially so with respect to CMCT to leg muscles (Figure

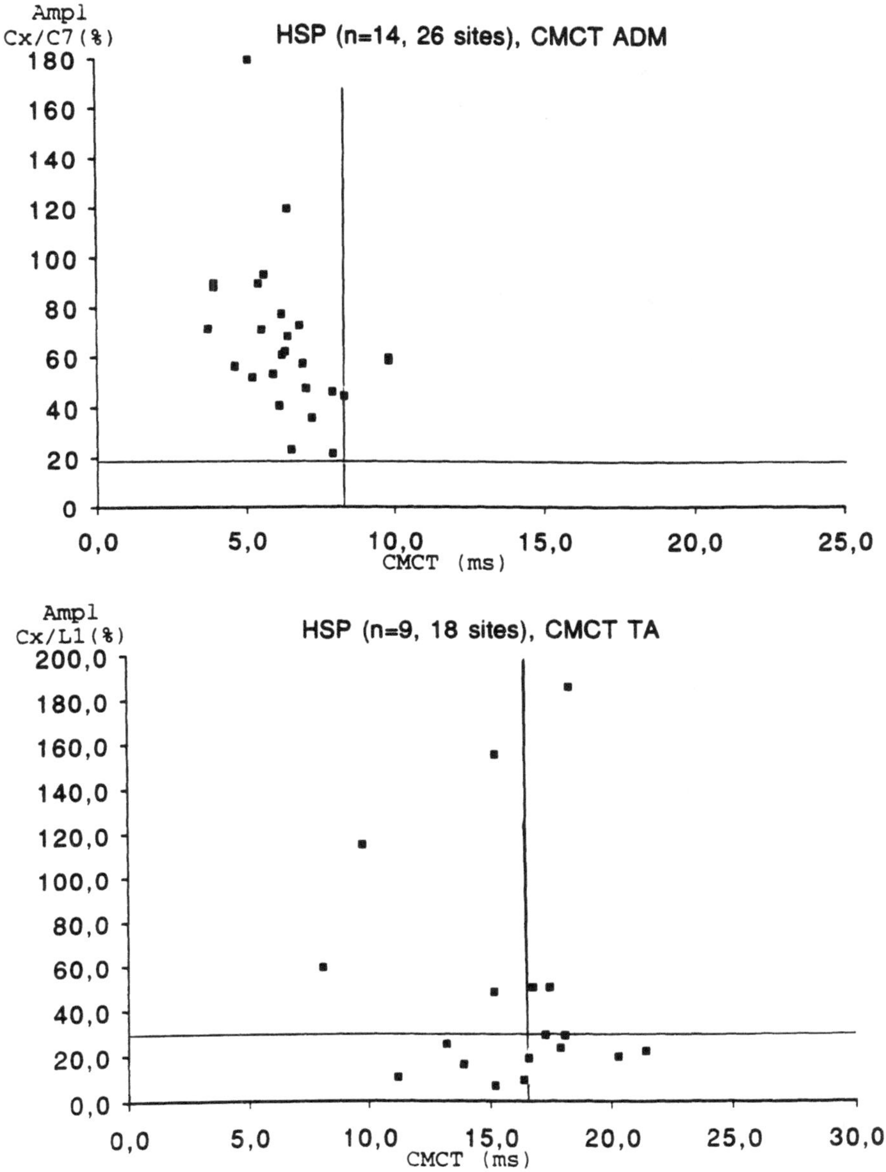

Figure 3 In HSP CMCT is almost always normal to upper extremities and abnormal to legs. Limits of norm marked by lines[6]

2). It is not a result of the segmental localization of cord compression, which was always at or above C7. Central conduction can be considerably delayed in cases with cord compression (up to 17 ms). Results are therefore not specific. Six patients without clear-cut central signs had abnormal CMCT to TA. Four of them had distinct cord compression in the MRI scan. Therefore, the method is able to detect subclinical lesions. In syringomyelia CMCT reflects impairment of central motor conduction, even in cases with radiculopathy and attenuated CMAPs after cervical stimulation. CMCT to TA was more sensitive than that to ADM. The relatively high sensitivity of CMCT to TA could probably be a result of the localization of descending neurons in the cord. Furthermore, the alteration of cervical motor roots may cause peripheral conduction disturbance which masks CMCT abnormality to upper extremities.

In HSP different investigators found degeneration of the corticospinal tract mainly in the lumbar and thoracic cord[8,9]. Later degeneration of the lateral pyramidal tracts extends from the lumbar to the upper cervical region[9]. Degeneration of Betz cells and atrophy of the central gyrus were also seen. In HSP CMCT is usually normal to upper extremities and abnormal to the legs. This is thought to be due to pronounced degeneration in the thoracic spinal cord, which is not part of the motor pathway investigated to cervical nerve roots[6]. CMCT results may help to prove the diagnosis at an early stage of the disease. They also describe the progression of the degenerative process. CMCT to ADM was also normal in spastic arms. CMCT sensitivity in HSP may be due to the morphology of the degenerative process which is different from ALS. Different from MS, results are symmetrical. Nevertheless, results are not specific and cannot exclude spinal space occupying lesions.

CMCT is a sensitive test of cord involvement in spinal disorders. It presents additional data in cases without clear-cut central paresis or spasticity. The investigation of CMCT to legs is believed to produce more sensitive results in spinal disorders than that to upper extremities. In syringomyelia CMCT is useful since peripheral motor neuron involvement makes central signs difficult to investigate. In HSP results are sensitive: they are, however, not specific. Thus, CMCT investigation cannot replace neuroradiology.

Acknowledgement

This work was supported by the Sander Foundation.

References

1. Claus D, Harding AE, Hess CW, Mills KR, Murray NMF, Thomas PK. Central motor conduction in degenerative ataxic disorders. A magnetic stimulation study. J Neurol Neurosurg Psychiatry. 1988;51:790–795.
2. Dvorak J, Herdmann J, Janssen B, Theiler R, Grob D. Motor-evoked potentials in patients with cervical spine disorders. Spine. 1990;15:1013–1016.
3. Masur H, Elger CE, Render K, Fahrendorf G, Ludolph AC. Functional deficits of central sensory and motor pathways in patients with cervical spinal stenosis: a study of SEPs and EMG responses to non-invasive brain stimulation. Electroenceph clin Neurophysiol. 1989;74:450–457.
4. Claus D. Central motor conduction: method and normal results. Muscle & Nerve. 1990;13:1125–1132.

5. Harding AE. Hereditary 'pure' spastic paraplegia: a clinical and genetic study of 22 families. J Neurol Neurosurg Psychiatry. 1981;44:871–883.
6. Claus D, Waddy HM, Harding AE, Murray NMF, Thomas PK. Hereditary motor and sensory neuropathies and hereditary spastic paraplegia: a magnetic stimulation study. Ann Neurol. 1990;28:43–49.
7. Chang CW, Lien IN. Estimate of motor conduction in human spinal cord: slowed conduction in spinal cord injury. Muscle & Nerve. 1991;14:990–996.
8. Behan WMH, Maia M. Stümpell's familial spastic paraplegia: genetics and neuropathology. J Neurol Neurosurg Psychiatry. 1974;37:8–20.
9. Strümpell A. Beiträge zur Pathologie des Rückenmarks. Arch Psychiatr Nervenkr. 1880;10:676–717.

35
Do neurogenic lesions of the spinal cord generate distinctive features of the epidurally recorded motor evoked potential?

V. DELETIS, K. KIPROVSKI, M. NEUWIRTH and G. ENGLER

INTRODUCTION

Since the introduction of intraoperative motor evoked potential (MEP) monitoring, MEPs have been recorded from many different patient populations. The best available method of monitoring MEPs in the operating room is recording from the epidural space of the spinal cord after transcranial electrical stimulation of the motor cortex[1]. Epidurally recorded MEPs are characterized by a D-wave, which represents the response following direct stimulation of the corticospinal tract. This wave may be followed by one or more I-waves, which represent the response following indirect (transynaptic) activation of the cortical motoneurons. The D-wave is the response of principal interest in intraoperative monitoring; it consists of a single, stable positive–negative component, exhibiting a similar morphology in most patients. In contrast, the I-wave complex shows characteristics which vary greatly across different patient populations; it may be absent or consist of one to four positive–negative components.

We present data showing I-wave attributes in different patient populations. We propose that these attributes represent the degree to which afferent activity of the spinal cord facilitates the motor cortex, and that patients with idiopathic scoliosis and patients with spinal cord tumours exhibit an increased amount of motor cortex facilitation.

PATIENTS AND METHODS

We recorded MEPs from 13 patients with idiopathic scoliosis, 13 patients with spinal cord tumours, and four patients with intracranial pathologies [two with arteriovenous malformations, one with cerebral aneurysms, and one with a metastatic tumour] from the epidural space of the upper thoracic spinal cord with an epidural catheter-electrode following transcranial electrical stimulation of the motor cortex (see Figure 1A). During surgery for the correction of scoliosis the catheter-electrode was inserted between the laminae through a flavectomy (Figure 1B). In the case of spinal cord tumours, MEPs were recorded from an electrode inserted rostral to the tumour following a laminectomy (Figure 1C). For patients with intracranial pathologies, the electrode was inserted percutaneously with a 14-gauge Tuohy needle (Figure 1D).

The patients with idiopathic scoliosis and intracranial pathologies had no sensory or motor deficits. Although some of the patients with spinal cord tumours did have deficits, the recordings reported here were recorded rostral to the level of the tumour.

The Digitimer D-180 (Digitimer, Welwyn Garden City, England) was used to deliver stimulation of 80–100% of stimulator output (100% output equals 750 V) to the scalp overlying the motor cortex. Three to seven responses were filtered between 100 and 1700 Hz and averaged. All patients were anaesthetized with a nitrous oxide/narcotic combination, with no halogenated anaesthetics above concentrations of 0.3%.

The data reported here were baseline recordings and were not used for clinical monitoring. I-waves (reflecting transynaptic activity) are influenced by halogenated and other anaesthetics, and are not suitable for monitoring. D-waves (recorded from the low thoracic region during surgery for correction of scoliosis and the removal of spinal cord tumours) were monitored throughout the surgical procedure to evaluate the functional integrity of the motor system.

RESULTS

Patients were categorized by the number of I-waves exhibited in their MEP and by their pathology. The distribution is displayed in Table 1. Note that more than 50% of the patients with idiopathic scoliosis showed MEPs with three or more I-waves. Only 32.5% of the patients with spinal cord tumours, and none of the patients with intracranial pathologies, had three or more I-waves.

Representative recordings from two of the patients with idiopathic scoliosis are presented in Figure 2, recordings from two of the patients with spinal cord tumours are presented in Figure 3, and recordings from two of the patients with intracranial pathologies are displayed in Figure 4.

DISCUSSION

These data represent the preliminary results of our study of I-wave characteristics in different patient populations. Although there are not enough patients in the study to achieve statistical significance, we note that there are recognizable differences in the number of I-waves observed in the three

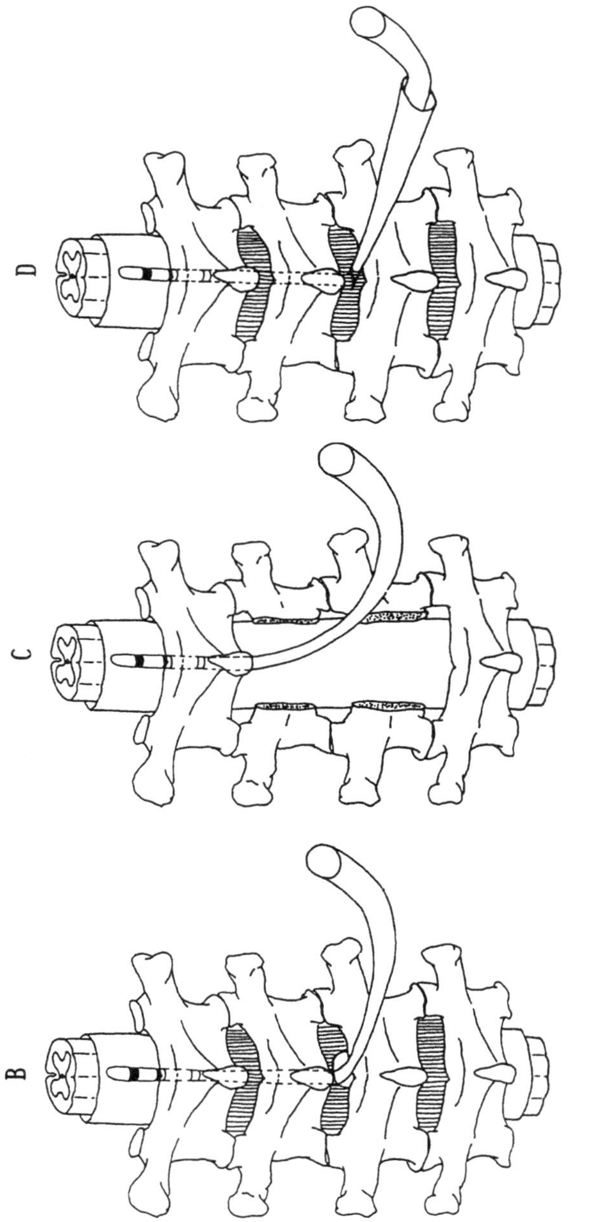

Figure 1 A. Electrode positions for transcranial electrical stimulation; anode at CZ and cathode 6 cm in front of CZ. Schematic diagram of placement of an epidural catheter-electrode: **B**. through a flavectomy; C. after a laminectomy; **D**. inserted percutaneously

Table 1 Number of I-waves observed in different patient populations

	Number of I-waves					
	0	1	2	3	4	*Total*
Idiopathic scoliosis	1	1	4	4	3	13
Spinal cord tumour	0	4	5	4	0	13
Intracranial pathology	1	3	0	0	0	4

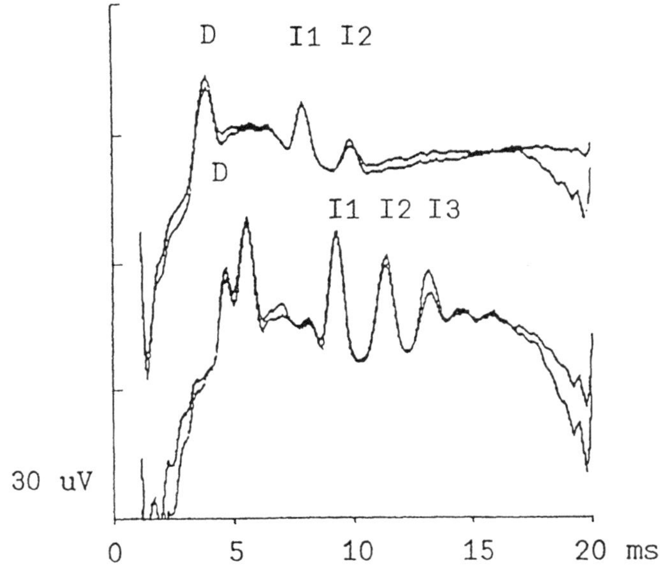

Figure 2 MEPs recorded from the epidural space of the thoracic spinal cord from patients with idiopathic scoliosis. Note the high amplitude and the large number of I-waves. [Positive is down]

groups studied. Specifically, the patients with idiopathic scoliosis showed greater numbers of I-waves than the patients with other pathologies.

We suggest that I-wave attributes reflect the degree to which spinal afferent activity facilitates the motor cortex as a continuous spino-cortical drive. The facilitatory effects of afferent activity have already been demonstrated in the MEP recorded over muscle following stimulation of the peripheral nerves[2,3]. Because patients with idiopathic scoliosis show large numbers of I-waves, we proposed that they have a greater degree of motor cortex facilitation than the patients with other pathologies. Patients with spinal cord tumours have variable degrees of facilitation and patients with intracranial pathologies have very little facilitation.

We propose that MEPs recorded from the epidural space of patients with idiopathic scoliosis do not represent the 'normal' response, as has been previously suggested[4], but reflect an increased amount of spino-cortical facilitation as the result of a spinal neurogenic lesion. The possibility of a

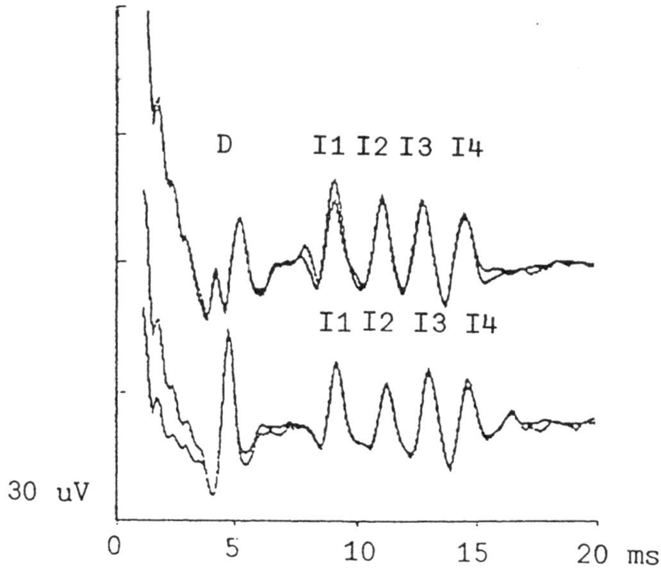

Figure 3 MEPs recorded from the epidural space of the thoracic spinal cord from patients with an intramedullary tumour. [Positive is down]

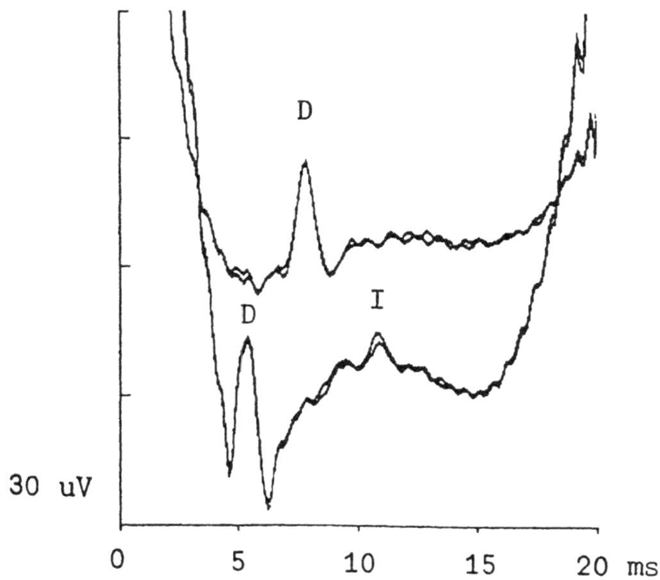

Figure 4 MEPs recorded from the epidural space of the thoracic spinal cord from patients with a cortical AVM. Note the low amplitude I-wave in one patient, and the complete absence of I-waves in the other. [Positive is down]

spinal neurogenic lesion was introduced by Trontelj et al.[5], who found an asymmetry in the stretch reflexes of patients with idiopathic scoliosis. They found that the stretch reflexes were diminished in the deep paraspinal muscles and increased in the superficial paraspinal muscles on the convex side of the curve, and suggested that the asymmetry in muscle activity is a result of the hypothesized segmental neurogenic lesion.

Other data supports the possibility of a segmental lesion in scoliotic patients. Zuk[6] and Redford et al.[7] observed an increase in the tonic EMG activity in the superficial paraspinal muscles of the convex side, relative to the concave side of the curvature. Hoogmartens and Basmajian[8] found that the tonic vibratory reflex is increased in the deep paraspinal muscles of the concave side of the curvature, and decreased in the deep muscles of the convex side of the curve.

There may be mechanisms other than spinal cord pathology that generate increased numbers of I-waves. Varying the stimulation geometry (by varying the electrode montage or magnetic coil orientation) or administering drugs which alter the threshold of cortical motoneurons may affect the number of I-waves generated by each stimulation of the motor cortex. Increased numbers of I-waves generated by these mechanisms may not reflect an increased amount of motor cortex facilitation.

We propose that the number of I-waves observed in the patients with spinal cord tumours also represents an increase in spino-cortical facilitation, resulting from chronic segmental irritation due to the tumour compression on the spinal cord. We believe that epidural MEPs recorded from patients with intracranial pathologies, who would be expected to have relatively unaffected spinal cords and spino-cortical drive, represent the 'normal' response. As it is unethical to record MEPs from the spinal epidural space of healthy subjects, we suggest that MEPs recorded from the epidural space of patients with intracranial pathologies (or pathologies not affecting the spinal cord) be considered the standard against which other recordings are compared.

References

1. Deletis V. Intraoperative monitoring of the functional integrity of the motor pathways. In: O Devinsky, A Beric, M Dogali, eds. Advances in Neurology: Electrical and magnetic stimulation of the brain. New York, NY: Raven Press; 1993: in press.
2. Deletis V, Schild JH, Beric A, Dimitrijevic MR. Facilitation of motor evoked potentials by somatosensory afferent stimulation. Electroenceph Clin Neurophysiol. 1992;85(5): in press.
3. Deuschl G, Michels R, Berardelli A, Schenck E, Inghilleri M, Lucking CH. Effects of electric and magnetic transcranial stimulation on long latency reflexes. Exp Brain Res. 1991;83: 403–410.
4. Burke D, Hicks RG, Stephen JPH. Corticospinal volleys evoked by anodal and cathodal stimulation of the human motor cortex. J Physiol. 1990;425:283–299.
5. Trontelj JV, Pecak F, Dimitrijevic MR. Segmental neurophysiological mechanisms in scoliosis. J Bone Joint Surg. 1979;61-B(3):310–313.
6. Zuk T. The role of spinal and abdominal muscles in the pathogenesis of scoliosis. J Bone Surg. 1962;44:102–105.
7. Redford JB, Butterworth, TR, Clements EL. Use of electromyography as a prognostic aid in the management of idiopathic scoliosis. Arch Phys Med Rehab. 1969;50:433–438.
8. Hoogmartens MJ, Basmajian JV. Postural tone in the deep spinal muscles of idiopathic scoliosis patients and their siblings. Electromyog Clin Neurophysiol. 1976;16:93–114.

36 Anaesthesia and the motor evoked potential

M. E. FENNELLY, B. A. TAYLOR and M. HETREED

INTRODUCTION

Sensory evoked potentials are widely used to monitor the integrity of the spinal cord during scoliosis surgery. Routine monitoring of the somatosensory evoked potential as recorded from the epidural space is relatively insensitive to routinely used anaesthetic agents.

Earlier prediction of impending damage to the spinal cord using intraoperative monitoring of the descending motor tracts is a desirable addition to the above. However almost all of the described techniques for monitoring the motor tracts are very anesthetic sensitive. The recently described double pulse stimulation of the cervico-thoracic spinal cord at the epidural space with recording the compound muscle action potential from tibialis anterior has proved no exception[1]. In this study we aimed to establish an anaesthetic technique compatible with recording stable maximal intraoperative compound muscle action potentials, and to demonstrate the effect of the nitrous oxide and isoflurane on the same. A recovery profile from propofol anaesthesia is also shown. Propofol has a short duration of action and is ideally suited for total intravenous anaesthesia in combination with an opioid[2].

METHODS – ANAESTHETIC

This study was approved by the local ethics committee and all patients, or their relatives, gave informed consent. All patients fell into the A.S.A. class I group and were in the 15–30 age group.

Pre-med

All patients received a papaveratum–hyoscine combination.

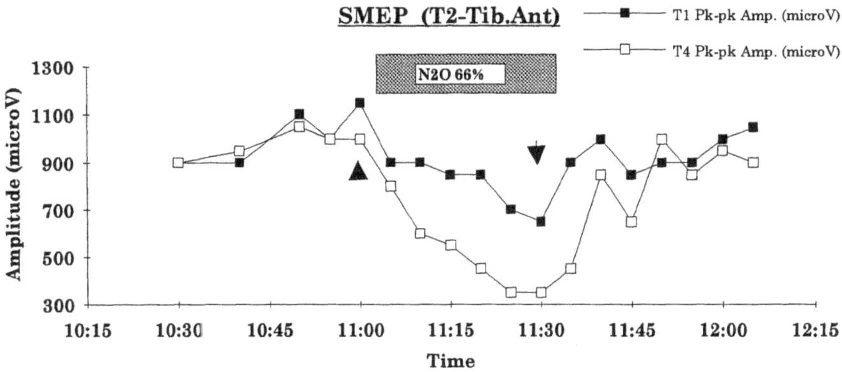

Figure 1 SMEP evoked by stimulating the epidural space at T2 and recording from tibialis anterior muscle. Stable baseline T1 and T4 recordings for 30 minutes before introduction of nitrous oxide 66%. Amplitude of both decreases over 25 minutes on N_2O and recovers to baseline 20 minutes after withdrawal. Note greater effect on T4 response. ↑ N_2O introduced; ↓ N_2O withdrawn

Anaesthetic

Anaesthesia was induced with propofol 1.5–2 mg/kg and alfentanil 100 μg/kg was then administrated slowly. Neuromuscular blockade was achieved with a bolus dose of vecuronium 0.1 mg/kg. For the purpose of the study no further muscle relaxation was given. Full recovery of neuromuscular function (assessed using a train of four stimuli from a peripheral nerve to muscle) signalled the start of the study period.

The patients were ventilated with an air–oxygen mixture. Adequate venous access and direct arterial pressure monitoring were established. Throughout surgery anaesthesia was maintained with a propofol infusion titrated to achieve steady state thirty to fifty minutes after induction. This was in conjunction with an alfentanil infusion 1–2 μg/kg/min.

METHODS – MONITORING

Patients having anterior spinal surgery had a percutaneous biopolar 3F gauge electrode sited at the C7–T1 or the T1–T2 interspace. Those undergoing posterior spinal surgery had a 4F gauge bipolar electrode sited peroperatively. Continuous somatosensory evoked potential monitoring was routine throughout all these cases. A constant voltage double pulsed stimulus ranging between 10 and 100 volts was applied with an inter-stimulus interval of 2 ms. Once a maximal stimulus was established, baseline readings were recorded. A train of four stimuli was also recorded across the cord. Motor evoked potentials were recorded from the tibialis anterior muscle using a silver/silver chloride surface electrode or from the gastronemius muscle using a needle electrode. In both study groups nitrous oxide 66% or isoflurane 1% were switched on and readings taken at five minute intervals for thirty

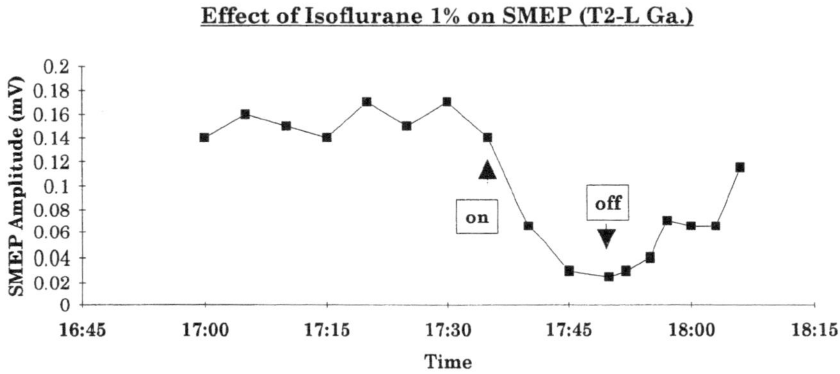

Figure 2 SMEP evoked by stimulating the epidural space at T2 and recording from gastrocnemius muscle. Stable baseline recordings for 30 minutes before introduction of isoflurane 1%. CMAP amplitude fell to 15% of baseline value after 20 minutes

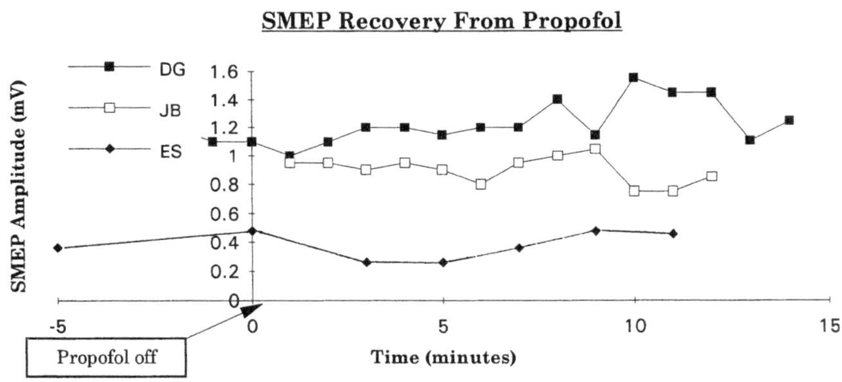

Figure 3 SMEP recovery profile from propofol anaesthesia in three cases. Recordings terminated when patients able to respond to command. No discernable increase in response amplitude suggests little influence of propofol anaesthesia on the SMEP

minutes. The inhalation agents were then switched off and readings taken for a further thirty minutes. The recovery profile from propofol was demonstrated at the end of three cases with recordings at 2 minute intervals from the time the propofol was switched off.

RESULTS

All patients studied showed a marked decrease in amplitude after twenty minutes of both nitrous oxide 66% and isoflurane 1%. The latency appeared unaffected. The motor evoked potential also returned to almost baseline values on discontinuation of the above agents. In other patients outside the study group we have found it impossible to record motor evoked potentials in the presence of nitrous oxide and isoflurane. Thiopentone would also appear to abolish the motor evoked potential recorded in this manner.

Somatosensory evoked potentials were recorded throughout the study period and remained stable and unchanged throughout. There was no evidence of a postoperative neurological deficit in any of these patients.

The average decrease in nasopharyngeal temperature was 0.5 to 1°C throughout surgery. Controlled hypotension was used in all patients and although there was a further average drop of 5 mm Hg mean arterial pressure on the introduction of nitrous oxide and or isoflurane, this was not felt to correlate with any changes seen in the motor evoked potential. All patients were ventilated to an end tidal carbon dioxide concentration of between 4.0 and 5.0 kPa. This tended to fluctuate with stimulation of the posterior tibial nerve for somatosensory recordings, but again did not contribute to any change in the recordings.

DISCUSSION

Monitoring the motor pathways per-operatively is a definite desirable addition to routine somatosensory evoked potential monitoring and using the technique described above, requires no additional invasive procedures. The sensitivity of this technique to routinely used anaesthetic agents, however, requires some anaesthetic manipulation. Total intravenous anaesthesia using propofol and alfentanil or fentanyl is now an established technique and in conjunction with controlled hypotension provides good operating conditions for scoliosis surgery.

Recent animal work has also shown nitrous oxide and the inhalational agents to have a deleterious effect on the amplitude of the spinal cord motor evoked potential[3]. Other studies also show similar effects on electrically stimulated motor evoked potentials in the epidural space[4] and on posterior tibial somatosensory cortical evoked potentials[5].

To date we have not been able to elicit a motor evoked potential when nitrous oxide and isoflurane are used in combination, and their individual effects are as above. As the technique of double pulse stimulation of the cord is still in the developmental phase, considerable further work is necessary to assess the effect of graded doses of nitrous oxide and isoflurane on the motor evoked potential. Also of interest is the idea of a central fatigue phenomenon induced by nitrous oxide on the T4 across the cord. This recovers as the nitrous oxide is switched off.

We have also found this anaesthetic technique to be compatible with eliciting reproducible motor evoked potentials from the tibialis anterior muscle per-operatively when using a double pulse stimulating magnet (Bistim).

In conclusion, it is possible to record supra maximal stable epidural motor evoked potentials per-operatively with a total intravenous technique.

References

1. Taylor, BA, Fennelly, ME, Taylor, A, Farrell J. Temporal summation and motor evoked potential spinal cord monitoring in man. In Jones SJ, et al., eds. Handbook of Spinal Cord Monitoring. Dordrecht: Kluwer; 1993:368–375.
2. Sebel PS, Lowdon J. Propofol, a new intravenous anaesthetic. Anesthesiology. 1989;71: 260–277.
3. Toda Y, Tamaki T. The effect of anaesthetic agents on the spinal cord motor evoked potential in cats. Japanese orthopedic research society, Kyoto, Sept 1991.
4. Loughnan B, Hall GM. Effects of halothane on electrically stimulated motor evoked potential in the extradural space. British Journal of Anaesthesia. 1989;62:297–300.
5. Kalkman CJ, Traast H, Zuurmond WWA, Bovill JG. Differential effects of propofol and nitrous oxide on posterior tibial somatosensory cortical evoked potentials during alfentanil anaesthesia. British Journal of Anaesthesia. 1991;66:483–489.

37
A method to assess the effect of neuromuscular blocking agents on the spinal motor evoked potential during spinal surgery

M. E. FENNELLY and B. A. TAYLOR

INTRODUCTION

The Spinal Motor Evoked Potential (SMEP) may be used to monitor spinal cord function during spinal surgery. To be certain that the cervico-dorsal, epidural stimulation of the spinal cord is not simply conducted antidromically in somatosensory pathways it is important to record the evoked Compound Muscle Action Potential (CMAP). Most anaesthetic techniques include muscular relaxation with neuromuscular blocking agents which significantly reduce the amplitude of the evoked CMAP. This is in contrast to per-operative routine monitoring of the spinal somatosensory evoked potential which is insensitive to routinely used anaesthetic agents, including muscle relaxants.

To quantify this effect the fatigue characteristics of the neuromuscular junction were assessed by determining the ratio of the amplitude of the fourth (T4) to the first (T1) response to a series of four stimuli at 2 Hz. This was initially accomplished using the established technique of stimulation of the ulnar nerve in the forearm with recording from the hypothenar muscles[1], but later incorporated into the SMEP technique by applying the stimuli to the epidural electrode and recording the CMAP from tibialis anterior muscle.

Experience of this method in ten patients quantified the effect of a non-depolarising neuromuscular blocking agent, vecuronium, on the intraoperative CMAP. In addition to the expected effect seen at the neuromuscular junction from a muscle relaxant, there seems to be a similar additional effect seen across the spinal cord with nitrous oxide.

METHODS

This study was approved by the local ethics committee and all patients or their relatives gave informed consent.

Anaesthetic

Pre-medication. All patients received a papaveretum/hyocine combination. Anaesthetic. The anaesthetic was a total intravenous one using propofol and alfentanil as described elsewhere in this volume[2].

Vecuronium 0.1 mg/kg was administered to facilitate intubation and full recovery of neuromuscular function occurred before the onset of the study period. Adequate venous access and direct arterial pressure monitoring were established.

Vecuronium was then given in incremental bolus doses of 0.01 mg/kg, 0.02 mg/kg and 0.04 mg/kg.

Stimulation

SMEP

An epidural bipolar 3F G electrode was sited per-operatively between the second and fourth thoracic interspace. A constant maximal voltage was established for each patient, usually between 25 and 50 volts. A dual 500 μs pulse with an interstimulus interval of 2 ms was then applied.

Neuromuscular junction stimulation

Surface electrodes were placed over the posterior tibial nerve and a constant voltage single 200 μs pulse applied.

Recording

Responses to the stimuli applied at both these sites were recorded from surface EMG electrodes over tibialis anterior. Amplification was set to 0.2 mV/cm, filter settings were 200–2000 Hz and the timebase was 5 ms/cm.

RESULTS

The effect of incremental doses of vecuronium on the peripheral neuromuscular junction shows an expected dose related reduction in EMG amplitude and a dose related reduction of T4:T1 ratio (Figure 1).

The effect on the spinal motor evoked potential shows a similar dose related reduction of SMEP amplitude, with a similar reduction in T1 to T4 ratio. There would also appear to be an exaggerated overshoot on recovery from each dose (Figure 1).

As described previously, changes in temperature and blood pressure were not felt to contribute to any change in the EMG. Also of interest is the fact that nitrous oxide appears to have a much more marked effect on the T4 response that on the T1 response (Figure 2).

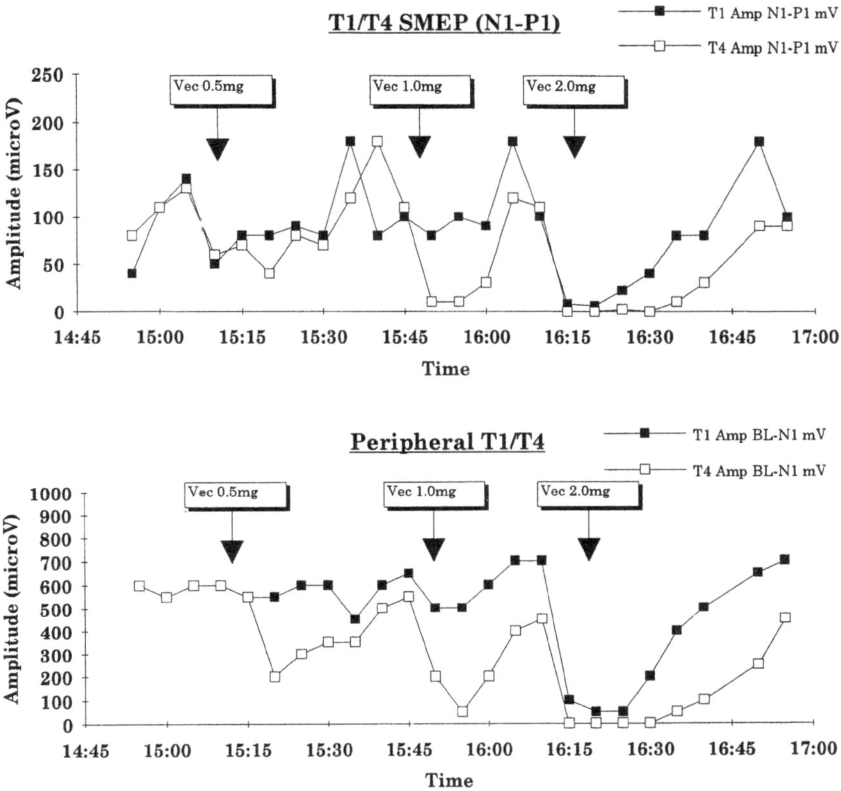

Figure 1 CMAPs recorded from tibialis anterior muscle in response to concurrent maximal epidural stimulation at T3 (upper graph), and posterior tibial nerve stimulation (lower graph). The effect of incremental bolus doses of vecuronium on the centrally and peripherally stimulated responses are shown, and the separately plotted T1 and T4 responses demonstrate the fade seen in the partially paralysed patient

DISCUSSION

The fade resulting from muscle relaxation in the spinal motor evoked potential pathway is similar to that seen at the neuromuscular junction alone. This suggests that monitoring of the SMEP pathway in this way provides adequate assessment of neuromuscular blockade in the pathway. It is now obviously possible to elicit a motor evoked potential in a partially paralysed patient, and this may be an advantage when curarization is deemed absolutely necessary. Reversal of neuromuscular blockade as expected also provides a return to pre-blocked values of the SMEP.

Also of note is the fact that there is a more pronounced unexplained overshoot on recovery in the SMEP than is seen in the peripherally stimulated EMG.

The phenomenon of a central fatigue across the cord with nitrous oxide

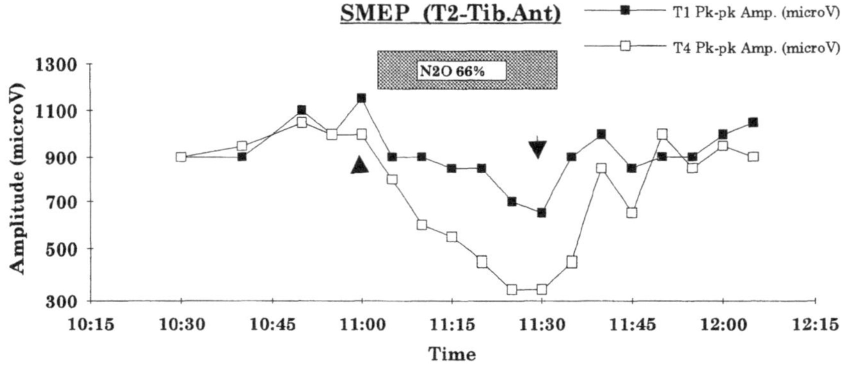

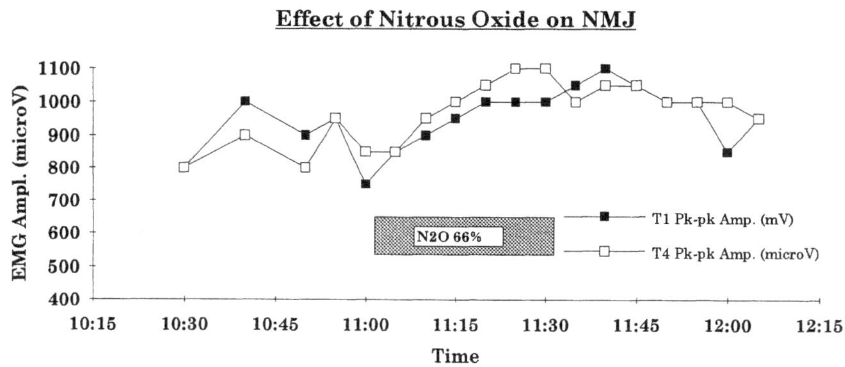

Figure 2 CMAPs recorded from tibialis anterior muscle in response to concurrent maximal epidural stimulation at T2 (upper graph), and posterior tibial nerve stimulation (lower graph). The introduction of 66% nitrous oxide (indicated by shaded bar) was followed by a reduction in both T1 and T4 response amplitudes over 30 minutes. The greater reduction in the T4 response, in the absence of any peripheral effect, suggests a 'central fatigue' in the cord analagous to the effect of muscle relaxants on the neuromuscular junction

is also of interest. This fatigue recovers as nitrous oxide is switched off.

In summary, the above described method provides a reliable non-invasive method of assessing neuromuscular blockade in the SMEP pathway in the anaesthetized patient.

References

1. Carter JA, Arnold R, Yate PM, Flynn PJ. Assessment of the Datex relaxograph during anaesthesia and Atracurium induced neuromuscular blocade. British Journal of Anaesthesia. 1986;58:1447–1452.
2. Fennelly ME, Taylor BA, Hetreed M. Anaesthesia and the motor evoked potential. In Jones SJ, et al., eds. Handbook of Spinal Cord Monitoring. Dordrecht: Kluwer; 1993:272–276.

38
Defining reliable anaesthetic protocols for intraoperative monitoring of transcranial magnetic motor evoked potentials

M. H. FRASER, B. SONI, G. NOBLE, J. H. WATT, R. CLAY and K. R. KRISHNAN

INTRODUCTION

Since the discovery by Merton and Morton in 1982[1] that transcranial motor evoked potentials (tcMEPs) could be recorded in limb muscles following stimulation of the motor cortex, there has been a great expansion in clinical applications for the tcMEP in the field of spinal surgery. But it is in the field of intraoperative monitoring that the advantages of non-invasive stimulation of descending motor pathways have opened up new alternative monitoring possibilities.

Barker et al.[2] in 1987 showed induction of the upper limb motor cortex to be possible with a tangentially placed 'double' coil but with the innovation of the 'double-cone' or 'butterfly coil' deeper cortical structures representing the lower limb could be stimulated more reliably (transcranial magnetic motor evoked potentials – tcMMEPs).

Authors such as Lesser et al.[3] and Ben-David[4] have pointed out the shortcomings of the Stagnarra wake-up test and the somatosensory evoked potential. It was Ghaly's work on primates in 1990 and 1991, with a variety of anaesthetic protocols that laid the foundation for human tcMMEP monitoring. He documented the recording of tcMMEPs under ketamine or etomidate anaesthesia with moderate potentiation occurring with 25–50% N_2O. However, midazolam and isoflurane prevented satisfactory monitoring. Herdmann[5] reported success with etomidate in 1991 although overall potentials were attenuated by 64% from preoperative recordings. Other workers have not been so successful, being unable to detect any responses with propofol, thiopentone, etomidate, isoflurane, Ethrane, fentanyl or 60% N_2O. The purpose of this study was to investigate the viability of a ketamine/N_2O or ketamine/etomidate/N_2O anaesthetic protocol for tcMMEP recording with attention to its reliability.

The main side-effects of ketamine were emergence delirium and peri-operative dreaming, preventable by adequate preoperative counselling and termination of the monitoring period with a benzodiazepine such as midazolam. Etomidate has been shown in some cases, to temporarily depress adrenocortical function in the immediate post-operative period; this can be overcome by the preoperative administration of hydrocortisone.

PATIENTS AND METHODS

An ethically approved protocol with informed consent was obtained to cover three categories of patients:

1. For neurologically intact controls who were not anaesthetized (20 cases).
2. For seven cases who had previously had a spinal cord injury but who had agreed to participate in the anaesthetic study whilst having routine surgery. In these cases electrodes were placed on the lowermost neurologically intact site.
3. There were two neurologically intact cases who were monitored during reduction and internal fixation of T12 or L1 fracture-dislocations.

Patients with cardiac pacers and an epileptic diathesis were excluded.

A Magstim 200 with a 'double-cone' head coil externally triggered a MS91 EMG recording unit. The power was set between 70 and 90%, enough to record a satisfactory baseline reading under stable anaesthesia and was left at this level for the period of monitoring. The coil was placed over the appropriate lateral pre-central gyrus for upper limb tcMMEPs or centrally over the coronal plane for lower limb tcMMEPs. The optimum position of stimulation was noted and the coil kept positioned with a 'G clamp' attached to the operating table. The repeatability of the teMMEPs was checked by averaging four recordings per measurement. The location of the optimum stimulation point on the motor cortex was critical and it was essential not to overstimulate the cortex within its recovery period (i.e. not more frequently than every 10–15 seconds) or post-excitation inhibition would lead to attenuation of the recordings.

Preoperative and operative tcMMEPs were recorded from thenar, extensor origin or anterior tibialis sites. After induction, intubation and stabilization with the appropriate infusion, baseline tcMMEPs were recorded and then at least one of the following anaesthetic protocols followed:

1. ketamine/N_2O
2. ketamine/etomidate/N_2O

with adjuncts such as alfentanil, fentanyl, isoflurane or propofol. Closure of the monitoring 'window' occurred with administration of midazolam or droperidol.

Table 1 Preoperative and operative latencies, amplitudes and site of measurement

Case	*Age*	*Preoperative*		*Operative*		*Site*
MP	21	41.0 ms	160 μV	37.4 $\pm$ 1.0 ms	72 $\pm$ 40 μV	R ant tib
KS	37	25.2 ms	660 μV	22.3 $\pm$ 1.7 ms	516 $\pm$ 273 μV	R apb
TC	36	4.3 ms	6.3 mV	4.5 $\pm$ 0.4 ms	3.4 $\pm$ 2.0 mV	R trap
JG	15	6.0 ms	1.1 mV	6.0 $\pm$ 0.7 ms	437 $\pm$ 335 μV	R stmd
PM	23	17.0 ms	500 μV	15.4 $\pm$ 1.1 ms	884 $\pm$ 320 μV	R ceo
PC	29	16.0 ms	1.6 mV	16.3 $\pm$ 0.5 ms	596 $\pm$ 170 μV	R ceo
PS	53	25.6 ms	3.2 mV	26.9 $\pm$ 1.3 ms	850 $\pm$ 285 μV	R apb
GT	40	20.6 ms	1.5 mV	22.1 $\pm$ 0.7 ms	302 $\pm$ 61 μV	R apb

ant tib = anterior tibialis, apb = abdnotor pollicis brevis, trap = trapezius, stmd = sternomastoid, ceo = common extensor origin.

Ketamine

Induction dose: 2 mg/kg

Maintenance: 150 μg/kg/min (30 min)
120 μg/kg/min (30 min)
100 μg/kg/min (remainder of procedure)

Etomidate

Induction dose: 100–300 μg/kg

Extended induction: 5–10 μg/kg/min

Alfentanil

Bolus: 250–500 μg

RESULTS

The mean latency for scalp to thenar tcMMEP was 20.7 + 1.4 ms and scalp to extensor hallucis brevis 36.2 + 2.2 ms in 20 normal controls. Table 1 gives the preoperative and operative latencies and mean amplitudes of the tcMMEP waveform for eight of the cases during the course of anaesthesia.

The maximum amplitude usually fell by up to 50% during anaesthesia. The amplitudes of recordings were higher in the larger and more proximal muscles. The standard deviation around the mean amplitude ranged from 20% to 60% for a particular case but the relative changes that occurred with respect to baseline were significantly related to anaesthetic events, e.g. the addition of etomidate bolus, fentanil or isoflurane. Examples of two patients' traces are shown.

tcMMEP amplitude variation during anaesthesia

P.S. after stabilizing on ketamine developed potentiation after the addition of N_2O but an etomidate bolus (given during patient movement) attenuated the response. Recovery of responses occurred 15 minutes later with the re-

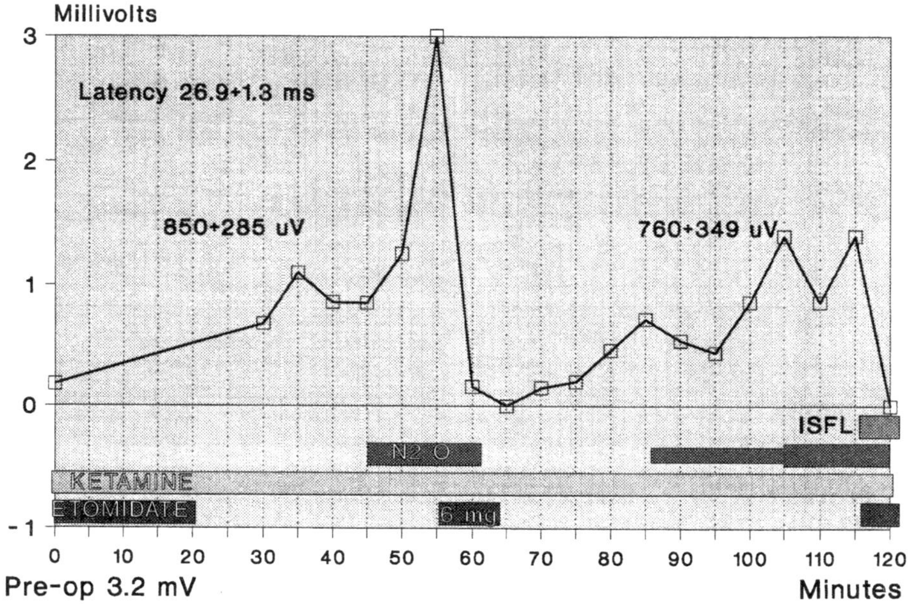

Figure 1 tcMMEPs against anaesthetic agent ketamine/N_2O. Case PS, 53 yr, R apb

introduction of N_2O. Isoflurane completely depressed the responses at the end of monitoring (Figure 1).

M.P. underwent Hartshill internal fixation of a T12 fracture-dislocation. Initially the potentials were reduced due to propofol/alfentanil and declined prior to surgical tightening 'T'. Changing over to ketamine led to a gradual increase in potentials up to 'T' but then a marked decline to 20 μV occurred after tightening (Figure 2). Fortunately this was short-lived, and without the need to release the fixation, there was continued recovery even when N_2O was added. There was no postoperative neurological deficit (Figure 2).

N_2O in up to 50% concentration was seen to potentiate the MMEP whereas etomidate as a bolus, isoflurane and propofol were seen to severely attenuate the response. Alfentanil was less depressant than fentanyl.

CONCLUSIONS

We would confirm that tcMMEPs can be measured under the following anaesthetic protocols and that there is sufficient consistency within an individual's record during the procedure to define a baseline. Repeatability of the protocol between different patients has been shown but more experience of monitoring internal fixations is needed for reliable inferences to be made of the amplitude attenuation in relation to the degree of cord compression which may be occurring. The following anaesthetic 'windows' have been found workable:

1. etomidate/ketamine/N_2O

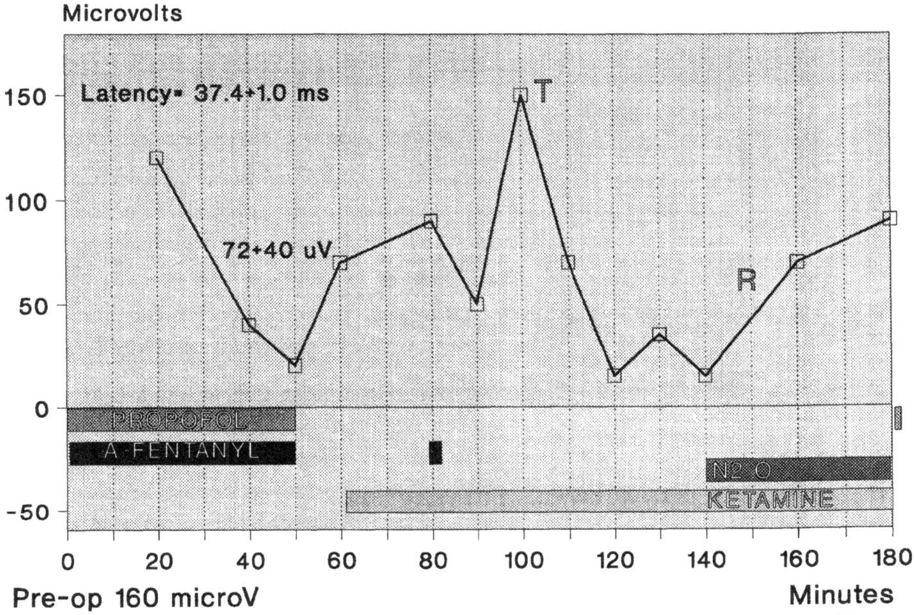

Figure 2 tcMMEPs against anaesthetic agent ketamine/N_2O. T12 fixation. Case MP, 21 yr, R ant tib

alfentanil not fentanyl as an adjunct

2. etomidate/N_2O
   disadvantage: adrenocortical suppression
3. ketamine/N_2O
   disadvantages: wide dose range, dreaming; depth of anaesthesia difficult to assess

Close cooperation between the surgical and anaesthetic teams is essential for these monitoring methods to be successful and therefore we acknowledge the support of Dr. J. H. Watt of the Regional Centre.

References

1. Merton PA, Morton HB, Hill DK et al. Scope of a technique for electrical stimulation of human brain, spain cord and muscle. Lancet 1982;2:597–600.
2. Barker AT, Freeston IL, Jalinous R, Jarratt JA, Magnetic stimulation of the human brain and peripheral nervous systems: an introduction and the results of an initial clinical evaluation. Neurosurgery. 1987;20:100–9
3. Lesser RP, Raudzens P, Luders H et al. Post-operative neurological deficits may occur despite unchanged intra-operative somatosensory evoked potentials. Ann Neurol. 1986;19: 22–25.
4. Ben-David B, Taylor PD, Haller GS. Posterior spinal fusion complicated by posterior column injury. A case report of a false negative wake-up test. Spine. 1987;12:540–43.
5. Herdmann J, Lumenta C, Dvorak J et al. Intra-operative use of the magneto-electric stimulation for motor pathway monitoring. Symp. Magnetic Brain Stimulation, Aachen, FRG Dec 6–7th 1991.

39
Transcranial magnetic stimulation for spinal cord monitoring

J. HERDMANN, M. KRZAN, F. SONNENSCHEIN and C. B. LUMENTA

INTRODUCTION

Intraoperative monitoring with somatosensory evoked potentials (SEPs) is a well established methodology. Despite well preserved SEPs severe postoperative motor disability may occur[1]. Therefore, monitoring of the motor pathways is of great importance. Initial experience with intraoperative MEP monitoring was gained using high-voltage electric stimulation during orthopaedic[2] and neurosurgical procedures[3]. Edmonds et al.[4] were the first to apply magnetic transcranial stimulation for intraoperative MEP monitoring during scoliosis surgery. Whereas good results were obtained during scoliosis surgery in neurologically intact patients, the same group reported inability to record MEPs following magnetic stimulation in neurologically impaired patients with spinal cord tumours during nitrous oxide-based anaesthesia[5].

Etomidate has been shown to enhance SEP responses[6] and it had a less depressant effect on MEPs than nitrous oxide[7] or benzodiazepines[8]. Using etomidate-based anaesthesia we have recorded MEPs in patients with existing motor impairment due to extramedullary tumours or spinal cord tumours.

METHODS

Transcranial magnetic stimulation was used for MEP monitoring in 11 patients aged 20–80 years (mean 43 years; five females, six males). They underwent spinal surgery for ependymoma ($n = 2$), astrocytoma ($n = 3$), meningioma ($n = 1$), and nerve root junction ($n = 1$). All but two patients showed preoperative motor impairment ranging from slight gait disturbance to inability to walk, with bladder and bowel dysfunction.

A MAGSTIM 200 magnetic stimulator (The MAGSTIM Company Limited, Whitland, GB) with a conventional circular coil generating a magnetic field with a peak intensity of 1.5 Tesla at the centre of the coil was used.

The surface EMG from limb muscles was recorded with DANTEC silver–silver chloride surface electrodes (DANTEC #13L20, DANTEC, Skovlunde, Denmark) using a BASIS 8000 EMG machine (ESAOTE BIOMEDICA, Florence, Italy). In some cases MEPs were recorded from the fibres of the cauda equina with a 1 mm diameter bipolar Vygon catheter electrode (VYGON Bi-Pacing-Cath #1124.098, VYGON, Aachen, Germany). The catheter which had an interelectrode distance of 5 mm was inserted into the epidural space through a 16 G Tuohy needle at the L3/L4 or L4/L5 interspinous level. Its tip was advanced to the L3 vertebral level.

Filter settings were 20 Hz–5 kHz for both surface EMG and cauda equina recordings. MEPs were analysed for peak-to-peak amplitudes and onset latencies.

For transcranial stimulation of the motor cortex the coil was centred over the vertex or slightly in front of it. Preoperatively a threshold stimulus intensity (the lowest stimulus intensity at which a MEP could be elicited in the awake and relaxing subject) was determined for each recording site. Maximum stimulator output intensity was always used thereafter. Following induction of anaesthesia and before any relevant surgical steps were performed, so-called 'baseline' traces were obtained from each recording site for comparison with later recordings. Three consecutive sweeps were recorded within 10 to 20 seconds. They were evaluated only if intersweep variability was low. In order to detect any changes as early as possible averaging techniques were not applied. We achieved a monitoring repetition rate of 5 per minute. The surgeon was alerted only if all three sweeps showed changes that exceeded the 'warning criteria'.

Warning criteria were defined in accordance with those given in literature: MEP amplitude changes were 'significant' when they exceeded 50% of baseline amplitude, while changes in latency were 'significant' when higher than 10% of baseline latency or when they exceeded the latter by more than 3 ms[9].

Where only motor pathways to lower limb muscles were endangered, control recordings could be obtained from hand muscles in order to distinguish technically determined or anaesthesia-related MEP changes from those induced surgically. Control recordings from the biceps brachii muscle in cases where hand muscles were relevant monitoring muscles were often disturbed by a large stimulus artefact.

Anaesthesia

Flunitrazepam (Rohypnol®, 2–4 mg) was used as premedication while continuous i.v. administration of etomidate (Hypnomidate®, 20–30 mg/h, i.e. 0.25–0.5 mg/kg/h) and fentanyl (Fentanyl-Janssen®, 0.05–0.1 mg/h, i.e. 0.6–1.5 μg/kg/h) was used for anaesthesia. Patients were ventilated with an oxygen–air mixture. Nondepolarizing neuromuscular blocking agents (vecuronium, pancuronium) were given for muscle relaxation which was monitored by recording the thenar M-wave after supramaximal median nerve stimulation using a DATEX Relaxograph (DATEX Instrumentation OY, Helsinki, Finland). Relaxation was about 80%, i.e. around 20% 'T1'.

RESULTS

The mean preoperative threshold stimulus intensity allowing for reliable MEP recordings after induction of anaesthesia was $60 \pm 13\%$ of the maximum stimulator output (mean ± SD; range: 40–80%; $n = 23$ recording sites). If threshold stimulus intensity was higher than 80% of maximum stimulator output no MEP could be obtained from this recording site after induction of anaesthesia.

Induction of anaesthesia with etomidate and fentanyl was accompanied by a decrease of MEP amplitudes by $64 \pm 34\%$ (mean ± SD) and an increase of MEP latencies by $7 \pm 8\%$ (1.9 ± 2.2 ms) compared with preoperative values.

In one patient improved intraoperative MEPs were associated with rapid postoperative improvement of gait disturbance. Two cases of irreversible intraoperative deterioration, or loss of MEPs, were associated with increased postoperative paresis. Reversible MEP changes or unchanged MEPs were seen in seven cases and correlated well with unchanged postoperative neurological findings. In one patient with a neurinoma from T4 to T11 no reproducible baseline could be recorded after induction of anaesthesia. In this case the preoperative threshold stimulus intensities were 90–100% of maximum stimulator output.

Case reports

A. Patient E. D. (61 years, male) was treated for grade 2 to 3 astrocytoma at C4 to C7. He showed quadriparesis (flaccid paresis of upper limbs and spastic paraparesis) confining him to a wheelchair. Intraoperatively MEP monitoring of the left abductor digiti minimi muscle was performed. Preoperative, intraoperative, and postoperative recordings are shown in Figure 1. During tumour removal MEP latencies suddenly increased by 14% (3.4 ms) and amplitudes decreased by 38% compared to baseline. Tumour dissection was interrupted and discontinued at this site. Within a few minutes latencies and amplitudes nearly returned to earlier values and at the end of the operation they were identical with the baseline values. The patient did not show any additional neurological deficit postoperatively.

B. Patient A. G. (52 years, female) with a ependymoma at T10 showed distinct spastic gait and reported slight bladder and bowel dysfunction. Intraoperative MEP monitoring was performed from both anterior tibial muscles, from the fibres of the cauda equina, and from the right abductor digiti minimi as control muscle. MEP amplitudes (Figure 2A) and latencies (Figure 2B) were plotted in relation to the respective baseline value indicated by the 'zero-line' as recorded at *B*. During tumour dissection (*D* to *O*) a

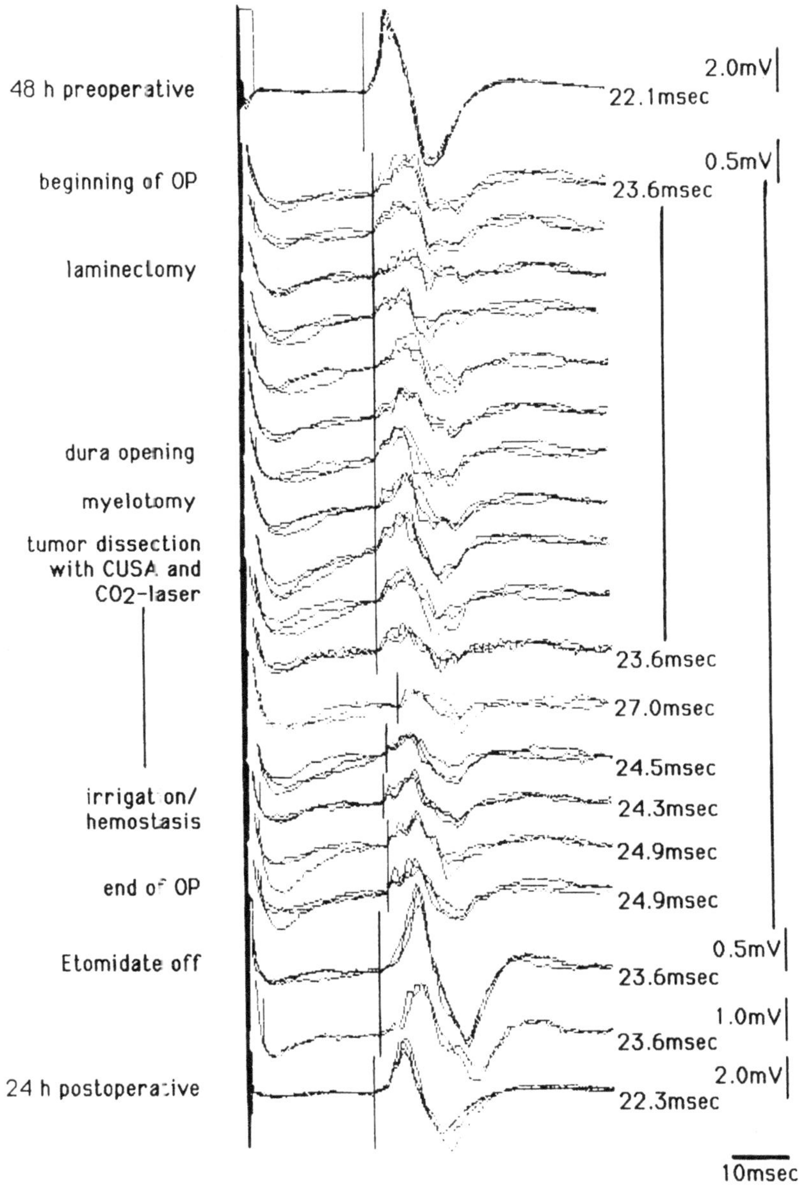

Figure 1 Astrocytoma at C4–C7 in a 61-year-old male. MEP monitoring was performed from the left abductor digiti minimi muscles. Surgical procedures, onset latencies and gain are indicated. During tumour dissection the amplitude of the MEPs critically dropped but regained earlier values after a short pause. The patient showed no additional motor deficit postoperatively

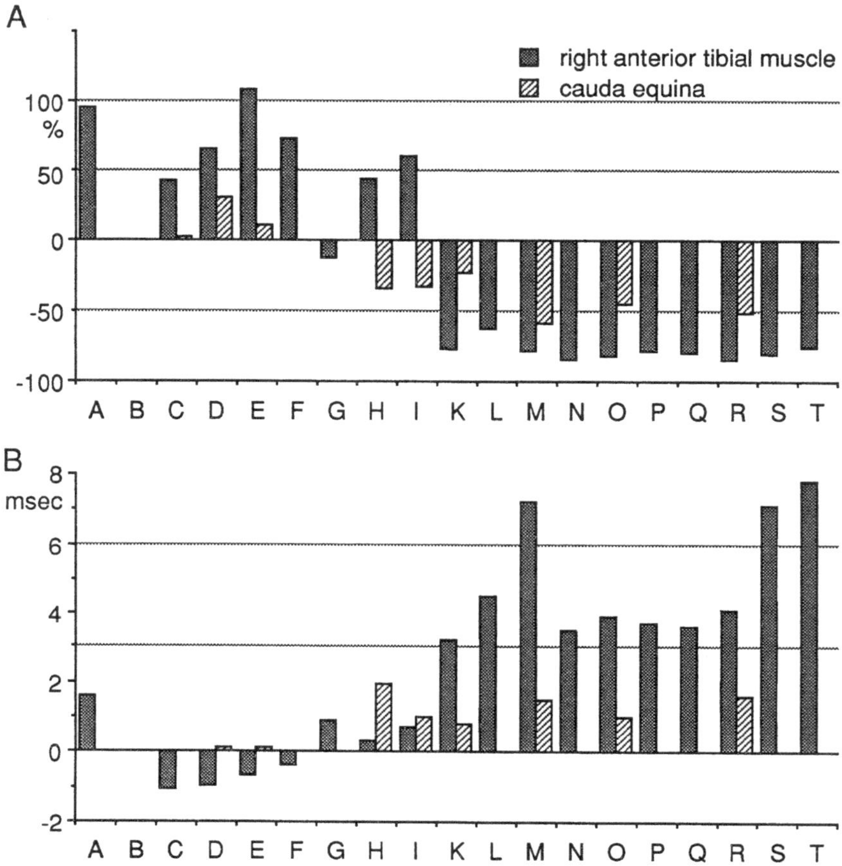

Figure 2 MEP amplitudes (A) and latencies (B) during surgery for an ependymoma at T10 (52-year-old female). MEPs were recorded from the right anterior tibial muscle (grey bar) and from the fibres of the cauda equina (hatched bars). A: preoperative; B: baseline; D–O: tumour dissection; S and T: 24 hours and 1 week postoperatively. MEPs from the left anterior tibial and from the right abductor digiti minimi muscle are not shown. They showed no significant intraoperative changes

decrease of MEP amplitude and an increase of latency was noted in the epidural recordings from the cauda equina (shaded bars, *H* and following) and in the recordings from the right anterior tibial muscle (gray bars, *K* and following). Whereas the amplitude and latency changes of the anterior tibial recordings distinctly exceeded the warning criteria, those from the cauda equina did not. Tumour dissection was immediately interrupted. During the consecutive intraoperative recordings (*L* to *R*) and postoperatively (*S* and *T*) neither amplitudes nor latencies of the right anterior tibial recordings recovered. Recordings from the left anterior tibial and right abductor digiti minimi muscles showed no significant changes. While left leg function remained unchanged compared to preoperative findings paresis of the right leg was markedly increased postoperatively but recovered within several weeks.

DISCUSSION

Evoked potential measurements critically depend on the signal-to-noise ratio which increases with signal amplitude and decreases with background noise. Anaesthetic drugs attenuate the evoked potential: signal attenuation is the more pronounced, the more synapses the evoked potential crosses between stimulating and recording electrodes. Appropriate anaesthesiological management, therefore, is of primary importance for successful intraoperative monitoring. Zentner[3] performed high-voltage electric MEP monitoring in nitrous oxide-based anaesthesia. Edmonds et al.[4] were able to perform magnetic MEP monitoring in 9 of 11 neurologically unaffected cases of scoliosis surgery also using nitrous oxide. They, however, reported major problems in neurologically impaired patients with spinal cord tumours or mechanical compression of the cord[5]. In another report MEP amplitudes decreased by 89 to 93% following induction of anaesthesia with nitrous oxide in 16 patients monitored using electric transcranial stimulation[7]. No data is available for MEP latency changes in nitrous oxide-based anaesthesia. Evoked potential monitoring was more successful during i.v. anaesthesia with ketamine[10], propofol[9], or etomidate[6,11]. In our study using an etomidate/fentanyl anaesthesia MEP amplitudes decreased by only $64 \pm 34\%$ (mean $\pm$ SD) and latencies increased by only $7 \pm 8\%$ (1.9 ± 2.2 ms) compared with preoperative findings. The favourable effect of etomidate on evoked potentials results from spinal as well as supraspinal mechanisms[12].

Controlled muscle relaxation with nondepolarizing neuromuscular blocking agents was compatible with MEP recordings from limb muscles[4]. We regularly observed mild muscle twitching following magnetic stimulation despite some 80% controlled muscle relaxation (20% T1). During microsurgical operations monitoring was therefore performed when compatible with the current surgical procedure.

Many anaesthetic drugs have their greatest effect at the anterior horn cell level so anaesthesia-related signal attenuation is less pronounced if MEPs are recorded from the spinal pyramidal tracts by means of epidural or subdural electrodes. Recordings of pyramidal tract activity following transcranial brain stimulation have hitherto only been performed using high voltage electric stimulation[13]. We recorded action potentials from motor root fibres of the cauda equina by means of epidural electrodes in the lumbar spinal canal following transcranial magnetic stimulation. The major disadvantage of these techniques is the difficulty in separating left/right information on pathway integrity because the unchanged waveforms of one side can conceal changes of the other side. In patient A. G. (see case report B) MEP changes did not exceed the warning criteria in the epidural recordings, whereas the changes were significant in the recordings from the anterior tibial muscle of the right leg, which postoperatively showed an increased paresis. No MEP changes or clinical signs of deterioration were noted in the left anterior tibial muscle. It therefore seems justified to define different warning criteria for techniques, which compromise information from both left and right spinal cord pathways. We propose that an amplitude decrease of more than 30% should then be regarded as 'significant'. The definition of warning criteria for latency changes appears arbitrary.

With the approach described above, intraoperative magnetic MEP monitoring was successful despite neurological impairment when preoperative MEP threshold stimulus intensity was lower than 80% of maximum stimulator output. Intraoperative MEP recordings could be obtained from 23 of 34 recordings sites rendering monitoring successful in ten of eleven patients. At three recording sites intraoperative MEPs could not be analysed due to large stimulus artefacts, while at two sites no MEPs could be recorded preoperatively. When MEPs were not obtainable from lower limb muscles or from the fibres of the cauda equina, MEP monitoring was performed from upper limb muscles, provided that the fibres to these muscles were at risk too. Lower success rates reported in cases of thoracic or lumbar spine surgery, for example in scoliosis surgery[4], are in part due to the fact that MEP monitoring had to be performed from lower limb muscles where threshold stimulus intensities are generally higher than in upper limb muscles.

In four patients with no intraoperative MEP changes and in three patients in whom the changes were transient and fully reversible, postoperative motor function was unchanged. In two patients with irreversible MEP deterioration an increased paresis was found postoperatively and in one patient with intraoperative MEP improvement following spinal cord decompression gait disturbance rapidly subsided postoperatively. These results indicate a clear correlation between intraoperative MEP findings and short term postoperative motor outcome. They are in favour of a high validity of magnetic MEP monitoring.

References

1. Dawson EG, Sherman JE, Kanim LEA, Nuwer MR. Spinal cord monitoring. Results of the Scoliosis Research Society and the European Spinal Deformity Society survey. Spine. 1991;16:S361–S364.
2. Boyd SG, Rothwell JC, Cowan JMA, Webb PJ, Morley T, Asselman P, Marsden CD. A method of monitoring function in corticospinal pathways during scoliosis surgery with a note on motor conduction velocities. J Neurol Neurosurg Psychiatry. 1986;49:251–257.
3. Zentner J. Noninvasive motor evoked potential monitoring during neurosurgical operations on the spinal cord. Neurosurg. 1989;24:709–712.
4. Edmonds HL, Paloheimo MPJ, Backman MH, Johnson JR, Holt RT, Shields CB. Transcranial magnetic motor evoked potentials (tcMMEP) for functional monitoring of motor pathways during scoliosis surgery. Spine. 1989;14:683–686.
5. Shields CB, Paloheimo MPJ, Backman MH, Edmonds HL, Johnson JR. Intraoperative use of transcranial magnetic motor evoked potentials. In: Chokroverty, ed. Magnetic stimulation in clinical neurophysiology, Boston, London: Butterworths; 1990:173–184.
6. Sloan TB, Ronai AK, Toleikis JR, Koht A. Improvement of intraoperative somatosensory evoked potentials by etomidate. Anesth Analg. 1988;67:582–585.
7. Zentner J, Kiss I, Ebner A. Influence of anesthetics – nitrous oxide in particular – on electromyograph response evoked by transcranial electrical stimulation of the cortex. Neurosurg. 1989;24:253–256.
8. Schönle PW, Isenberg C, Crozier TA, Dressler D, Machetanz J, Conrad B. Changes of

transcranially evoked motor responses in man by midazolam, a short acting benzodiazepine. Neuroscience Let. 1989;101:321–324.

9. Jellinek D, Jewkes D, Symon L. Noninvasive intraoperative monitoring of motor evoked potentials under propofol anesthesia: effects of spinal surgery on the amplitude and latency of motor evoked potentials. Neurosurg. 1991;29:551–557.
10. Ghaly RF, Stone JL, Aldrete JA, Levy WJ. Effects of incremental ketamine hydrochloride doses on motor evoked potentials (MEPs) following transcranial magnetic stimulation: a primate study. J Neurosurg Anesthesiol. 1990;2:79–85.
11. McPherson RW, Sell B, Traystman RJ. Effects of thiopental, fentanyl and etomidate on upper extremity somatosensory evoked potentials in humans. Anesthesiology. 1986;65: 584–589.
12. Meinck HM, Möhlenhof O, Kettler D. Neurophysiological effects of etomidate, a new short-acting hypnotic. Electroencephalogr Clin Neurophysiol. 1980;50:515–522.
13. Deletis V. Intraoperative monitoring of the functional integrity of the motor pathways. In: Devinsky O, Beric A and Dogali M, eds. Electrical and magnetic stimulation of the brain. New York: Raven Press; 1992: (in press).

40
Evaluation of motor pathways in patients with cervical myelopathy by magnetic stimulation: pre-, intra-, and postoperative study

T. IIZUKA, S. IIDA, T. IMAI and H. AZUMA

INTRODUCTION

Measurement of motor evoked potentials (MEP) induced by transcranial magnetic stimulation, a technique introduced by Barker et al. in 1985, has revolutionized electrodiagnosis within the central motor system[1]. Since 1987, we have examined more than 200 patients with neurological disorders including 100 with cervical myelopathy by this noninvasive method[2,3]. The purpose of this study was to evaluate the reliability and usefulness of this technique in the assessment of pre-, intra-, and postoperative motor function in-patients with cervical myelopathy.

MATERIALS AND METHODS

The study population comprised 68 surgical patients with cervical myelopathy (45 males and 23 females; mean age, 59 years; mean height, 157 cm). Sixty-six healthy volunteers (30 males and 36 females; mean age, 34 years; mean height, 161 cm) served as controls. The diagnosis was confirmed in all cases by plain X-rays of the cervical spine, computed tomography, myelography, and magnetic resonance imaging.

The operative procedures were expansive laminoplasty in 50 cases and anterior decompression and fusion in 18. Follow-up periods were six to 48 months. Cervical myelopathy was assessed according the scoring system of the Japanese Orthopaedic Association ('JOA score'). In essence, motor function of the upper and lower limbs is graded on a scale from 0 to 4: in which 4 denotes normal; 3, slightly impaired; 2, moderately impaired; 1, severely impaired. A score of 0 indicates inability to eat with a spoon or to walk, even with support.

Electrophysiological methods

Transcutaneous magnetic stimuli were delivered over the scalp, cervical, and lumbar regions with a magnetic stimulator (MAGSTIM 200, UK). Surface electrodes were placed over the bilateral abductor digiti minimi (ADM) and abductor hallucis (AH) muscles and MEP were recorded simultaneously. The band-pass filter of the amplifier was set at 20–2000 Hz (Neuropack 4-mini, JPN or Nicolet Viking, USA). MEPs were recorded while the subjects were relaxed and also as they voluntarily contracted the target muscles (facilitation). The onset-latency, duration, and peak-to-peak amplitude of the MEP were measured. The central motor conduction time (CMCT) to the spinal cord segments innervating arm or leg muscles as calculated by subtracting the latency after cervical (or lumbar) stimulation from the latency after cortical stimulation. F wave studies were also performed for measurement of the latent period from anterior horn cells of the spinal cord to the hand or foot muscles, calculated according to the formula: $(F + M - 1)/2$. The CMCT(F) was also calculated in the manner described above. The criteria for an abnormal response were: no response, or the occurrence of MEP latency and CMCT values exceeding the upper confidence limit (mean + 2.5 SD) of the control population. Student's t test was used for statistical analysis. In eight cases, intraoperative transcranial magnetic stimulation and epidural recordings were performed under general anaesthesia. Postoperative studies were performed in all cases. A new indirect method for measuring the spinal cord motor conduction velocity (Sp-MCV) was also introduced.

RESULTS

A new indirect method for measuring the Sp-MCV

Figure 1 illustrates the non-invasive, indirect method for measuring the Sp-MCV by magnetic stimulation and F wave studies. Sp-MCV is obtained by the formula: Sp-MCV = Distance/CMCT(T). On the right are magnetic MEP and F waves of a 51-year-old myelopathy patient with cervical ossification of posterior longitudinal ligament. In all four limbs, the latencies of all MEP were significantly prolonged relative to the control values. This patient's Sp-MCV was 31.1 m/s, while the mean control Sp-MCV was 51.6 m/s.

Latency without or with facilitation (Table 1)

The mean latencies of cortical MEP, CMCT, and CMCT(F) obtained from the relaxed hand and foot muscles of the patients were significantly increased compared with those of normal controls.

Those obtained from the slightly contracted hand and foot muscles of the patients were also significantly longer than those of controls. The mean

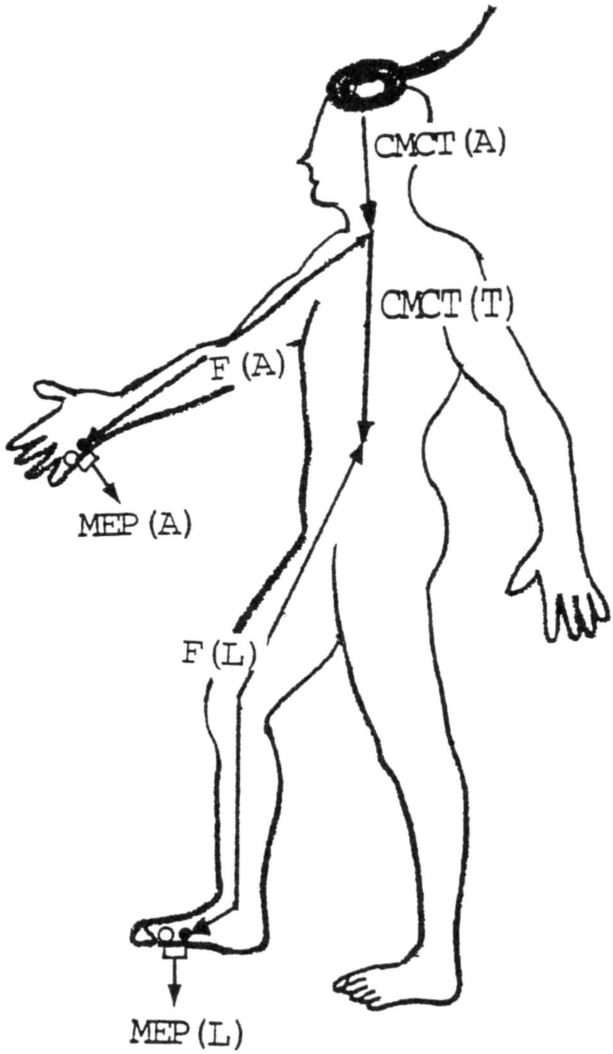

Figure 1 A new indirect method for measuring the Sp-MCV. Left shows schematic illustration of the measurement. Sp-MCV = Distance (C7–Th12 spine) / CMCT(T); CMCT: central motor conduction time, A: arm, T: trunk, L: leg, F: F value; latency from anterior horn cell to target muscle, MEP: latency of MEP. On the right are MEP and F waves of a patient with myelopathy. Sp-MCV in this patient was 31.1 ms

latency of cortical MEP was shortened by a few milliseconds on facilitation in both controls and myelopathy patients.

Motor function and CMCT (Table 2)

The mean CMCT increased according to the degree of motor function

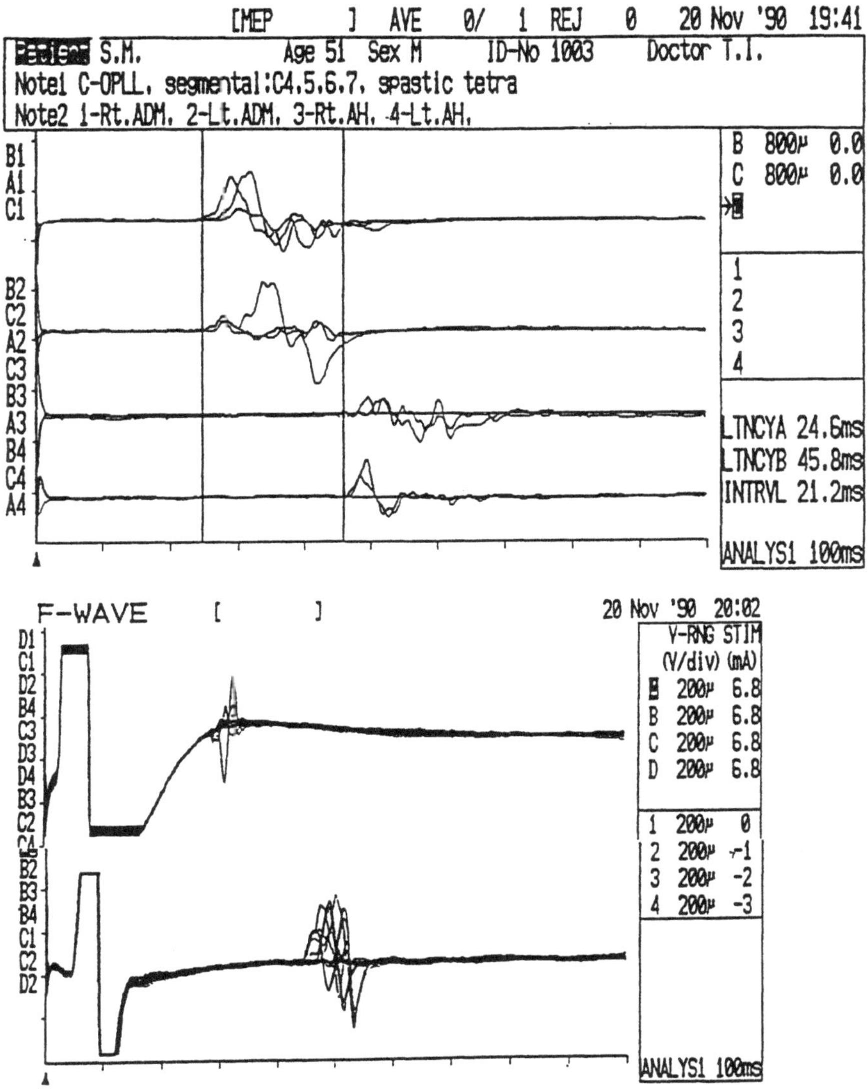

Figure 1 *Continued*

impairment, as reflected by JOA scores, in both upper and lower limbs. That is, the more severe the motor dysfunction, the longer the CMCT.

Intraoperative study

We successfully recorded intraoperative transcranial MEP in seven of eight patients with myelopathy. As for the spinal cord monitoring of the patient without intraoperative magnetic MEP, we used conventional electrical

Table 1 Latency without and with facilitation

	(mean ± SD, ms)			
	Without facilitation		*With facilitation*	
Target muscle	*controls*	*myelopathy*	*controls*	*myelopathy*
ADM	$n = 66$	$n = 68$	$n = 66$	$n = 68$
Head	21.1 ± 1.5	26.7 ± 4.7**	18.6 ± 1.4	25.0 ± 4.6**
Cervical	12.0 ± 1.1	12.6 ± 1.1	12.0 ± 1.1	12.6 ± 1.1
(F + M − 1)/2	12.8 ± 1.0	13.8 ± 1.4	12.8 ± 1.0	13.8 ± 1.4
CMCT	8.8 ± 1.4	14.0 ± 4.4**	6.2 ± 1.4	12.5 ± 4.5**
CMCT(F)	8.2 ± 1.3	12.8 ± 4.4**	5.3 ± 1.2	11.4 ± 4.3**
AH				
Head	37.1 ± 2.3	46.7 ± 9.4**	35.0 ± 2.2	43.8 ± 5.4**
Lumbar	20.8 ± 2.2	21.9 ± 2.1	20.8 ± 2.2	21.9 ± 2.1
(F + M − 1)/2	22.9 ± 1.3	25.0 ± 2.2*	22.9 ± 1.3	25.0 ± 2.2*
CMCT	17.0 ± 1.5	24.6 ± 8.4**	14.9 ± 2.0	22.2 ± 4.8**
CMCT(F)	14.4 ± 2.0	21.2 ± 8.9**	11.9 ± 1.7	18.9 ± 4.7**

*$p < 0.05$, **$p < 0.01$, ABM: Abductor digiti minimi, AH: Abductor hallucis, CMCT: central motor conduction time, F: latency of F wave, M: latency of M wave

Table 2 Motor function and CMCT(c)

JOA Score (*Motor*)	*CMCT(U/E)* *Mean ± SD* (n)	*CMCT(L/E)* *Mean ± SD* (n)
0 (unable)	18.2 ± 1.4 (2)	26.3 ± 5.5 (6)
1 (severe)	15.5 ± 4.0 (14)	24.6 ± 5.3 (17)
2 (moderate)	12.6 ± 4.8 (28)	22.8 ± 4.6 (31)
3 (mild)	11.7 ± 3.7 (20)	19.2 ± 2.7 (9)
4 (normal)	6.2 ± 1.4 (66)	14.9 ± 2.0 (66)

c: with facilitation, U/E: upper extremity, L/E: lower extremity, JOA: Japanese Orthopaedic Association

epidural stimulation, which we have been using for over 10 years.

Figure 2 presents the intraoperative MEP from the C6 and C3 epidural spaces obtained with transcranial magnetic stimulation during general anaesthesia with a muscle relaxant in a patient with cervical myelopathy.

Postoperative motor function and CMCT

Postoperatively, the mean JOA score improved by about 1.1 point in the upper limbs and by 0.8 point in lower limbs. The mean CMCT were shortened by 2.7 ms (head to neck) and 3.7 ms (head to lumbar region).

DISCUSSION

There have been only a few reports describing the perioperative changes in magnetic MEP in cervical myelopathy[2–4]. Preoperatively, we found significant delays or even absence of MEP responses in the 68 patients with cervical

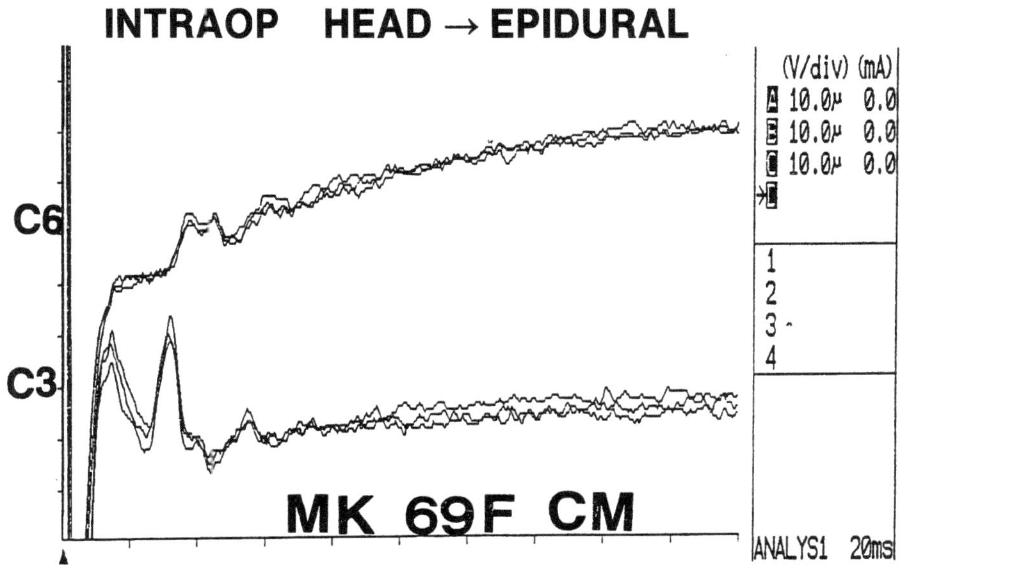

Figure 2 Intraoperative magnetic MEP from epidural space. Intraoperative MEP from C6 and C3 epidural spaces were obtained with transcranial magnetic stimulation during general anaesthesia in a 69-year-old female patient with cervical myelopathy

myelopathy compared with the results from 66 normal controls. These differences were observed regardless of the site of recording (hand or foot), the condition of the target muscle (with or without facilitation), or the method of calculation of the CMCT (cervical and lumbar stimulation or F wave study). The mean CMCT and clinical motor function were well correlated. The fact that abnormal MEP were recorded in 93% of the patients indicates that magnetic MEP induction is highly sensitive in detecting motor pathway disturbance in this disease.

Our intraoperative results showed the potential for monitoring of central motor pathways during surgery. In our experience, this monitoring system has proved useful in surgery at the site of the cranio-cervical junction. However, the origins of MEP from the epidural space have not yet been determined, so further investigation is necessary.

Postoperative data also indicate the reliability of this method in the assessment of motor pathways in cervical myelopathy.

This technique shares with conventional magnetic stimulation the disadvantage of poor localization of the stimulated site. However, use of double or spherical coil promises to improve this situation[5].

No serious complications of magnetic transcranial stimulation have been reported, and no side-effects were noted in this study. Although this method appears to be very safe, further prospective studies should be performed to definitively establish its safety.

CONCLUSIONS

1. CMCTs were significantly prolonged in 68 cervical myelopathy patients compared with 66 healthy controls.
2. CMCT and the degree of motor disturbance were well correlated.
3. The mean postoperative CMCT was shorter, by a few milliseconds, than the preoperative latency.
4. A new direct method of measuring spinal cord MCV was introduced.
5. Intraoperative spinal cord monitoring of motor tract was successfully performed with magnetic MEP.
6. This method is noninvasive and reliable in assessing motor function in cervical myelopathy.

References

1. Barker AT, Freestone IL, Jalinous R, Merton PA, Morton HB. Magnetic stimulation of the human brain. J Physiol. 1985;369:3p.
2. Iizuka T, Tanaka H, Sugioka H, Kondo T. Assessment of the motor pathway in cervical myelopathy by transcutaneous magnetic stimulation. Spine & Spinal Cord. 1989;2:611–616.
3. Iizuka T, Tanaka H, Sugioka H, Kondo T, Azuma H. Magnetic transcutaneous stimulation of the motor pathway in spinal cord disorders. In: Shimoji K, Tamaki T, Kurokawa T, Jones SJ, eds. Spinal cord monitoring and electrodiagnosis. Berlin/Heidelberg: Spring-Verlag; 1991:253–261.
4. Eisen A, Shtybel W. Clinical experience with transcranial magnetic stimulation. Muscle & Nerve. 1990;13:995–1011.
5. Ueno S, Tashiro T, Harada K. Localized stimulation of neural tissues in the brain by means of a paired configuration of time-varying magnetic fields. J Appl Phys. 1988;64:5862–5864.

41
An attempt at noninvasive mapping of the human motor cortex using transcranial magnetic stimulation

T. IZUMI, Y. NISHIJIMA, M. OKADA, Y. OHTA, N. TODA and Y. YAMAZAKI

INTRODUCTION

In recent years, motor evoked potentials (MEPs) following magnetic stimulation have been introduced into clinical assessment of central motor pathways. One disadvantage of magnetic stimulation is that focal stimulation has been considered impossible. However, a certain type of circular coil, in which the electromagnetic field is most intense at the centre, makes this possible. Another problem is that the direction of current flow in the coil has some effect on the potentials elicited. Our purpose in this study was to provide data for mapping the human motor cortex using such a circular coil and to clarify the effect of the direction of current flow in the coil on MEPs with transcranial magnetic stimulation.

SUBJECTS

The experimental group consisted of 24 healthy adult volunteers (18 males and 6 females). Their mean age was 32.6 ± 14.4 years (range: 23–66 years), and their mean height was 167.8 ± 5.7 cm (range: 155–173 cm). They had no evidence of neurological disease.

THE PROPERTIES OF THE COIL

The vertical, transverse and composite components of the magnetic field created by a coil manufactured by the DANTEC Corporation were investigated. The maximal vertical magnetic field of the stimulation coil was at its centre (Figure 1). This characteristic differed from that of other types of coils which had their maximal magnetic field at their edges.

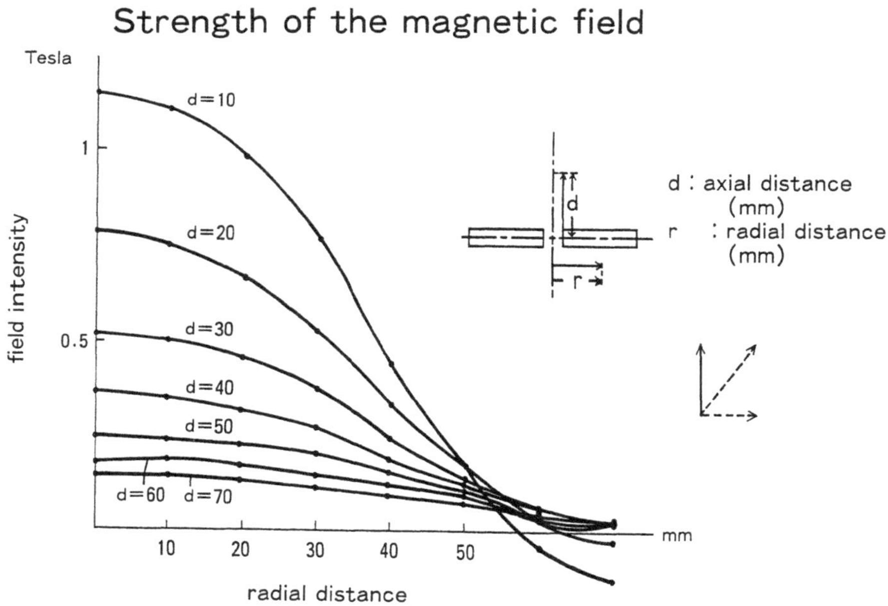

Figure 1 The vertical component of the magnetic field. The centre of the stimulation coil has a maximal vertical magnetic field

METHODS

Measurement was performed with the subjects in a quiet room maintained at a controlled temperature. The subjects sat with their eyes closed. Stimulation was provided to Erb's point, the dorsum of the 5th cervical vertebra and the motor cortex with the DANTEC Corporation magnetic stimulator, and was performed with the stimulating coil placed flat over the sites of the scalp stimulated. Erb's point and the 5th cervical vertebra were stimulated at 80% (1.44 Tesla) of maximum output and the motor cortex at 100% (1.8 Tesla). The sites stimulated on the scalp over the motor cortex were Cz, C3′ and C4′ (1 cm anterior and 3.5 cm lateral to Cz) and F3′ and F4′ (2 cm anterior and 7 cm lateral to Cz). Both clockwise and anticlockwise stimuli were delivered at each site of stimulation.

Recordings were made using an EVOMATIC 8000 electromyograph, with the analysis time set at 60 ms, filter band width at 20 Hz–2 kHz, and sensitivity at 5 mV. The MEPs were recorded over both relaxed and voluntarily contracted deltoid (D), biceps brachii (BB), abductor pollicis brevis (APB) and abductor hallucis (AH) muscles. MEPs were recorded 2 to 3 times to confirm reproducibility. Latency was measured as the time elapsed between stimulus onset and the onset of the first negative peak, and amplitude was determined as the difference between the first negative peak and the subsequent positive peak.

The frequency of MEPs recorded from each muscle was evaluated at the different sites stimulated at Erb's point, C5 and on the scalp. The effects of clockwise and anticlockwise stimulation at each site of stimulation were investigated with regard to frequency, latency and amplitude.

RESULTS

The MEPs elicited following motor cortical stimulation had higher frequency, shorter latencies and higher amplitudes in voluntarily contracted muscles than in relaxed muscles, but no relation was shown between the degree of the muscle contraction and the changes of MEPs. The MEPs recorded after stimulation at Erb's point and C5 did not show any changes between relaxed and facilitated muscles. The range of standard deviations of MEP latency was limited, whereas the range of MEP amplitudes was rather wide. Shorter latencies and higher amplitudes were obtained as the output of the magnetic stimulator was increased. The latencies of the MEPs were linearly correlated with subject height, but not correlated with age.

The frequency of MEPs recorded over each muscle following stimulation of the different sites on the scalp is shown in Figures 2, 3 and 4.

When stimulated at Cz, the AH muscles were most frequently excited. Stimulation at C3′ or C4′ elicited MEPs from the deltoid and biceps brachii muscles more often than the APB or AH muscles. The APB muscles were predominantly excited following F3′ or F4′ stimulation. Based on these data, we considered that the most effective sites of stimulation to evoke MEPs from the AH muscle were over Cz, from the BB and D muscles, over C3′ and C4′ and from the APB muscle, over F3′ and F4′.

The effects of direction of current flow in the coil on MEPs were also investigated at all sites stimulated. With the exception of Erb's point, clockwise stimulation tended to evoke MEPs in muscles on the left side and anticlockwise stimulation to evoke MEPs in muscles on the right side. However, when stimulated at the Cz site, there were almost no differences between clockwise and anticlockwise stimulation in the frequency of MEPs recorded from the facilitated AH muscles, and neither clockwise nor anticlockwise stimulation at sites of F3′ and F4′ made much difference in the frequency of MEPs elicited from relaxed and facilitated APB muscles.

Clockwise stimulation at C5 produced MEPs clearly on the left, regardless of each stimulus site (on the left, on the centre, on the right), and anticlockwise stimulation on the right.

Both clockwise and anticlockwise stimulation at Erb's point on the left produced MEPs in all of the muscles on the left of the subjects, while delays in latency and decreases in amplitude were observed in MEPs recorded after clockwise stimulation in comparison with anticlockwise stimulation. Similarly, delays in latency and decreases in amplitude of MEPs evoked following anticlockwise stimulation of Erb's point on the right were observed in comparison with clockwise stimulation.

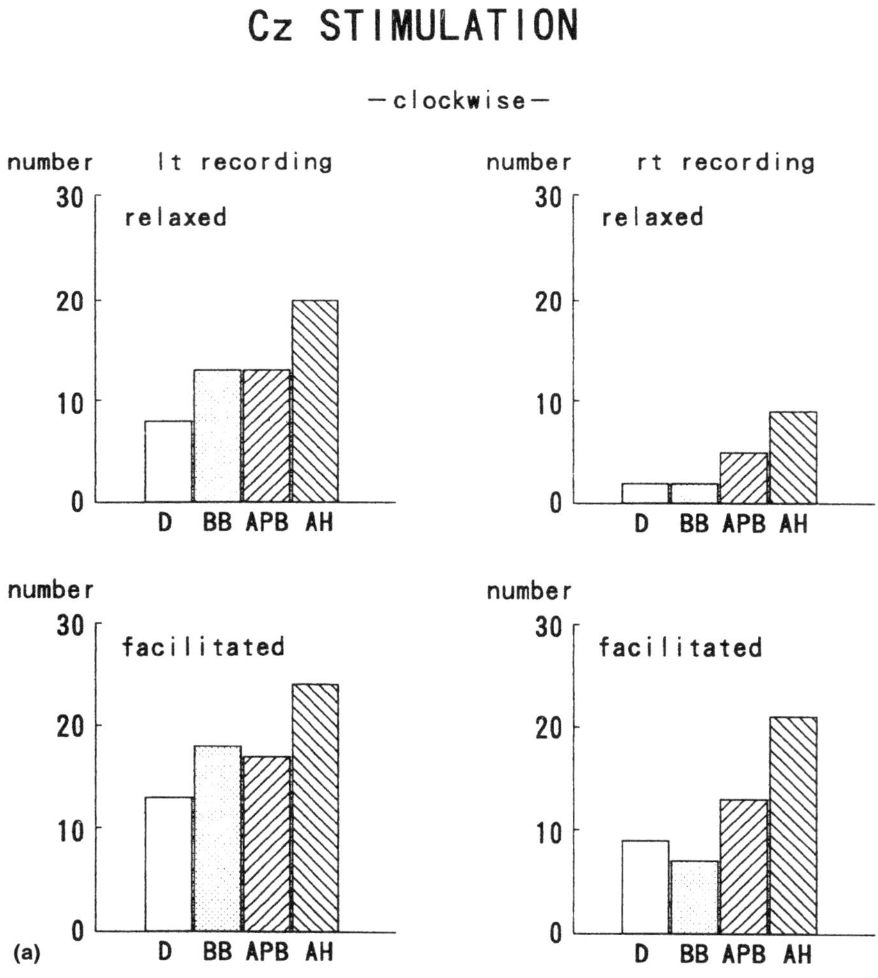

Figure 2a, b The frequency of MEPs recorded from each muscle following stimulation at Cz. (**a**) Clockwise stimulation. The AH muscles on the left are most frequently excited. (**b**) Anticlockwise stimulation. The MEPs are best evoked from the AH muscles on the right

CASE PRESENTATIONS

Case 1 (Figure 5): The subject is a 24-year-old healthy male volunteer. The MEPs recorded from both AH muscles following Cz stimulation are shown. In relaxed AH muscles, clockwise stimulation produced MEPs only on the left, and anticlockwise stimulation on the right. In the facilitated condition, MEPs were recorded from both muscles, regardless of the direction of current flow in the coil.

Case 2 (Figure 6): The subject is a 25-year-old healthy male volunteer. MEPs were recorded from both APB muscles after C5 stimulation. Clockwise

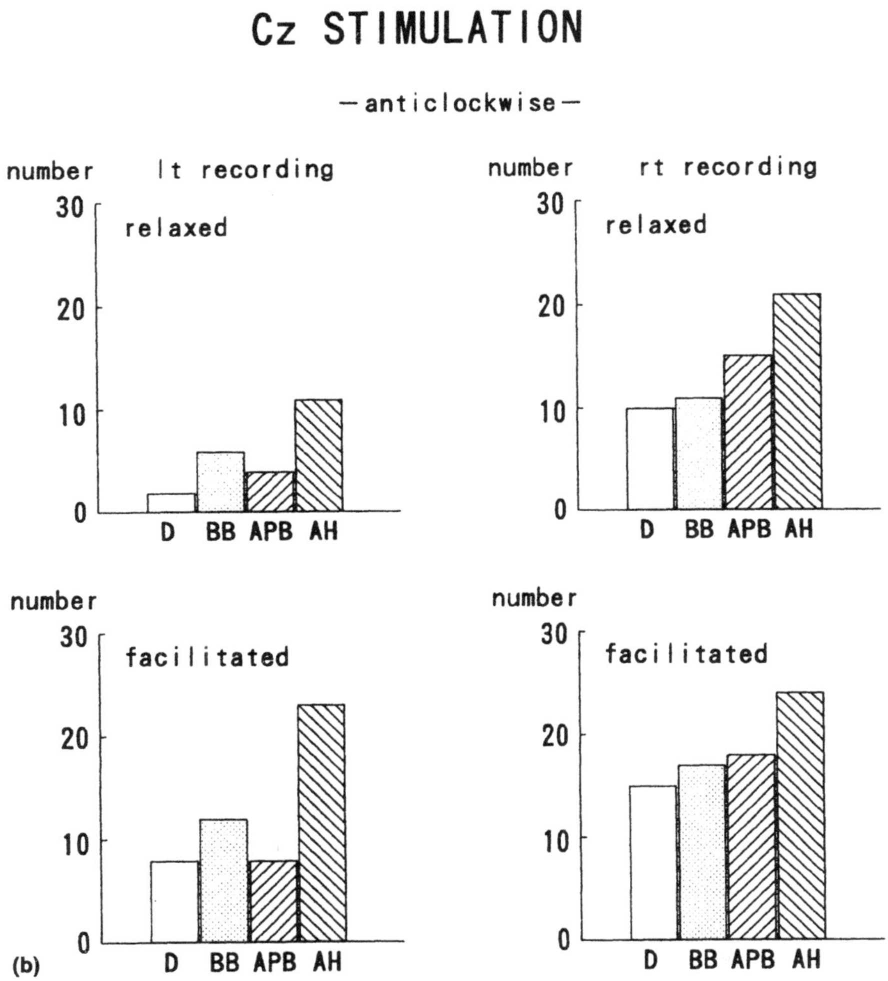

Figure 2 *Continued*

stimulation evoked MEPs clearly on the left, and anticlockwise stimulation on the right.

DISCUSSION

Barker et al.[1] described a new technique for noninvasive magnetic stimulation of the human motor cortex, and Barker et al.[2] and Hess et al.[3,4] reported that it was possible to study central motor conduction times using magnetic stimulation. The main advantage of this technique is that it is less painful than electrical stimulation, and thus more suitable for clinical purposes. One

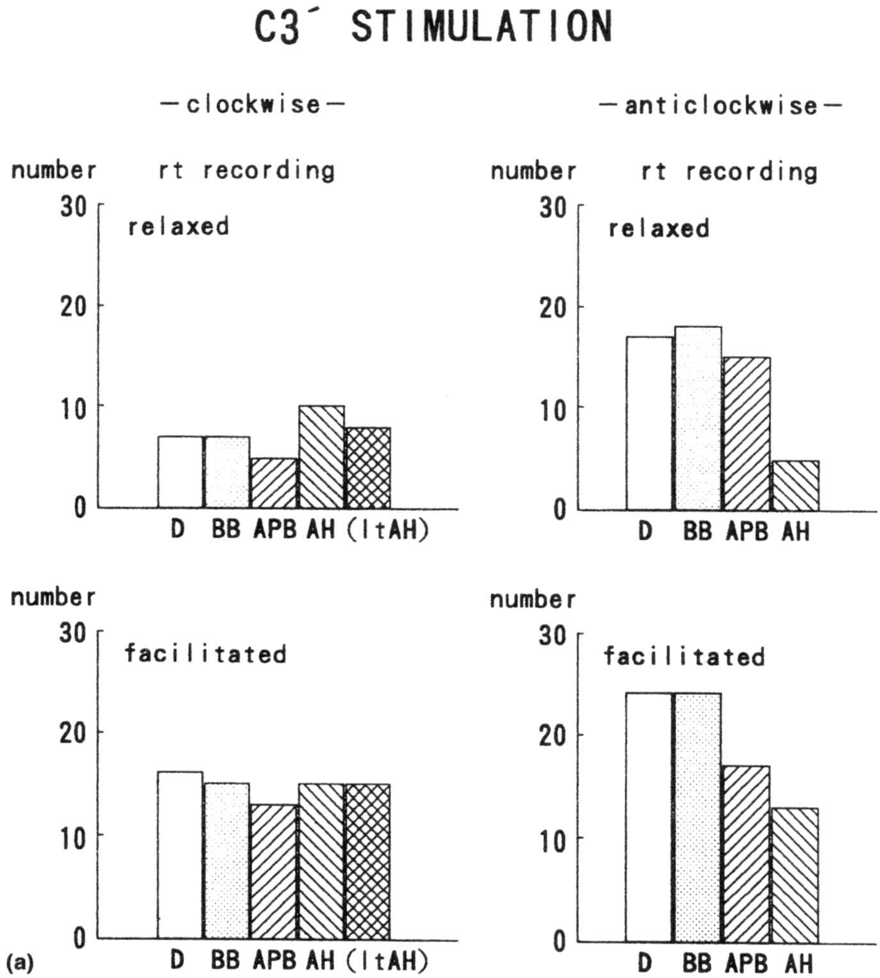

Figure 3a, b The frequency of MEPs recorded from each muscle following stimulation at C3′ or C4′. (**a**) C3′ stimulation. Anticlockwise stimulation elicits MEPs from the deltoid and biceps brachii muscles more often than APB or AH muscles. (**b**) C4′ stimulation. Clockwise stimulation best evokes MEPs from the deltoid and biceps brachii muscles compared with the other muscles

disadvantage is that focal stimulation appears to be impossible.

Cohen et al.[5] experimented with noninvasive mapping of the human motor cortex by transcranial magnetic stimulation. They have stated that their study allowed them to formulate a technique for noninvasive mapping of the human motor cortex using magnetic stimulation with magnetic coils available at the time, and the coil was kept tilted at a constant angle of 30° with the distal portion pointing anteriorly in contact with the scalp and the handle pointing posteriorly. Hand motor maps so defined were positioned slightly in front of the site on the scalp where the N20 component of median

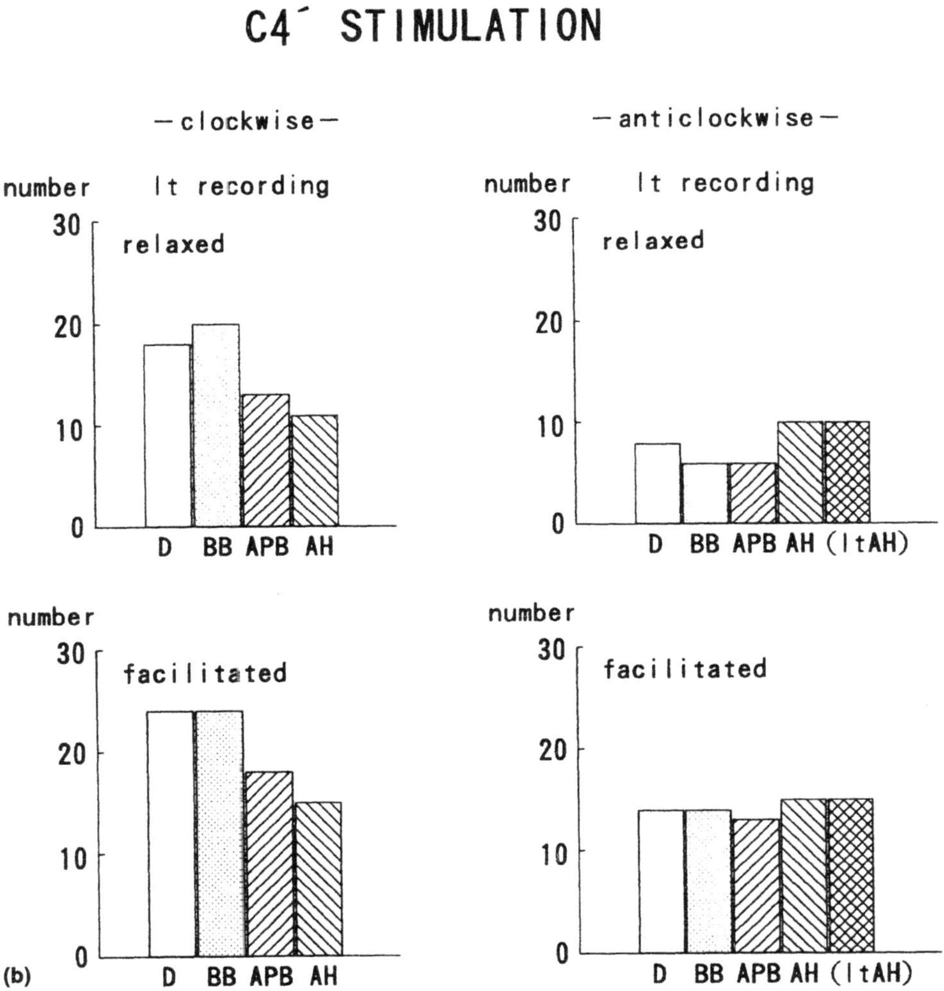

Figure 3 *Continued*

nerve somatosensory evoked potentials was maximal.

However, the coil we used (inner diameter, 27 mm: outer diameter, 137 mm) had its maximal vertical magnetic field in the centre of the coil. Keeping the centre of the stimulating coil on the sites of the scalp stimulated with a flat orientation enabled us to achieve focal stimulation. In this way, we were able to map appropriate sites of stimulation for the different muscles on the scalp, that is, AH muscles at the midline close to Cz, D and BB muscles over C3′ and C4′ and APB muscles over F3′ and F4′. It is also important to be attentive to the direction of current flow through the coil. Clockwise stimulation on the scalp tended to evoke MEPs from the muscles on the left side, and anticlockwise stimulation on the right side. Thus, to obtain MEPs

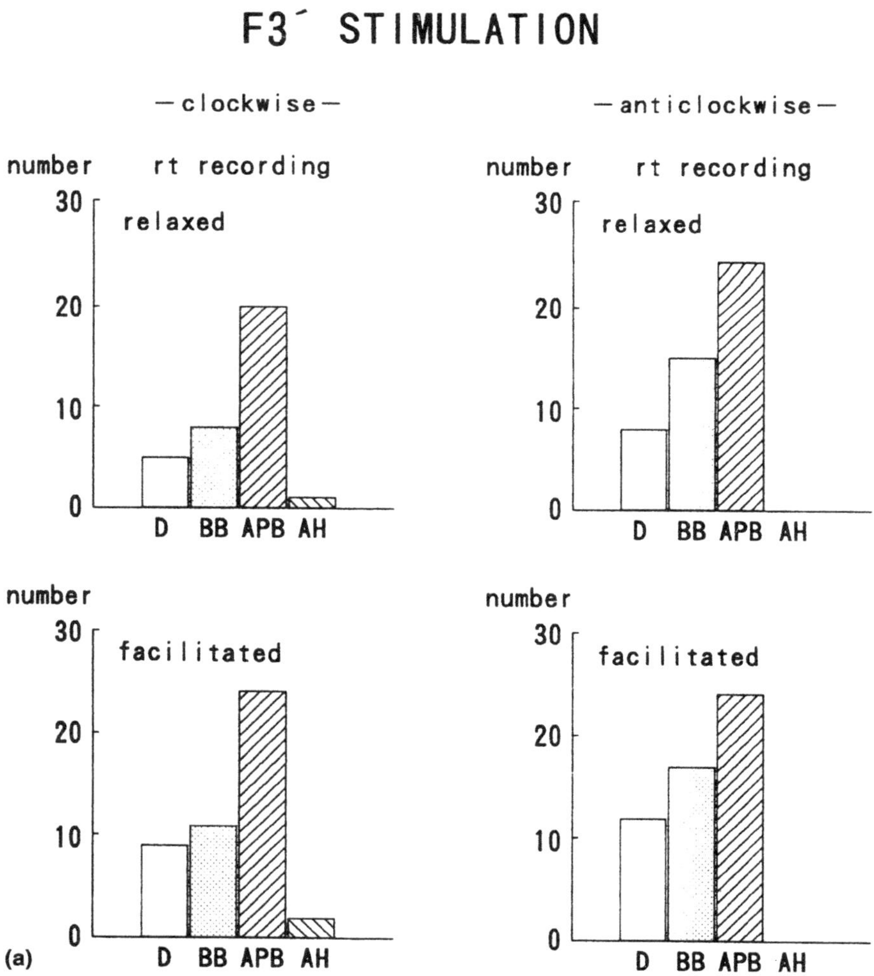

(a)

Figure 4a, b The frequency of MEPs recorded from each muscle following stimulation at F3′ or C4′. (**a**) F3′ stimulation. The APB muscles are predominantly excited following anticlockwise stimulation. (**b**) F4′ stimulation. The APB muscles are best excited after clockwise stimulation

from the muscles on the left, clockwise stimulation of the motor cortex should be used. Conversely, anticlockwise stimulation of the motor cortex should be applied in order to evaluate MEPs elicited from muscles on the right. Each stimulus site on the scalp should be stimulated in facilitated muscles as well as in relaxed muscles to obtain high frequency of MEPs elicited.

An effect of direction of current flow in the coil on MEPs was also observed when stimulation of the fifth cervical vertebra was performed[6]. Maximal MEPs in APB muscles on the left were obtained with clockwise stimulation

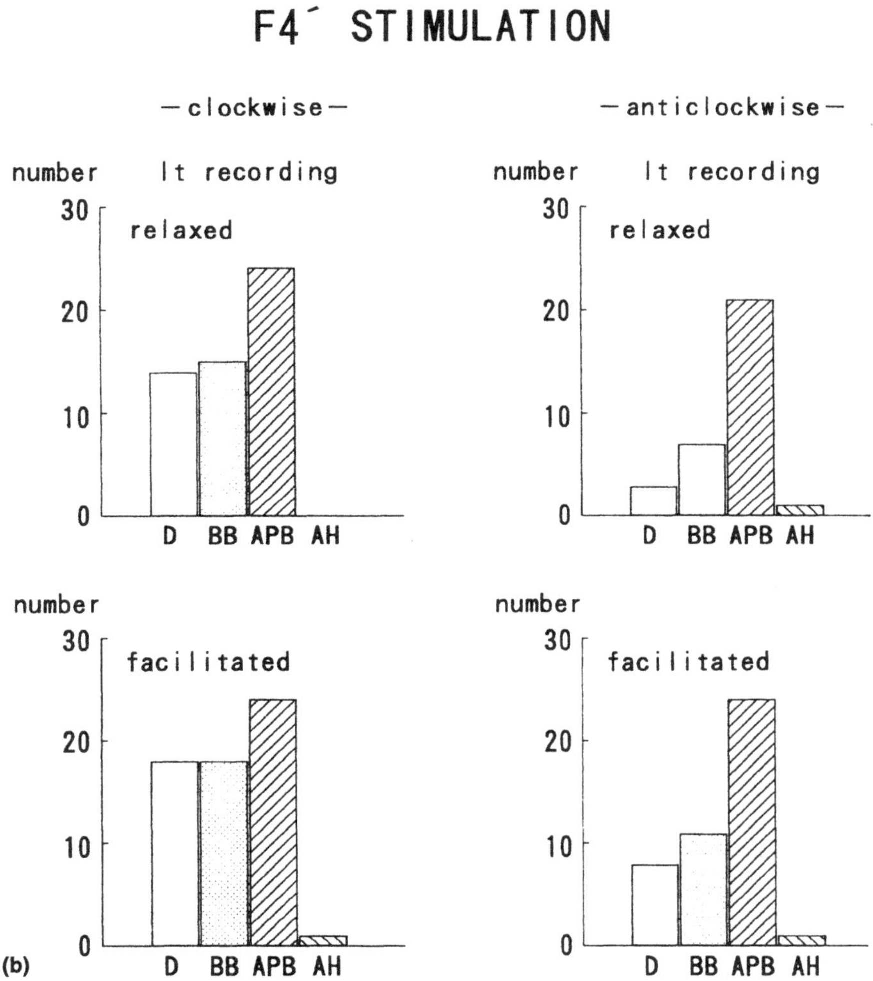

Figure 4 *Continued*

of the fifth cervical vertebra on the left. On the other hand, maximal MEPs in the APB muscles on the right were obtained by anticlockwise stimulation of the fifth cervical vertebra on the right.

The precise mechanism explaining these findings remains unknown, but maximal muscle excitation occurs when the anatomical orientation of the nerve tissue supplying the muscles recorded precisely coincides with the direction of the current induced.

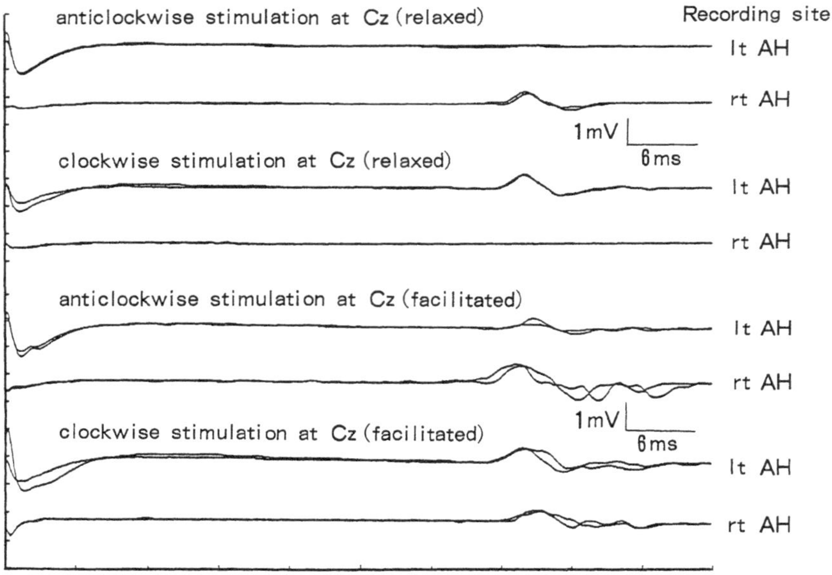

Figure 5 Case 1. A healthy 24-year-old male volunteer, 168 cm (Cz stimulation)

CONCLUSIONS

Based on our findings, we conclude that, with the exception of Erb's point, clockwise stimulation tends to evoke MEPs in muscles on the left side and anticlockwise stimulation in muscles on the right side. The coil we used makes focal stimulation of the motor cortex possible, that is, the left APB muscle maps maximally over F4′ and the right APB muscle over F3′, the left D and BB muscles map over C4′, the right D and BB muscles over C3′, and the left and right AH muscles map at the midline close to Cz.

References

1. Barker AT, Jalinous R, Freeston IL. Noninvasive magnetic stimulation of human motor cortex. Lancet. 1985;2:1106–1107.
2. Barker AT, Freeston IL, Jalinous R, Jarratt JA. Clinical evaluation of conduction time measurements in central motor pathways using magnetic stimulation of the human brain. Lancet. 1986;1:1325–1326.
3. Hess CW, Mills KR, Murray NMF. Measurement of central motor conduction in multiple sclerosis by magnetic brain stimulation. Lancet. 1986;2:355–358.
4. Hess CW, Mills KR, Murray NMF, Schriefer TN. Magnetic brain stimulation: Central motor conduction studies in multiple sclerosis. Ann Neurol. 1987;22:744–752.
5. Cohen LG, Hallett M, Lelli S. Noninvasive mapping of human motor cortex with transcranial magnetic stimulation. In: Chokroverty S, ed. Magnetic stimulation clinical neurophysiology. Stoneham, MA: Butterworth; 1990:113–119.
6. Nishijima Y, Okada M, Izumi T, Yamazaki Y, Toda, N. The standard methods of magnetic stimulation to the skull and neck to elicit compound muscle action potentials. SEIKEIGEKA, 1992;43:571–576 (Japanese).

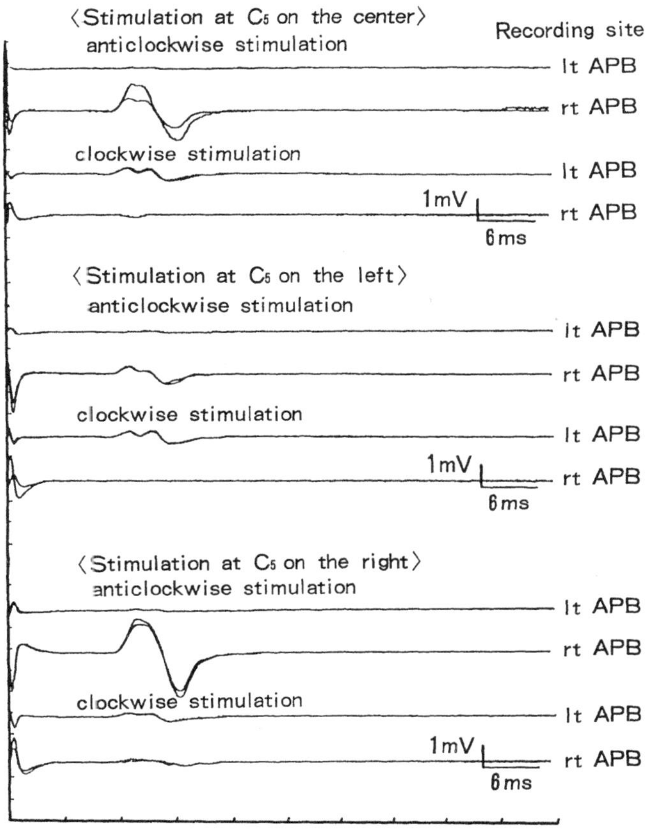

Figure 6 Case 2. A healthy 25-year-old male volunteer, 165cm (C5 stimulation)

42 Separation of upper and lower extremity components of the corticospinal MEP (D-wave) recorded at the cervical level

Y. KATAYAMA, T. TSUBOKAWA, T. YAMAMOTO, T. HIRAYAMA and S. MAEJIMA

INTRODUCTION

It has been demonstrated in experimental animals that synchronous impulses occurring in efferent axons of corticospinal tract (CST) neurons in response to stimulation of the motor cortex can be recorded from the spinal cord[1]. The response consists of an initial D-wave and a later sequence of volleys termed I-waves. The D-wave reflects impulses arising from direct activation of the axons of CST neurons, and is induced most efficiently with anodal stimulation of the surface of the motor cortex.[1].

Responses identical to the D-wave can be recorded in humans from the epidural space of the spinal cord following stimulation of the motor cortex exposed for intracranial surgery[2,3]. There are many similarities between the responses recorded in humans and the D-wave reported in experimental animals[3-6]. Among these, the following features are especially important for concluding that these responses represent impulses arising from direct activation of the axons of CST neurons. First, it has been demonstrated that the response is capable of following double pulse stimulation with interstimulus intervals of more than 500 Hz[3]. Second, the response is limited to stimulation of restricted areas corresponding to the location of the motor cortex. Furthermore, disappearance of the response is seen only when surgical damage is caused to these areas.[2].

Devastating deficit in motor function as a result of surgical damage to the motor cortex or CST is a major concern in operations on lesions located in close proximity to these structures. Intraoperative monitoring of electromyographic responses to stimulation of the motor cortex (corticomuscular MEP) has been employed for the purpose of monitoring the conductivity of CST fibres. Changes in the corticomyographic MEP are not, however, specific to disruption of the conductivity of CST fibres, since the amplitude is strongly affected by changes in the excitability of spinal motoneurons which are under the influence of a number of neural pathways other than the CST. We have been using the human D-wave (corticospinal MEP) for identifying the localization of the motor cortex and monitoring the conductivity of CST fibres during intracranial surgery. The corticospinal MEP meets the following requirements for the signal to be monitored for such a purpose: (1) the signal is capable of being recorded rapidly and constantly; (2) anaesthesia and other drugs exert minimal influences; (3) the intraoperative changes are highly specific to changes in CST function; and (4) the intraoperative changes are well correlated with the postoperative deficits in motor function[2,6].

It is uncertain, however, which components of the CST fibres are represented by the corticospinal MEP recorded at the epidural space of the spinal cord. In contrast to the corticomuscular MEP, changes in the corticospinal MEP do not provide information regarding the extent of the deficits of motor function. While disrupted components of the CST fibres may be estimated from differences in the changes in the corticospinal MEP between different stimulation sites of the motor cortex, this procedure requires multiple recordings and is not practical for the purpose of intraoperative monitoring. In order to overcome this disadvantage of the corticospinal MEP, we have been applying a dual recording technique. This report summarizes our experience with this technique, which enables the site of stimulation and the disrupted component of CST fibres to be estimated from a single recording.

METHODS

We analysed corticospinal MEPs obtained from 10 patients with brain tumours without any motor deficits preoperatively. The patients and their families gave informed consent for the procedures described below to be carried out. The present study was approved by the Committee for Clinical Trials and Research in Humans of our university and by the Japanese Ministry of Health and Welfare as part of an advanced medical care programme. Patients who demonstrated clinical or electroencephalographic evidence of seizure activity were excluded.

During the course of intracranial surgery, the exposed motor cortex is directly stimulated with electrode arrays having four plate electrodes (M-3586, Medtronic Inc., Minneapolis, Minn., USA). The four electrodes are 5 mm in diameter and spaced 5 mm apart. The location of the stimulation site is estimated roughly from bony landmarks by means of a method usually employed for intracranial surgery. The location of the motor cortex, however, could not be defined precisely by any method other than recording of the MEP. Attention is therefore always focused on identifying areas from which the corticospinal MEP is recorded with the lowest threshold.

Under fluoroscopic control, a pair of flexible platinum electrodes (M-8483, Medtronic Inc., Minneapolis, Minn., USA) were inserted into the epidural space from the upper thoracic or lower cervical levels, and one was advanced rostrally to the C2 or C3 level and the other to the C7 or Th1 level. The corticospinal MEP was recorded monopolarly from the epidural electrodes with a reference electrode placed at the paravertebral muscles, in response to motor cortex stimulation[3,7]. Multiple recordings were performed with stepwise withdrawal of the rostrally placed epidural electrodes, if necessary[7]. Signals from the electrodes were fed into an amplifier with a bandpass range of 5 Hz to 5 kHz and 32–64 sweeps were averaged using a signal processor.

For the purpose of delineating the extent of the motor cortex, monopolar stimulation naturally provided better spatial resolution. However, the response to monopolar stimulation during intracranial surgery is often unstable in its amplitude and threshold during the course of surgery and is sometimes associated with uncontrollable artefacts, presumably due to large changes in impedance of the current pathways induced by various surgical procedures. While it is difficult to obtain a clear response using bipolar stimulation with an interpolar distance of less than 10 mm for stimulation, bipolar stimulation with a longer interpolar distance of more than 10 mm clearly produced the D-wave. The lowest intensity for evoking the D-wave is as low as 2 mA, when the stimulation electrodes are appropriately placed on the motor cortex. When both poles were placed on the motor cortex, the amplitude remained approximately the same after reversing the polarity of stimulation and stimulation artefacts could be removed by a signal averaging technique by applying anodal currents alternately to one and then the other pole of the stimulation electrodes[3].

The present study analysed the data from alternating bipolar stimulation employing the most distant two plate electrodes of the electrode array (interpolar distance, 3 mm). Two electrode arrays are placed on the motor cortex in the mediolateral orientation parallel to the central fissure; one on the trunk and hand areas placing the two plate electrodes 5 and 8 cm lateral from the midline (Pair 1), and the other on the thigh and trunk areas placing the two electrodes 1 and 4 cm lateral from the midline (Pair 2). This technique (Figure 1) is practically most useful for intraoperative monitoring of the conductivity of CST fibres. The locations of the stimulation electrode arrays are not altered, once the monitoring has been initiated. Stimuli are applied as monophasic square wave pulses of 0.2–0.5 ms duration delivered at 4 Hz. Delivery of the stimuli is triggered by electrocardiograms.

Stimulation intensity was limited to submaximal level since supramaximal stimulation does not appear to be applicable safely. The incapability of employing supramaximal stimulation in recording the corticospinal MEP is another reason for uncertainty regarding the site of stimulation. It is always uncertain to what extent the motor cortex and the subcortical white matter is being activated by given stimulation intensity. It is therefore important to establish a method to determine the extent of activation, such as the technique analysed in the present study.

RESULTS

The corticospinal MEP, consisting of a D-wave sometimes followed by one or more I-waves, was clearly recorded in all cases. The D-wave demonstrated

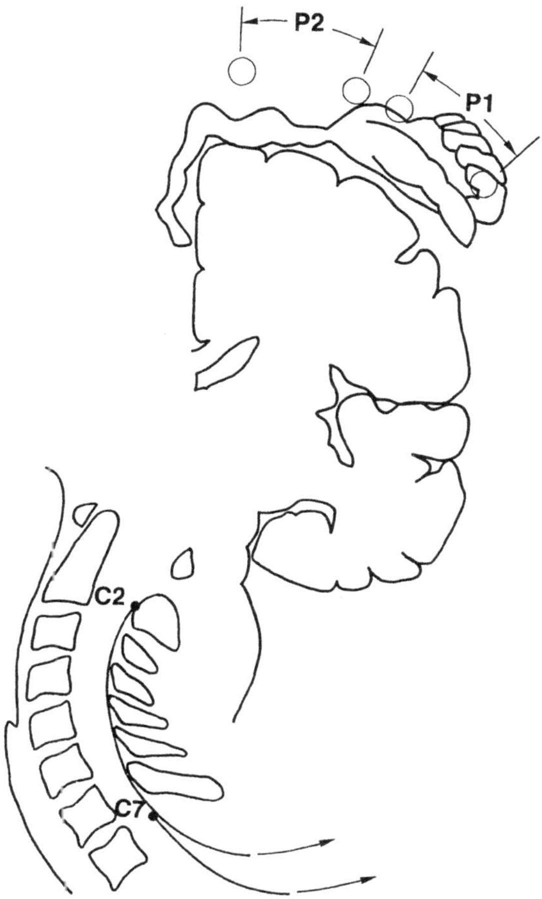

Figure 1 Schematic representation of dual recording technique. The motor cortex is stimulated by one of two pairs of stimulation electrodes (P1 and P2). The corticospinal MEP was recorded from two levels of the spinal cord (C2 and C7)

minimal variations and was capable of following repetitive stimulation at a frequency above 500 Hz so that recordings could be updated as fast as every 15–30 s. Since the D-wave is relatively large, reaching up to 7–20 μV, averaging of 32 sweeps is sufficient for clear recording. This response is recorded regardless of whether the recording electrodes are placed dorsolaterally to the spinal cord.

As the recording site was moved caudally, the latency of the response increased and the amplitude decreased (Figure 2). The difference in peak latency of the negative peak of the response over the known distance between two recording sites is usually used for calculation of the conduction velocity. The calculated conduction velocity lies within the range 50 to 75 m/s[2,7]. The width of the response also increased as the recording site was moved caudally,

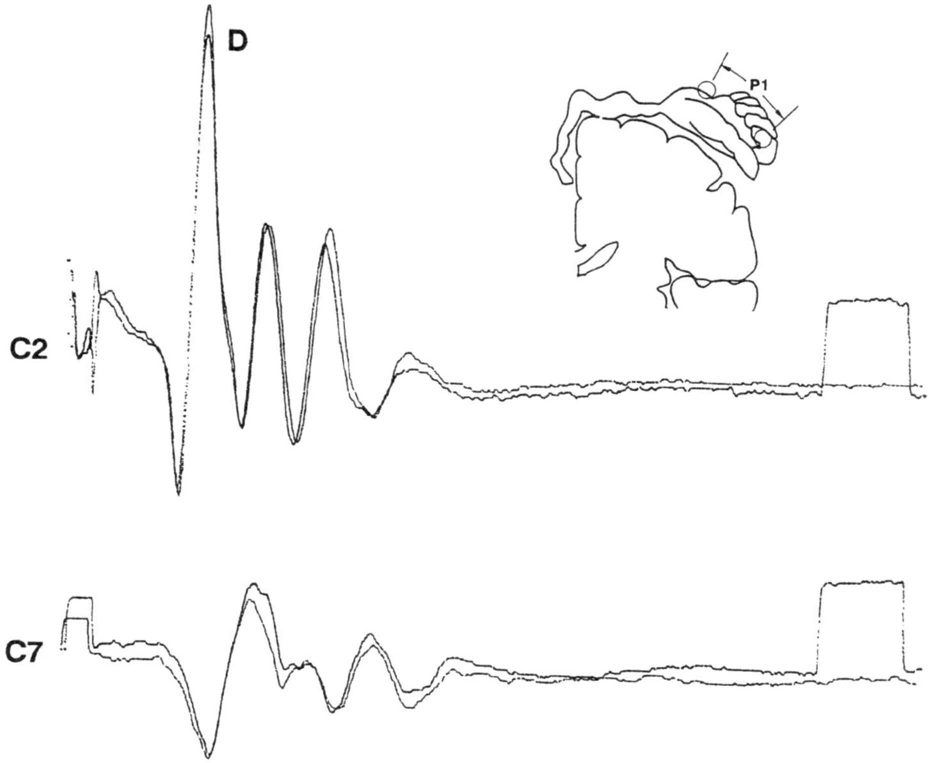

Figure 2 Representative examples of corticospinal MEPs recorded at C2 and C7 levels in response to stimulation of the lateral part of the motor cortex. Calibration, 5 μV and 2 ms. Upwards, negative. A large decrease in amplitude of the D-wave between these two recording sites indicates that a large component of the corticospinal tract fibres leaves the spinal cord at the cervical level and innervates the upper extremity. The amplitude at the C7 level represents impulses which do not leave the spinal cord at the cervical level and mostly innervate the lower extremity. Thus, the ratio of the amplitudes between C7(–Th1) and C2(–3) reflects the relative contributions of the upper and lower extremity (and/or trunk) components to the recorded D-wave

suggesting temporal dispersion.

The decrease in amplitude at the caudal recording site was, however, too large to be accounted for only by temporal dispersion. The amplitude at the C2–3 and C7–Th1 levels varied depending on the stimulation site. The amplitude at the C2–3 level was slightly larger when the Pair I electrodes were used for stimulation, whereas the amplitude at the C7–Th1 level was clearly larger when the Pair 2 electrodes were used. Thus, the ratio of the amplitudes between C7–Th1 and C2–3 (LU ratio) was larger when the Pair 2, rather than Pair 1 electrodes, were used. The LU ratio was sometimes smaller than 0.1 when the motor cortex was stimulated by the Pair 1 electrodes. In contrast, the LU ratio was as large as 1.0 (Figure 3).

In the ten cases investigated in the present study, the motor cortex as

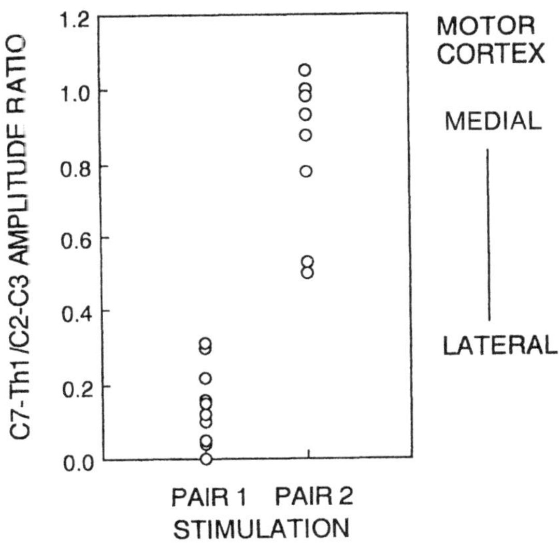

Figure 3 The amplitude ratio of the D-wave between C7–Th1 and C2–3 levels (LU ratio) in relation to the stimulation site (PAIR 1 and PAIR 2, see Figure 1). The LU ratio reflects (C7–Th1/C2–3) the relative contributions of the upper and lower extremity (and/or trunk) components to the recorded D-wave. The larger the ratio, the more the lower extremity (and/or trunk) component is being activated. The smaller the ratio, the more the upper extremity component is contributing to the recorded D-wave

identified from the corticospinal MEP was left undamaged during surgery. However, a decrease in amplitude was observed more or less in all cases during resection of the tumours involving the subcortical white matter. Three cases demonstrated moderate motor deficits in the upper extremity, two in the lower extremity, and two in both extremities postoperatively.

In the cases demonstrating motor weakness in the upper extremity postoperatively, the amplitude decreased at the C2–3 level, especially when the Pair 1 electrodes were used for stimulation. In contrast, the amplitude at the C7–Th1 level was affected less markedly. Thus, the LU ratio increased during subcortical resection of the brain tumours in these cases (Figure 4, left). The increase in ratio was more clearly observed when the Pair 1 electrodes were used.

In the cases in which muscle weakness of the lower extremity was induced postoperatively, the amplitude decrease at the C7–Th1 level, especially when the Pair 2 electrodes were used for stimulation. The amplitude decreased at the C2–3 level also but less markedly. A decrease in the LU ratio was therefore observed during subcortical resection of the tumours in these cases (Figure 4, right). The decrease in ratio was clearer when the motor cortex was stimulated by the Pair 2 electrodes.

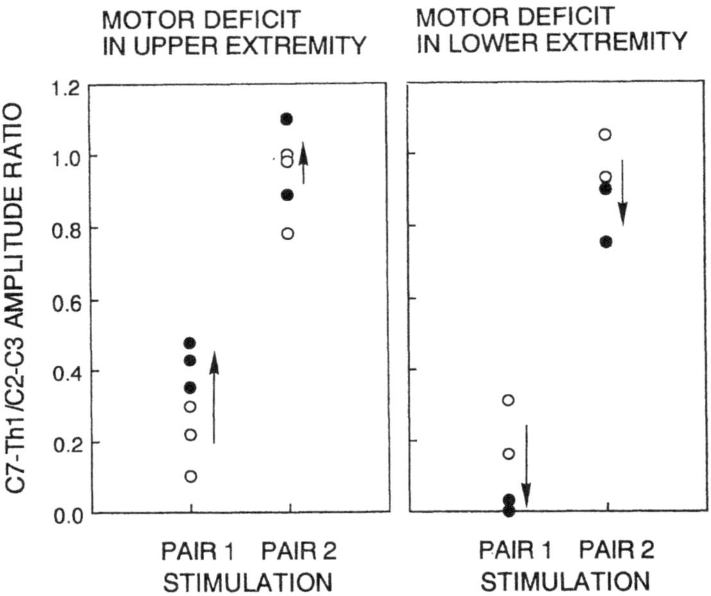

Figure 4 Changes in the amplitude ratio of the D-wave between C7–Th1 and C2–3 levels (LU ratio) in relation to postoperative motor deficits (left, deficit in upper extremity; right, deficit in lower extremity). The LU ratio increased when motor deficit was observed in the upper extremity component, whereas the LU ratio decreased when motor deficit was observed in the lower extremity. Even if the location of the stimulation site is not completely certain, an increased LU ratio indicates predominant disruption of the upper extremity component of CST fibres and a decreased LU ratio indicates predominant disruption of the lower extremity (and/or trunk) component

DISCUSSION

The dual recording technique at the C2–3 and C7–Th1 levels provides information concerning which components of the CST fibres contributed to the recorded D-wave. A rapid decrease in amplitude of the D-wave between these two recording sites indicates that a large component of the CST fibres leaves the spinal cord at the cervical level and innervates the upper extremity. The amplitude at the C7–Th1 level represents impulses occurring in CST fibres which do not leave the spinal cord at the cervical level and mostly innervate the lower extremity. Thus, the ratio of the amplitudes between C7–Th1 and C2–3 reflects the relative contributions of the upper and lower extremity (and/or trunk) components to the recorded D-wave. The larger the ratio, the more the lower extremity (and/or trunk) component is being activated. The smaller the ratio, the more the upper extremity component is contributing to the recorded D-wave.

The LU (C7–Th1/C2–3) ratio was clearly smaller when the Pair 1 electrodes were used, implying that the upper extremity component is predominantly activated when the stimulation is delivered to the lateral part of the motor cortex. When the Pair 2 electrodes were used, the LU ratio was larger, indicating that components which do not leave the spinal cord at the cervical level appear to arise in the more medial part of the motor cortex. This observation is clearly consistent with the topographic organization of the motor cortex.

While information from completely normal subjects is unavailable, the normal LU ratio may not differ greatly from the values obtained in the patients analysed in the present study, since no appreciable motor deficits were observed preoperatively. The location of the stimulation site sometimes remains uncertain because of the limited extent of craniotomy. In addition, because of the incapability of employing supramaximal stimulation in recording the corticospinal MEP, it is always uncertain to what extent the motor cortex and the subcortical white matter is activated by given stimulation intensity. The LU ratio is useful in such cases for estimating the location of the stimulation site insofar as the preoperative motor deficits are minimal. The LU ratio basically varies between 0 to 1.0. When the LU ratio is very small, the lateral part of the motor cortex is stimulated (see Figure 3). In contrast, a large LU ratio indicates that the more medial part of the motor cortex is being stimulated.

It is important to verify that both upper and lower extremity components are reflected in the corticospinal MEPs to be monitored. The LU ratio should be carefully checked prior to initiation of resection of the brain tumour. Otherwise, false negative findings may result. The present study confirmed that the disrupted component of CST fibres can be estimated from the LU ratio. When the upper extremity component is disrupted, the amplitude of the D-wave recorded at the C2–3 level decreases following stimulation with the Pair 1 electrodes. Even if the location of the stimulation site is not completely certain, an increased LU ratio indicates predominant disruption of the upper extremity component. When the lower extremity (and/or trunk) component is disrupted, the amplitude of the D-wave recorded at the C7–Th1 level following stimulation with the Pair 2 electrodes decreases and the LU ratio decreases.

Theoretically, the LU ratio should not be changed by disruption of the lower extremity component when lower extremity component is specifically activated. Since the LU ratio with stimulation by the Pair 2 electrodes varied from 0.5 to 1.0, the arrangement of the Pair 2 electrodes in the present study appeared to have activated a considerable portion of the upper extremity component together with the lower extremity (and/or trunk) component.

It is not known directly to what extent the lower extremity (and/or trunk) component is activated by the Pair 2 electrodes. The present results suggested, however, that decreases in amplitude of the D-wave at the C7–Th1 level by more than 50% predicted an incapability to walk. Since the walking capability involves not only muscles of the lower extremity but also those of the trunk, it is possible that the decrease largely represents a disruption of the CST fibres arising from the trunk area of the motor cortex. The corticospinal MEP recorded by the Pair 2 electrodes and the LU ratio appears to be practically useful for predicting the quality of the postoperative motor deficits.

CONCLUSION

It is important in monitoring the corticospinal MEP to verify that both upper and lower extremity components are appropriately reflected in the D-wave to be monitored. The dual recording technique at the C2–3 and C7–Th1 levels provides information concerning which components of the CST fibres contributed to the recorded D-wave.

References

1. Patton HD, Amassian VE. Single- and multiple-unit analysis of cortical stage of pyramidal tract activation. J Neurophysiol 1954;17:345–363.
2. Tsubokawa T. Clinical significance of the centrifugal evoked potentials to cortical surface or scalp stimulation for evaluation of motor function during surgery. Neurosurgeons (Tokyo) 1985;5:111–122.
3. Katayama Y, Tsubokawa T, Maejima S, Hirayama T, Yamamoto T. Corticospinal direct response in humans: identification of the motor cortex during intracranial surgery under general anesthesia. J Neurol Neurosurg Psychiatry. 1988a;51:50–59.
4. Levy WJ, York DH, McCaffrey M, Tanzer F. Motor evoked potentials from transcranial stimulation of the motor cortex in humans. Neurosurgery. 1984;15:214–227.
5. Boyd S, Rothwell JC, Cowan JMA, Webb PJ, Morley T, Asselman P, Marsden CD. A method of monitoring function in corticospinal pathways during scoliosis surgery with a note on motor conduction velocities. J Neurol Neurosurg Psychiatry. 1986;49:251–257.
6. Tsubokawa T. Intraoperative monitoring of corticospinal motor evoked potentials in neurosurgical operations. In: Schramm J, Moller AR eds. Intraoperative Neurophysiologic Monitoring in Neurosurgery. Berlin, Germany: Springer; 1991:106–126.
7. Katayama Y, Tsubokawa T, Yamamoto T, Hirayama T, Maejima S. Preoperative determination of the level of spinal cord lesions from the killed end potential. Surg Neurol. 1988b;29:91–94.

43 Changes in the corticospinal MEP (D-wave) during microsurgical removal of intramedullary spinal cord tumours: experience in 16 cases

Y. KATAYAMA, T. TSUBOKAWA, T. YAMAMOTO, T. HIRAYAMA and S. MAEJIMA

INTRODUCTION

It has become increasingly clear that complete surgical excision is the best therapeutic approach for the majority of intramedullary spinal cord tumours. There is, however, always a risk in the radical excision of intramedullary spinal cord tumours that a greater postoperative neurological deficit may be inflicted. Extensive dissection can cause damage to the well-functioning cord tissue and/or the anterior spinal artery and penetrating branches. In order to minimize neurological deficits occurring as a result of such damage, the somatosensory evoked potential (SEP) has often been employed as an intraoperative monitor of cord function. The SEP is, however, not an ideal monitor since it reflects primarily the conductivity of the dorsal column. There are many cases in which SEP monitoring has missed motor deficits following damage to the corticospinal tract (CST), which is more devastating than any other neurological deficits[1,2]. These cases emphasize the need for a direct monitor specific to the CST integrity.

Classical research on experimental animals has demonstrated that impulses mediated by CST neurons in response to activation of the motor cortex, termed the D-wave, can be recorded from the spinal cord (Amassian et al.[3] for review). We have shown that responses analogous to the D-wave can be recorded from the epidural space of the human spinal cord following stimulation of the motor cortex exposed for intracranial surgery[4,5]. This signal, which is termed the corticospinal motor evoked potential (MEP), is efficacious in the intraoperative identification of the motor cortex[4,5]. Furthermore, by employing this method, surgical damage of the CST was detected intraoperatively and unnecessary damage was efficiently avoided[6]. Use of the D-wave as a monitor of the CST integrity during certain forms of spinal surgery has been described previously by various authors[7–9]. This report summarizes our experience of monitoring the corticospinal MEP in patients who underwent radical excision of intramedullary spinal cord tumours.

METHODS

Intraoperative monitoring of the corticospinal MEP was performed in a consecutive series of 16 patients with intramedullary spinal cord tumours. All the patients gave informed consent for all the procedures described below to be carried out. During surgery, a flexible platinum electrode (Medtronic Inc., M8483, Minneapolis, Minn., USA) was inserted into the epidural space from the caudal edge of the laminectomized area, advanced 5 cm caudally and placed on the midline of the cord. The corticospinal MEP was recorded monopolarly from the epidural electrode with a reference electrode placed at the paravertebral muscles, in response to motor cortex stimulation. Signals from the electrodes were fed into an amplifier with a bandpass of 5 Hz to 5 kHz and 16–32 sweeps were averaged using a signal processor.

The motor cortex can be stimulated transcranially[7–9]. While the motor cortex in children can be stimulated with a relatively low intensity current because of their thin cranial vault, a higher intensity current proved necessary for stimulating the motor cortex through the thick skull of adults. This non-invasive method of motor cortex stimulation often leads to wide current spreading over the scalp which renders the MEP recorded from the cervicothoracic region obscure due to large stimulation artefacts. We therefore employed epidural stimulation through a small burr hole in adult cases and transcranial stimulation in one child's case. The burr holes were made in the skull overlying the motor cortex at 30 mm off the midline and electrode arrays having four plate electrodes each separated for 10 mm (M-3586, Medtronic Inc., Minneapolis, Minn., USA) were inserted into the epidural space. The most medial electrode was placed close to the midline. The stimuli were applied bipolarly using the most medial and most lateral electrodes (interpolar distance, 30 mm). An anodal current was delivered to both electrodes alternately, so that subcortical CST fibres originating from the trunk and upper leg areas of the motor cortex were stimulated. The corticospinal MEPs in response to stimulation of the right and left hemispheres were alternately recorded. The stimuli were applied bilaterally in one case with anodal current given monopolarly through the most medial electrode. In the one child case in which transcranial stimulation was employed, the motor cortex was stimulated bilaterally with the anodal current delivered through a plate electrode placed at the vertex. The stimuli were applied as monophasic square wave pulses of 0.2–0.5 ms duration delivered at 4 Hz and triggered by electrocardiograms.

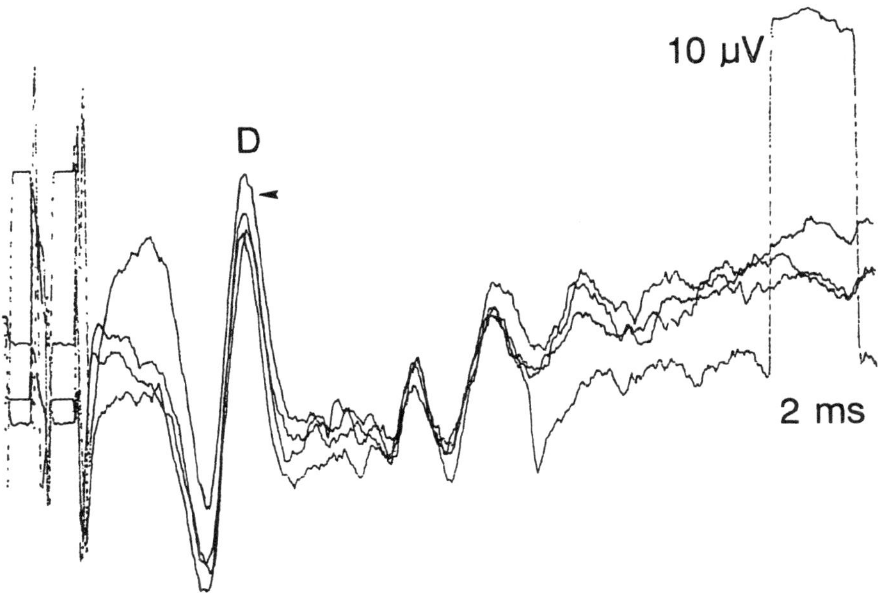

Figure 1 Representative example of the corticospinal MEP in a patient with intramedullary spinal cord tumour. Upwards correspond to negativity. Three MEP traces repeatedly recorded every 30 s before dural incision are superimposed. One more trace superimposed (indicated by arrow head) is a response recorded immediately after dural incision

RESULTS

A corticospinal MEP was clearly recorded in all of the 11 patients who were capable of walking: an initial biphasic positive–negative response, termed a D-wave, sometimes followed by small additional responses, termed I-waves, was observed (Figure 1). The amplitude and latency were not significantly different from those in other series of patients without lesions in the CST. The duration of the D-wave from the onset of positivity to the end of negativity in these patients, however, appeared to be prolonged.

In four patients who were not capable of walking preoperatively, no detectable D-wave was recorded. In addition to the severity of motor deficits, the level of the caudal pole of the tumour appeared to be related to undetectable D-wave. Thus, three of five cases in which the lowest level of the tumour was at Th6 or below revealed no detectable D-wave, whereas all of six cases in which the lowest level of the tumour was at Th6 or below revealed no detectable D-wave, whereas all of six cases in which the lowest level of the tumour was at C7 or above showed a clearly detectable D-wave. It appears that the lower the site of recording, the more easily does the D-wave become undetectable.

The D-wave in the present cases was often greater than 10 μV in amplitude and was clearly demonstrated with average of 16–32 sweeps. The MEP

Table 1 Intraoperative MEP changes and postoperative motor deficits

Postoperative motor deficits† *(lower extremity)*	*n*	*Intraoperative MEP findings (D-wave)*		
		Improved	*Unchanged**	*Deteriorated*
Improved	5	4 (80%)	1	
Transiently improved	2		2 (100%)	
Unchanged	8		6 (75%)	2 (25%)
Deteriorated	1		1	

†3 months after surgery. *includes cases whose D-wave remained undetectable.

Table 2 Intraoperative MEP changes and tumour types

Histological classification	*n*	*Intraoperative MEP findings*		
		Improved	*Unchanged*	*Deteriorated*
Ependymoma	7	4 (57%)	3 (43%)	
Astrocytoma	4		3 (75%)	1 (25%)
Haemagioblastoma	3		3 (100%)	
Metastatic germinoma	1		1	
Intramedullary neurinoma	1			1

traces were therefore updated as fast as every 15–30 s. Repeated recordings of the corticospinal MEP every 30 s before intramedullary surgical procedures revealed minimal variations in amplitude and latency. There was a semi-real-time stream of information with feedback of the data from the MEP monitoring to the ongoing surgical procedures. Changes in the D-wave were induced in response to certain surgical procedures, such as excessive tension applied to the pial traction suture or haemostatic compression of the cord. The D-wave otherwise displayed minimal variations.

Improvement of the MEP, i.e. an increase in amplitude and decrease in latency or the appearance of a previously undetectable D-wave, was noted in four patients during surgery and was invariably associated with improvement of the motor deficits (Table 1). The increase inûamplitude and decrease in peak latency of the D-wave were often observed in these cases when the dura was opened or the cyst fluid was drained. In the present series, such improvements were encountered more frequently in ependymoma than astrocytoma (Table 2). Unchanged MEPs in six cases were associated with a benign postoperative course. In four cases, the D-wave remained undetectable at the end of surgery, which naturally did not predict the surgical outcome. One of these cases revealed obvious deterioration of motor deficits postoperatively. Moderate but persistent deterioration of the D-wave, i.e. decrease in amplitude or increase in latency of the D-wave, was noted during surgery in two cases. These patients showed no appreciable deterioration of weakness at 3 months postoperatively, presumably because further resection was abandoned.

DISCUSSION

The following characteristics are required for the recorded signal when monitoring the conductivity of CST fibres: (1) the signal should be rapidly and constantly recordable so that feedback of information to the ongoing surgery is readily possible; (2) the intraoperative changes are highly specific to changes in the integrity of the CST; and (3) anaesthesia and other drugs induce minimal influences. The present study confirms that the corticospinal MEP meets these requirements as a monitor of the CST integrity.

Corticospinal MEPs were clearly recorded in all patients who were capable of walking preoperatively. The only abnormality which seemed to be present in these patients was a prolonged duration of the D-wave. This change probably represents a temporal dispersion of the D-wave due to an increased variability in the conduction velocity of the CST fibres. In contrast, the D-wave was never recorded in patients who had been incapable of walking. Thus, complete disappearance of the D-wave during surgery appears to predict severe postoperative motor deficits. It should also be noted, however, that the lower the site of recording, the more easily does the D-wave become undetectable, presumably because of the lower amplitude of the D-wave at lower levels.

The improvement of the D-wave observed in four cases during surgery indicates that the conductivity of the CST in these patients had been suppressed functionally. Favourable effects of surgery may thus be expected when improvements in the corticospinal MEP are observed. Nevertheless, moderate deterioration of the corticospinal MEP does not necessarily indicate a poor outcome, insofar as resection is not continued further. Since the corticospinal MEP is recorded with minimal averages of sweeps, the trace can be updated and communicated to the operating surgeon quickly. This allows the surgeon to identify the manipulation responsible for any observed deterioration.

There is now evidence to indicate that, in most cases of intramedullary spinal cord tumours, better results can be obtained by radical excision of the tumour rather than by limited surgery with radiation therapy. Radical excision is always associated with a risk of inflicting a greater neurological deficit as a result of extensive dissection within the functioning cord tissue. The corticospinal MEP appears to represent an important surgical adjunct for minimizing such a risk.

References

1. Ginsburg HH, Shetter AG, Raudzens PA. Postoperative paraplegia with preserved intraoperative somatosensory evoked potentials. J Neurosurg. 1985;63:296–300.
2. Lesser RP, RaudzensP, Luders H, Newer MR, Goldie WD, Dinner DS, Klem G, Hahn JF, Shetter AG, Ginsburg HH, Gurd AR. Postoperative neurological deficits may occur despite unchanged intraoperative somatosensory evoked potentials. Ann Neurol. 1986;19:22–25.
3. Amassian VE, Stewart M, Quirk GJ, Rosenthal JL. Physiological basis of motor effects of a transient stimulus to cerebral cortex. Neurosurgery. 1987;20:74–93.
4. Tsubokawa T. Clinical significance of the centrifugal evoked potentials to cortical surface or scalp stimulation for evaluation of motor function during surgery. Neurosurgeons (Tokyo). 1985;5:111–122.
5. Katayama T, Tsubokawa T, Yamamoto Y, Maejima S. Spinal cord potentials to direct stimulation of the exposed motor cortex in humans. Comparison with data from transcranial motor cortex stimulation. In: Rossini PM, Marsden CD, eds. Non-invasive stimulation of

brain and spinal cord function. Fundamentals and clinical applications, New York: Alan R Liss, Inc; 1988:305–311.
6. Tsubokawa T. Intraoperative monitoring of corticospinal motor evoked potentials in neurosurgical operations. In: Schramm J, Moller AR, eds. Intraoperative neurophysiologic monitoring in Neurosurgery. Berlin, Germany: Springer; 1991:106–126.
7. Levy WJ, York DH, McCaffrey M, Tanzer F. Motor evoked potentials from transcranial stimulation of the motor cortex in humans. Neurosurgery. 1984;15:214–227.
8. Boyd S, Rothwell JC, Cowan JMA, Webb PJ, Morley T, Asselman P, Marsden CD. A method of monitoring function in corticospinal pathways during scoliosis surgery with a note on motor conduction velocities. J Neurol Neurosurg Psychiatry. 1986;49:251–257.
9. Katayama Y, Tsubokawa T, Maejima S, Hirayama T, Yamamoto T. Corticospinal direct response in humans: Identification of the motor cortex during intracranial surgery under general anesthesia. J Neurol Neurosurg Psychiatry. 1988;51:50–59.

44
Ketamine anaesthesia allows recording of muscle responses to magnetic cortical stimulation: an electrophysiological study in man

K. KOTHBAUER, U. D. SCHMID, S. LIECHTI and K. M. RÖSLER

INTRODUCTION

Transcranial magnetic cortex stimulation (magStim) is a routinely used diagnostic tool in neurology. It provides a non-invasive means to assess motor pathway function[1]. Efforts to apply this method to intraoperative neurophysiological monitoring in neurosurgical procedures compromising the motor pathways have been hindered by the inhibitory effect of anaesthetics on magnetically evoked muscle responses. In order to establish a feasible anaesthesia regimen which allows recording muscle responses to magStim during surgery, various anaesthetic agents have been extensively investigated on their effect on MEPs to magStim. Unfortunately those agents that did not suppress the responses did not produce sufficient anaesthesia for surgery (fentanyl, nitrous oxide). Those that did abolished the responses (propofol, midazolam, isoflurane, pentothal)[2]. Recent studies in monkeys suggested that ketamine may be a more promising choice[3]. In this study MEPs to CortStim were recorded during anaesthesia-induction with ketamine hydrochloride.

METHODS

Patients

Five patients (one female, four males) who were at low risk for anaesthesia (ASA I, American Society of Anesthesiologists physical status[4]) and aged between 31 and 57 years (mean: 40.6 years) took part in this study. Their average body weight was 80.4 kg (range: 75–90 kg). They were scheduled to undergo lumbar disc surgery. The study protocol has been used successfully and without adverse effects in the past, and it is approved by the local ethical committee. Subjects gave written informed consent to participate in the study.

Stimulation and recording

The magnetic stimulator had a total capacitance of 800 μF, and a maximum charging energy of 2500 joules. A stimulating coil composed of 17 concentric circular turns of copper wire with a mean diameter of 8.1 cm and stimulus intensities between 1.8–2.1 T was used. At a maximum output of 2.5 kV, the peak magnetic field was 2.3 Tesla (T) at the centre of the coil, with a rise time of 115 μs and a virtually exponential decay. The coil was placed tangentially over the vertex and the current flow was counter-clockwise (conventionally defined) as seen from behind to preferentially stimulate the pathways to the right-sided target muscle[1]. Coil position, orientation and stimulus intensity were left unchanged throughout the experiment.

Muscle responses (CMAPs) were recorded with surface electrodes from the belly and tendon of the M. abductor digiti minimi (ADM) of the right hand. The grounding electrode was placed at the right wrist. High gain audio monitoring was used to ascertain the target muscle's relaxed state during stimulation. Conventional EMG-equipment (Medelec Sensor 94a) interfaced to a personal computer (Apple II GS, Apple Computers Inc.) and an x–y-plotter (HP 7470, Hewlett-Packard) was used. Filter bandpass was 30–5000 Hz and the amplitude gain 1.5 mV/division.

Anaesthesia procedure

Patients were premedicated with nicomorphine-HCl 7.5 mg, midazolam 2.5 mg and atropine 0.5 mg 45 minutes before induction of anaesthesia. An intravenous infusion of 0.9% NaCl solution was established at the left arm, i.e. contralateral to the site of recording. The experiments were terminated after induction of anaesthesia was completed.

Vital parameters such as electrocardiogram, arterial blood pressure, pulse and respiratory rate were monitored. Patients breathed 100% oxygen spontaneously through a mask unless a respiratory depression or apnoea necessitated artificial mask ventilation to keep O_2 saturation within 97–100% as measured by pulse oximetry.

Depth of anaesthesia was assessed clinically and classified according to guidelines provided by Guedel[5] as: stage I (awake, or drowsy, but responsive); stage II (excitation: unconscious, abolished eyelid reflex, dilated but reactive pupils, eye gaze divergent or oscillating, elevated muscle tone, irregular respiration; or stage III (surgical anaesthesia: unconscious, abolished eyelid reflex, contracted pupils, gaze concentrically fixed, depressed or absent voluntary ventilatory effort). Baseline was established as 'stage 0', i.e. after premedication.

Induction experiments

For induction of anaesthesia, patients were placed in a comfortable supine position on the operating table in a quiet induction room and were draped with warm towels. Approximately 45 minutes after the patient had been premedicated, baseline values were obtained to 12 magStim. Ketamine hydrochloride (Ketalar®, Parke-Davis) was then administered in steps (Figure 1) until the patient reached stage III (i.e. surgical) anaesthesia, or, independent of the stage of anaesthesia, until the maximal dosage for that agent had been given.

The time interval between successive doses was 90 seconds. The stage of anaesthesia was assessed 60 seconds prior to administration of the next dose; four single magnetic magStim were given 30 seconds later.

Analysis of data

Onset latencies and peak to peak amplitudes of the first negative component of the CMAP were measured. The mean and standard deviations of values were calculated for each subject and for an overall calculation of all five subjects. Paired differences between baseline and step-dosage levels of amplitudes and latencies were assessed using the Wilcoxon-test.

RESULTS

Effect of ketamine on consciousness and anaesthesia stage

All five patients were comatose with 1.5 mg/kg BW ketamine and stage III anaesthesia was reached after 2.0 mg/kg BW.

Effect of ketamine on amplitudes and onset latencies of CMAPs

Overall onset latencies of the CMAPs were 22.8 ($\pm$1.6) ms at baseline and 1 mg/kg BW and only slightly longer at 2 and 3 mg/kg BW (23.4 $\pm$ 1.4 ms and 23.5 $\pm$ 1.1 respectively). Amplitudes dropped from 101% ($\pm$29) at 1 mg/kg to 78% ($\pm$26) at 2 mg/kg and 57% ($\pm$39) at 3 mg/kg (baseline = 100% ($\pm$25) after premedication). These differences turned out to be statistically insignificant. Furthermore, the evoked CMAPs were easily reproducible at a constant amplitude gain of 1.5 mV/U with a constant stimulus intensity in each experiment (Figure 1).

Assessment of latencies and amplitudes in correlation to consciousness and anaesthesia stage

Latencies were minimally prolonged in the unconscious patient in relation to the baseline value awake (Figure 2). Amplitudes were diminished to about

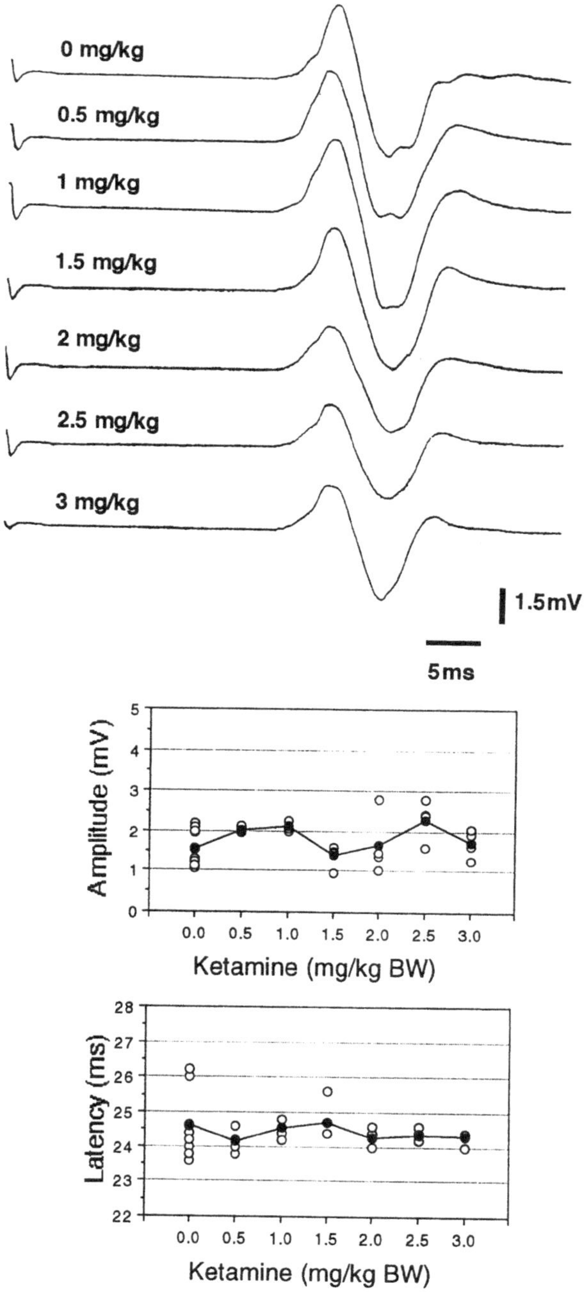

Figure 1 Original recordings of one patient. Each curve represents an average of four single recordings at each step of 0 mg/kg BW–3 mg/kg BW. Tables show mean values of amplitude and latency with twelve recordings at baseline and four recordings at each dosage step

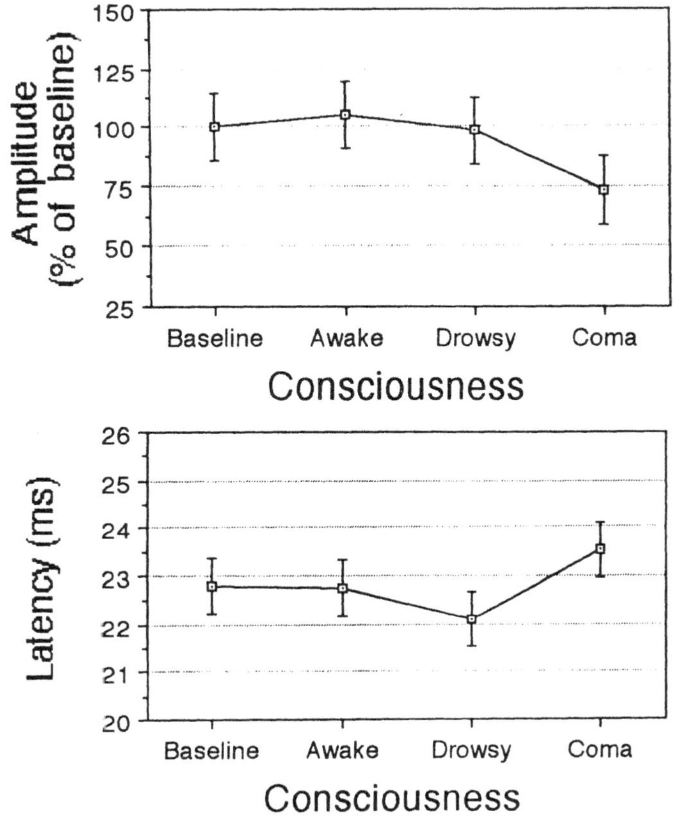

Figure 2 Correlation of the assessed state of consciousness to amplitudes and latencies of recorded muscle responses to magStim

three-fourths of the baseline value. Similar alterations of latencies and amplitudes are found in correlation to anaesthesia stage (Figure 3).

DISCUSSION

As can be seen in several other contributions in this volume somatosensory evoked potential-monitoring is a well-established technique in spinal cord monitoring for orthopaedic and neurosurgical procedures that may pose a risk to spinal cord function. However, with SSEP-monitoring only the sensory pathways are under the scrutiny of the monitoring team and this allows only indirect inference on the state of the motor pathways, i.e. the pyramidal tracts. Therefore, many efforts have been made to develop a means of effective intraoperative electrophysiological monitoring of the motor system. In the last decade electrical and especially magnetic transcranial cortex stimulation have emerged as diagnostic tools in neurology. While electrical stimulation is a painful procedure, transcranial magnetic cortex

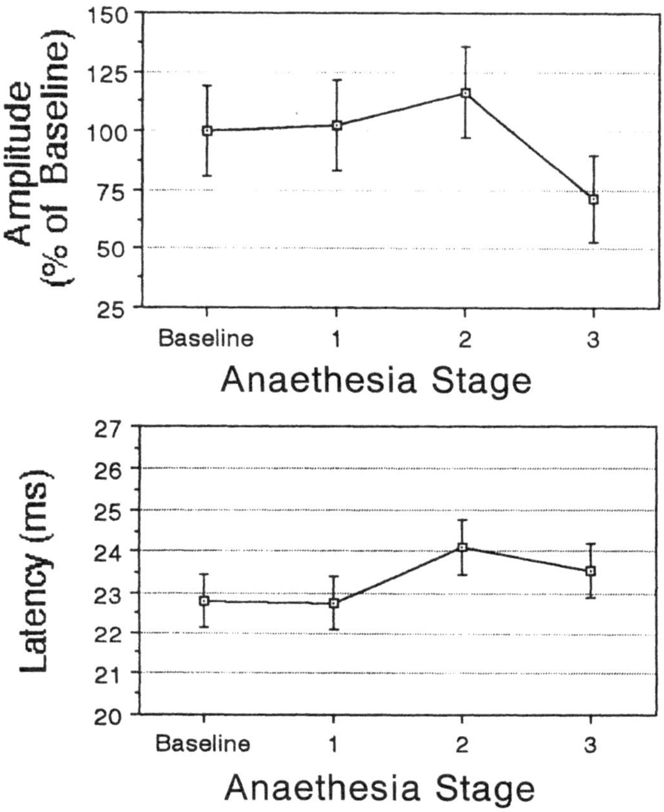

Figure 3 Correlation of anaesthesia stage to amplitudes and latencies of recorded muscle responses to magStim

stimulation is non-invasive and easily applicable. Thus several studies have been undertaken to establish a motor system monitoring with the use of magStim. After promising early reports[6] the application of the technique in a surgical environment has turned out to be more difficult than expected. Several studies show now that those anaesthetic agents such as nitrous oxide or fentanyl that allowed recording of muscle responses to cortex stimulation did not produce sufficient anaesthesia for surgery. Those that did, like midazolam, propofol, barbiturates or halothane, abolished the responses. The aim of this investigation was to contribute to the establishment of a feasible anaesthesia regimen in spinal surgery that will allow motor-system-monitoring with magStim.

Using an already well-defined protocol[2] we recorded muscle responses in the right hypothenar muscles to magStim during anaesthesia induction with ketamine hydrochloride as a single agent. Ketamine is a cyclohexanone that induces what is called 'dissociative' anaesthesia[7]. As a short-acting, potent drug it is frequently used in emergency medicine or intensive care units.[8].

With the administration of 2 mg/kg body weight, ketamine produced surgical anaesthesia in all five patients. Compound muscle action potentials in the right hypothenar were constantly recorded up to a maximum dosage of 3 mg/kg. Amplitudes and latencies changed only minimally. Reliable recordings were possible with constant stimulation energy and a constant recording amplification. These data obtained in humans are in line with previous results of primate studies that also suggested ketamine as a promising agent for anaesthesia in spinal procedures that require motor system monitoring.

The result of this investigation may be an important step towards the definition of a feasible anaesthesia regimen for spinal surgery that requires motor pathway monitoring. Ketamine hydrochloride may be the anaesthetic agent of first choice for these procedures.

Acknowledgements

The authors wish to thank the OR and anaesthesia staff of the neurosurgical clinic, Inselspital, Bern, for their cooperation. Cooperation and support as well as valuable stimulation and discussion came from RW Seiler, Dept. of Neurosurgery, Bern Switzerland; D. Thompson, Dept. of Anaesthesiology, Bern, Switzerland, and RF Ghaly, Cook County Hospital, Chicago, Illinois.

References

1. Hess CW, Mills KR, Murray NMF. Responses in small hand muscles from magnetic stimulation of the human brain. J Physiol (Lond). 1987;338:397–419.
2. Schmid UD, Boll J, Liechti S, Schmid J, Hess CW. Influence of some anesthetic agents on muscle responses to transcranial magnetic cortex stimulation: A pilot study in humans. Neurosurgery. 1992;30:85–92.
3. Ghaly RF, Stone JL, Aldrete JA, Levy WJ. Effects of incremental ketamine hydrochloride doses on motor evoked potentials (MEPs) following transcranial magnetic stimulation: a primate study. J Neurosurg Anesthesiol. 1990;2:79–85.
4. Atkinson RES, Rushman GB, Lee JA, eds. A synopsis of anaesthesia, 2nd edn. Bristol: Wright: 1987; pp 107–118.
5. Guedel AE. Inhalation anaesthesia: a fundamental guide, 2nd edn. New York: Macmillan; 1951;pp.1–43.
6. Edmonds HL, Paloheimo MPJ, Backm MH, Johnson JR, Holt RT, Shields CB. Transcranial magnetic motor evoked potentials for functional monitoring of motor pathways during scoliosis surgery. Spine. 1989;14:683–686.
7. Winters WD, Ferred-Allado T, Guzman-Flores C. The cataleptic state induced by ketamine: a review of the neuropharmacology of anesthesia Neuropharmacology. 1972;11:303–315.
8. Godman-Gilman A et al, eds. Goodman and Gilman's: The pharmacological basis of therapeutics, 8th edn New York; 1990.

45
Long-term and quantitative assessment of spinal cord motor function by motor evoked potentials

B. MEYER, V. ROHDE and J. ZENTNER

INTRODUCTION

As has been shown previously both in animal experiments with spinal cord trauma and ischaemia, as well as in humans, MEP have proven to be a valuable diagnostic tool for detection of lesions along the spinal cord[1–8]. However, reports addressing the significance of MEP for quantitative evaluation of spinal function are still lacking. Therefore, this study was designed to assess whether MEP (1) allow quantitative assessment of the actual motor function and (2) follow the clinical development of motor status. To achieve this goal, we correlated the clinical with the electrophysiological findings in 110 patients who had various lesions along the spinal cord and compared postoperative clinical and electrophysiological follow-up examination of 42 patients.

PATIENTS AND METHODS

A total of 110 patients (68 males, 42 females) with ages ranging from 28 to 83 (mean age 59 years) were included in this study. The lesion was tumorous in 39 and non-tumorous in 71 patients. Localization of disease was cervical in 59, thoracic in 37 and thoracolumbar in 14 cases. In all 110 patients, MEP were elicited by electrical, in 50 of them additionally by magneto-electric, stimulation of the motor cortex, while the peripheral conduction time was determined by electrical stimulation of the lumbar nerve roots.

Electrical stimulation was performed with voltage constant condenser discharges using a Digitimer D 180 which delivers a maximum output of 750 V with a time constant of 50 or 100 μs. For stimulation of the motor leg area the anode was placed over the bregma and the cathode 6 cm behind it in the midline. When the lumbar nerve roots were stimulated, the anode was

Table 1 Classification of clinical findings

Category	*Motor status*
0	Unremarkable
1	Normal gait, elevated muscle tone, sinking of legs
2	Spastic gait, ambulant without support
3	Ambulant with support
4	Able to stand with support
5	Minimal movements, frustrated contractions
6	Complete paralysis

positioned at the intervertebral space of D11–12 and the cathode one segment below, each in the midline. Magneto-electric excitation of the motor leg area was achieved by positioning the coil over Fz with the current running clockwise for activation of the right side and counterclockwise for the left side (Magstim 200, 1.5 T). Electromyographic responses were recorded from the anterior tibial muscle using EMG-electrodes in a belly/tendon fashion. While electrically evoked responses were recorded with the target muscle at rest, magneto-electrically elicited potentials were facilitated by voluntary contraction of the target muscle with about 10% of maximum muscle strength. This was controlled by audio-EMG. Stimulus strength was gradually increased until a clear EMG response was obtained or the absence of any response was documented despite the maximum stimulus strength that could be tolerated by the individual patient. The time base was 100 μs with a gain ranging from 100 μV to 1 mV per division. Filter settings ranged from 20 Hz to 3 kHz. Only single stimuli were applied. The time interval between the stimuli was 3–5 s. At least three potentials were obtained from each recording site. The central motor conduction time (CMCT) was determined by subtraction of the peripheral latency from the total latency (electrophysiological system Compact 4, Nicolet).

We decided upon the following criteria for pathological MEP changes: Absence of any response on one or both sides, CMCT cortex-D12 > 14 ms (electrical stimulation) or > 16.3 ms (magneto-electric stimulation). These limits for acceptable CMCT result from our normative date plus 2.5 SD as obtained from 50 healthy volunteers.

In order to determine whether unilateral clinical accentuation correlates with MEP, we compared side differences in CMCT with the clinical findings. Statistical analysis was performed by calculating the mean side differences in CMCT and standard deviations for patients with and without accentuation of motor deficits using the Student's *t*-test.

For quantitative correlation of electrophysiological and clinical findings, MEP and the results of neurological examination were divided into seven categories, respectively. Category 0 of MEP means normal latencies and category 6 absence of any response. MEP category 1 represents a borderline increase in CMCT of up to 1 ms, while categories 2–4 comprise patients with increase of CMCT in steps of 5 ms. Category 5 of MEP represents patients with a CMCT of more than 30 ms (electrical) and 32 ms (magneto-electric) (Table 1, 2). Correspondingly, the clinical motor status was divided into seven categories as defined by the extent of the motor deficit. Categories

Table 2 Classification of MEP findings

Category	*CMCT ms (Electric)*	*CMCT ms (Magneto-electric)*
0	<14.0	<16.3
1	14.0–15.0	16.3–17.0
2	15.0–20.0	17.0–22.0
3	20.0–25.0	22.0–27.0
4	25.0–30.0	27.0–32.0
5	>30.0	>32.0
6	no potentials	no potentials

mean CMCT (electric) 21.9 ms SD 4.7
mean CMCT (magneto-electric) 22.8 ms SD 4.8

Table 3 Central motor conduction time (CMCT) in patients with unilateral ($N = 43$) and symmetrical ($N = 67$) motor deficits

CMCT (msec) / Stimulation	MOTOR DEFICIT: Unilaterally Accentuated N = 43			Symmetrical N = 67		
	Mean	Range	S. D.	Mean	Range	S. D.
Electrical	4.3	2.6 - 13.0	2.8	2.7	0 - 5.6	2.1
Magnetoelectric	4.2	1.2 - 14.2	2.1	1.4	0.4 - 3.2	1.1

ranged from normal motor status to complete paralysis (Table 1, 2). Based on the categories, MEP were correlated with the clinical motor status. Statistical evaluation was performed by means of a one-way analysis of variance.

The initial clinical and electrophysiological status was defined preoperatively. Short-term follow-up was performed in all patients ($N = 42$) within the first 2 postoperative weeks. Additionally, 9 patients were examined 8–15 months postoperatively.

RESULTS

Clinically, both sides were equally affected in 67 patients while motor deficits were unilaterally pronounced in 43 cases. In patients with symmetrical motor deficits, differences in CMCT right/left in response to electrical stimulation ranged from 0 to 5.6 ms (mean 2.7 ms, SD 2.1 ms). Corresponding values for magneto-electric stimulation were 0.4 and 3.2 ms, respectively (mean 1.4 ms, SD 1.1 ms). In all of the 43 patients with unilaterally pronounced motor deficits, MEP changes were more severe on the clinically more affected side. With electrical stimulation, differences in CMCT ranged from 2.6 to 13.0 ms (mean 4.3 ms, SD 2.8 ms), and from 1.2 ms to 14.2 ms (mean 4.2 ms, SD 2.1 ms) with magneto-electric stimulation. These results were highly significant (Student's *t*-test; alpha = 0.05) (Table 3). Statistical evaluation of our clinical

Table 4 Correlation of clinical (C0–C6) and MEP (M0–M6) categories in 110 patients. By means of one-way analysis of variance no statistically significant correlation could be found ($F = 0.42$)

	M 0	M 1	M 2	M 3	M 4	M 5	M 6	Σ
C 0	14	2	5	3	1			25
C 1		3	3	4	2	1		13
C 2		1	3	5	3	1		13
C 3		1	3	4	6			14
C 4		2	3	3	2	2		12
C 5		1	4	2	2	2	4	15
C 6							18	18
Σ	14	10	21	21	16	6	22	

and electrophysiological results, each divided into seven categories, by means of a one-way analysis of variance shows that there was no correlation between the extent of the clinical motor deficit and the prolongation of CMCT (one-way analysis of variance; $F = 0.42$). Only the absence of electromyographic responses usually correlated with complete paralysis. These results were valid both for electrical and magneto-electric stimulation. Within the first two postoperative weeks the motor status in $N = 19$ patients (45.2%) was unchanged compared to preoperative, in $N = 16$ cases (38.1%) better and in $N = 7$ cases (16.7%) worse. MEP however were unchanged in $N = 35$ cases (80.3%). Electrophysiological improvement was never observed. In contrast all seven patients (16.7%), who deteriorated clinically showed corresponding deterioration of MEP. A long-term follow-up of nine patients (for 8–15 months postoperatively) showed in three cases (33.3%) an improved, deteriorated and unchanged clinical motor status respectively. Six patients (66.7%) – all clinically unchanged or improved – had unchanged MEP findings.

Improvement in MEP was never observed. Three patients (33.3%) showed a parallel clinical and electrophysiological deterioration.

DISCUSSION

Comparing prolongation of the CMCT right/left we found differences of up to 5.6 ms (electrical stimulation) and 3.2 ms (magneto-electric stimulation) in patients without clinical evidence of unilaterally accentuated deficits. However, statistical analysis of side differences of CMCT in patients with and without unilaterally accentuated motor deficits was highly significant (alpha = 0.05; Student's *t*-test). Thus, changes in MEP are usually more severe on the clinically more affected side. The main question however, was whether MEP allow quantification of the clinical motor deficit. Our results suggest that this is not the case. We did not observe significant correlation between the extent of the clinical deficit and the prolongation of CMCT as documented by statistical analysis. Only complete paralysis usually correlated with the absence of electromyelographic responses. On the other hand, absence of MEP does not necessarily indicate complete paralysis in every case, since this was also observed in patients showing residual motor function. Here we found the only difference between results obtained with the electrical and magneto-electric stimulation techniques. With electrical stimulation, MEP were overwhelmingly pathological due to prolonged CMCT, while responses were absent in only a few patients. With magneto-electric stimulation, the rate of absent responses was much higher.

We also found that in no case was an improvement of motor status accompanied by a corresponding electrophysiological (MEP) development. In all these – improved – cases MEP were unchanged. In contrast each clinical deterioration went parallel with electrophysiological worsening.

In conclusion, examination of the descending pathways by means of MEP is a useful test for electrophysiological assessment of lesions along the spinal cord. Pathological MEP indicate damage even in cases when clinical examination produces ambiguous results. Moreover, MEP may indicate the clinically more affected side. However, MEP fail to provide quantification of the clinical motor deficit, since there is no linear correlation between the clinical motor status and the electrophysiological changes. In our opinion, changes in MEP may be related more closely to the type of spinal cord damage, e.g. demyelination vs. axonotmesis, and the duration of the disease than to the actual clinical motor function. However, further studies are necessary to define more clearly the essential factors influencing latencies of motor responses and to provide evidence as to whether additional evaluation of other variables, e.g. configuration and/or amplitudes of the potentials, could improve the correlation between clinical and electrophysiological findings.

Furthermore we conclude that only a deteriorating clinical motor status can be assessed reliably by motor evoked potentials, whereas this is not the case with the clinical improvement. This limits the usefulness of MEP as an objective tool for monitoring the development of spinal cord motor function.

References

1. Abbruzzese G, Dall'Agata D, Morena M, Simonetti S, Spadavechia L, Severi P, Andriolo GC, Favale E. Electrical stimulation of the motor tracts in cervical spondylosis. J Neurol Neurosurg Psychiat. 1988;51:796–802.
2. Cusick JF. Monitoring of cervical spondylotic myelopathy. Spine. 1987;13:877–880.
3. Fehlings MG, Tator CH, Linden RD, Piper IR. Motor evoked potentials recorded from normal and spinal cord injured rats. Neurosurgery. 1987;20:125–130.
4. Hess CW, Mills KR, Murray NMF, Schriefer TN. Magnetic brain stimulation: central motor conduction studies in multiple sclerosis. Ann Neurol. 1987;2:744–752.
5. Patil AA, Nagaray MP, Mehta R. Cortically evoked motor action potentials in spinal cord injury research. Neurosurgery. 1985;16:473–476.

6. Rossini PM, Caramia MD, Zarola F. Mechanisms of nervous propagation along central motor pathways: noninvasive evaluation in healthy subjects and in patients with neurological disease. Neurosurgery. 1987;20:183–191.
7. Simpson RK, Baskin DS. Corticomotor evoked potentials in acute and chronic blunt spinal cord injury in the rat: Correlation with neurological outcome and histological damage. Neurosurgery. 1987;20:131–137.
8. Tsubokawa T, Katayama Y, Maejima S, Hirayama T, Yamamoto T. Assessment of spinal cord injury with multimodality evoked spinal cord potentials. Part 2: Correlation with neurological outcome in clinical spinal cord injury. Neuro-Orthopedics. 1987;3:82–89.

46
Correlation of clinical, radiological and electrophysiological (MEP) findings in cervical myelopathy

B. MEYER and J. ZENTNER

INTRODUCTION

As has been shown previously both in animal experiments as well as in humans, motor evoked potentials (MEP) have proven to be a sensitive electrophysiological tool for detecting lesions along the spinal cord even in the subclinical condition[1–6]. As there is still some controversy regarding the indication for surgery in cervical spondylotic disease, this study was designed to clarify whether MEP can provide additional information especially in cases in which clinical and radiological findings are ambiguous. To achieve this goal, we correlated MEP with radiological and clinical results in 62 patients with suspected or proven cervical myelopathy.

PATIENTS AND METHODS

A total of 62 patients (41 males, 21 females) aged from 27 to 73 years (mean age 57 years) were included in this study. All were admitted for evaluation of cervical spine syndrome and examined clinically with respect to local-cervical, radicular and myelopathic symptoms. Radiological findings were categorized as follows: 1. Evidence of myelomalacia encompassed cases with intramedullar hyperintensity on T2-weighted MRI or intramedullar contrast depot on delayed myelo-CT. 2. Evidence of spinal cord impairment was indicated by oval deformation of the cord on myelo-CT or complete stop of the contrast medium on myelography. 3. Borderline signs of cord impairment included incomplete stop of contrast medium on myelography or consumption of ventral subarachnoid space on myelo-CT[7,8]. Based on these criteria the patients were categorized into three groups: Group I encompassed patients with clinical and radiological evidence of cord. Patients in group II had radiological but no clinical evidence of cord impairment. Group III consisted of patients without clinical myelopathic signs and with no or only borderline radiological cord impairment.

In all 62 patients MEP were elicited by electrical and magneto-electric stimulation of the motor leg area. The peripheral conduction time was determined by electrical stimulation of the lumbar nerve roots. Electrical stimulation was performed with voltage constant condenser discharges using a Digitimer D 180 which delivers a maximum output of 750 V with a time constant of 50 or 100 μs. For central stimulation the anode was placed over the bregma and the cathode 6 cm behind it on the midline. The lumbar nerve roots were stimulated by positioning the anode at the intervertebral space of D 11–12 and the cathode one segment below, each on the midline. Magneto-electric excitation of the motor leg area was achieved by positioning the coil over Fz with the current running clockwise for activation of the right side and counter-clockwise for the left side. (Magstim 200, 1.5 T). Electromyographic responses were recorded from the anterior tibial muscle using EMG-electrodes in a belly/tendon fashion. While electrically evoked responses were recorded with the target muscle at rest, magneto-electrically elicited potentials were facilitated by voluntary contraction of the target muscle with about 10% of the maximum muscle strength. This was controlled by audio-EMG. Stimulus strength was gradually increased until a clear EMG response was obtained, or the absence of any response was documented despite a maximum stimulus strength as tolerated by the individual patient. The time base was 100 ms with a gain ranging from 100 μV to 1 mV per division. The bandwidth was 20 Hz to 3 kHz. Only single stimuli were applied with an inter-stimulus interval of 3 to 5 s. At least three potentials were recorded at each site. The central motor conduction time (CMCT) was determined by subtraction of the peripheral latency from the total latency (electrophysiological system Compact 4, Nicolet). We decided upon the following criteria for pathological MEP changes: Absence of any response on one or both sides, CMCT cortex-D12 > 14 ms (electrical stimulation) or > 16.3 ms (magneto-electric stimulation). These limits for acceptable CMCT result from our normative data (including 2.5 S.D.) obtained from 50 healthy volunteers. Based on these criteria, MEP were correlated with the clinical and radiological findings in our 62 patients.

RESULTS

Thirty-two patients belonged to group I. Thirteen patients had sensory and 19 motor signs of cervical myelopathy. Fifteen of them showed radiological evidence of myelomalacia and 17 of cord impairment. Eleven of the 18 patients of group II had radicular and seven local symptoms. Radiological findings indicated myelomalacia in three patients and cord impairment in the remaining 15 cases. In group III consisting of 12 patients, five had radicular and seven had local symptoms. Six of them showed either borderline cord impairment or were inconspicuous radiologically (Table 1).

Comparing MEP with the three different radiological categories regardless of clinical findings, we obtained the following results: In category I with

Table 1 Results of MEP in the three groups of patients as categorized according to clinical and radiological findings

Patients group	*Clinical examination*	*Radiological findings*	*Pathological MEP*
I $N = 32$	Myelopathy sensory $N = 13$ motor $N = 19$	Myelopathy myelomalacia $N = 15$ myelon impairment $N = 17$	100%
II $N = 18$	Local/radicular symptoms local $N = 7$ radicular $N = 11$	Myelopathy myelomalacia $N = 3$ myelon impairment $N = 15$	100%
III $N = 12$	Local/radicular symptoms local $N = 7$ radicular $N = 5$	Borderline/inconspicuous borderline $N = 6$ inconspicuous $N = 6$	 50% 25%

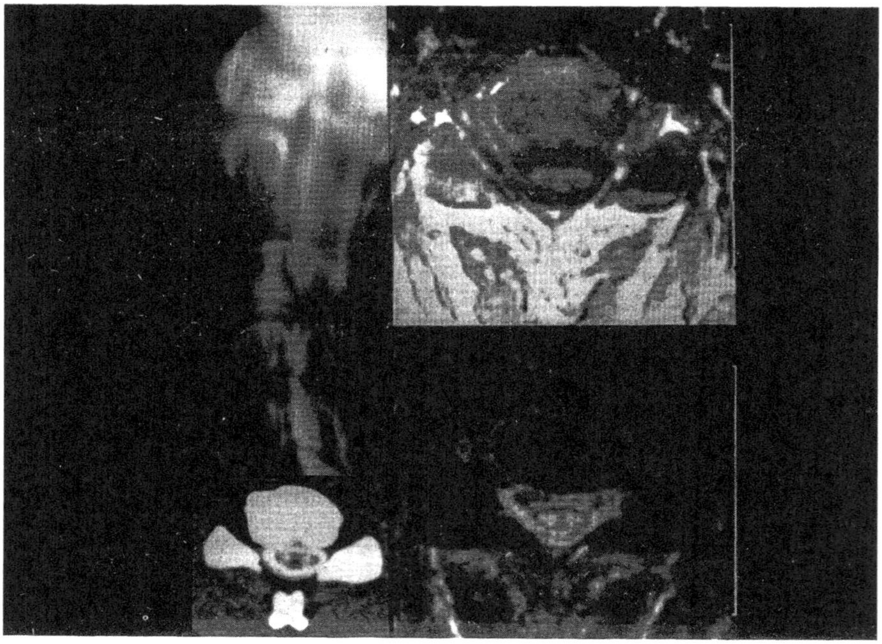

Figure 1 A 62-year-old patient with radiological evidence of myelomalacia. Conventional X-rays and CT show a hard disc at C3–4. Myelography demonstrates incomplete stop of the contrast medium at C3–4. Myelo-CT and MRI show intramedullar contrast depots and intramedullar hyperintensities, respectively. MEP were pathological in every case of this category

evidence of myelomalacia ($N = 18$) we observed pathological MEP in all cases. This is also true for category II with evidence of cord impairment ($N = 32$). In category III ($N = 6$) with borderline radiological signs MEP were abnormal in 50% of the patients. Figures 1–3 show typical examples.

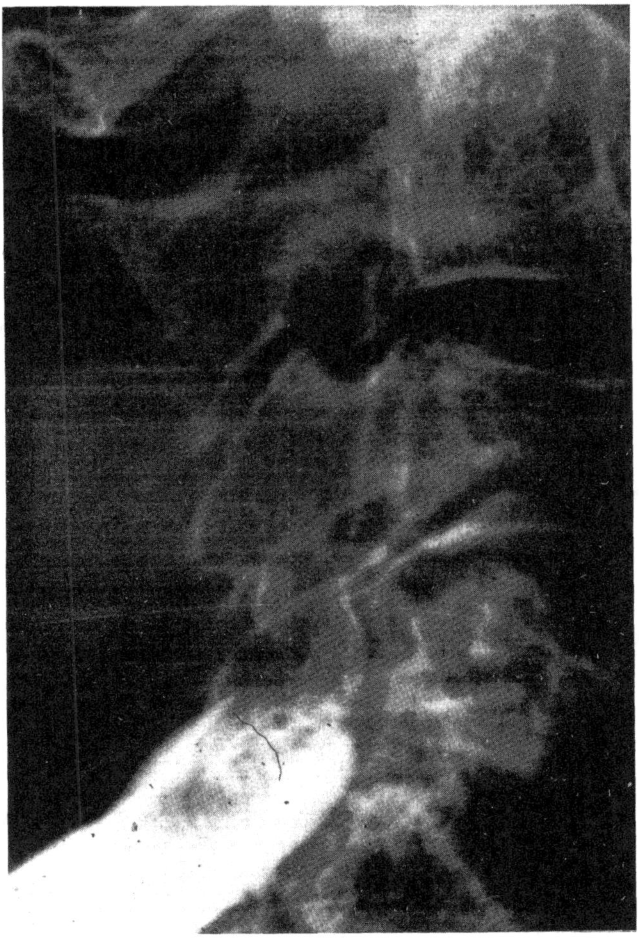

Figure 2 Radiological evidence of myelon impairment as demonstrated by complete stop of the contrast medium on myelography at C4–5. MEP were pathological in every case of this category

DISCUSSION

Cervical spondylosis is a frequent disease in elderly patients. Consensus exists about the indication for surgery in patients demonstrating myelopathy or severe radicular deficits when the lesion has been confirmed by radiological studies. However, there is still controversy about the indication for surgery in patients with only minor radicular and/or local symptoms, but with gait disturbance which, however, cannot be unequivocally related to myelopathy and may also result from other disorders. Therefore, it was the aim of our work to find out whether the electrophysiological assessment of the descending pathways by means of MEP might provide additional information with

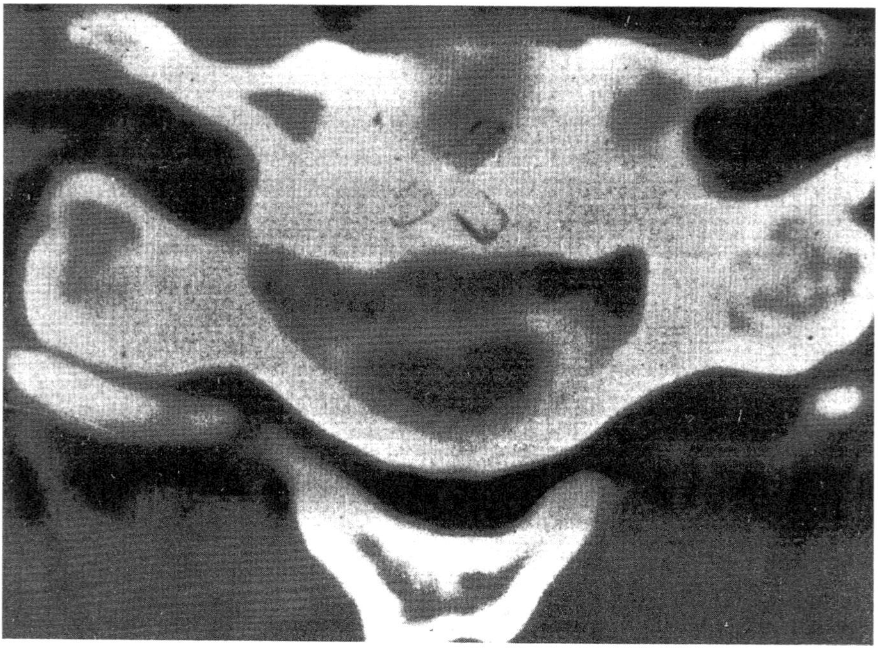

Figure 3 Borderline signs of myelon impairment with consumption of the ventral subarachnoid space on myelo-CT. There were pathological MEP in 50% of the patients belonging to this category

respect to the indication for surgical treatment in these cases. To achieve this goal, MEP findings were correlated with the clinical and radiological results of our 62 patients.

Our results clearly show that clinically evident cervical spondylotic myelopathy is always accompanied by pathological changes in MEP. Thus, MEP are sensitive for detection of clinically evident spinal cord lesions. In our experience, MEP are also able to indicate subclinical lesions of the descending pathways[6]. Therefore, MEP can provide useful information in patients in whom clinical examination is not conclusive for myelopathy and may support the indication for surgery in these cases. This is especially true when there is radiological evidence of spinal cord involvement (myelomalacia or cord impairment). All of these patients in our series had electrophysiological signs of myelopathy. In cases where results of clinical examination are ambiguous and where radiologic studies show only borderline cord impairment, follow-up examinations with MEP may provide useful information, allowing neurological deterioration to be detected at an early stage. Patients could then be referred for surgical treatment at an opportune time.

In conclusion, it is not our intention to suggest that motor evoked potential studies are necessary in every patient with cervical spondylotic disease, or that MEP findings alone may justify surgical treatment. However, we believe that MEP may be a helpful addition to clinical and radiological studies in making decisions about surgical or conservative management.

References

1. Abbruzzese G, Dall'Agata D, Morena M, Simonetti S, Spadavechia L, Severi P, Andriolo GC, Favale E. Electrical stimulation of the motor tracts in cervical spondylosis. J Neurol Neurosurg Psychiat. 1988;51:796–802.
2. Cushick JF. Monitoring of cervical spondylotic myelopathy. Spine. 1987;13:877–880.
3. Fehlings MG, Tator CH, Linden RD, Piper IR. Motor evoked potentials recorded from normal and spinal cord injured rats. Neurosurgery. 1987;20:125–130.
4. Levy WJ. Clinical experience with motor and cerebellar evoked potential monitoring. Neurosurgery. 1987;20:169–182.
5. Patil AA, Nagaray MP, Mehta R. Cortically evoked motor action potentials in spinal cord injury research. Neurosurgery. 1985;16:473–476.
6. Zentner J, Rieder G. Diagnostic significance of motor evoked potentials in space-occupying lesions of the brainstem and spinal cord. Eur Arch Psychiatr Neurol Sci. 1990;239:285–289.
7. Al-Mefty O, Harkey LH, Middleton TH, Smith RR, Fox JL. Myelopathic spondylotic lesions demonstrated by magnetic resonance imaging. J Neurosurg. 1988;68:217–222.
8. Larsson EM, Holtas S, Cronquist S, Brandt L. Comparison of myelography, CT myelography and magnetic resonance imaging in cervical spondylosis and disc herniation. Pre- and postoperative findings. Acta Radiol. 1989;30:233–239.

47

Recording of myogenic motor evoked potentials (mMEP) under general anaesthesia using a train of transcranial electric stimuli

U. PECHSTEIN, M. TANIGUCHI, J. NADSTAWEK and J. SCHRAMM

INTRODUCTION

Recording of motor evoked potentials is of interest for the neurosurgeon during surgery in the vicinity of the central region, in and around the brain stem and in surgery of the spinal cord.

Previous authors showed that myogenic motor evoked potentials after single pulse transcranial electric stimulation are very susceptible to general anaesthesia[1,2]. Only extraordinary strong stimulating intensities allow recording of mMEP under sufficient anaesthetic levels. Zentner's group used a constant voltage stimulation, which is associated with motion artefacts, so that this type of stimulation cannot be used during microsurgical procedures.

Two approaches to this problem are possible: by changing the method of anaesthesia or by changing the method of stimulation. We focused our interest on the latter one.

High frequency direct cortical electric stimulation was shown to allow recording of myogenic motor evoked potentials under general anaesthesia[3]. This study aims to show that it is also possible to record myogenic motor evoked potentials under general anaesthesia by applying repetitive high frequency electric stimuli to the scalp over the motor cortex using a stronger than usual but conventional type of stimulation.

PATIENTS AND METHODS

Fourteen patients were examined with repetitive high frequency transcranial electric stimulation. Nine of them received balanced anaesthesia and five total intravenous anaesthesia. In three patients we were not successful.

Stimulating electrodes were placed on Cz (according to the 10–20 system)

Table 1 Stimulating and recording parameters for mMEP. Range of tested parameters in brackets

Stimulation:	
Pulses per stimulus	(1)–5–(10)
Duration	(200)–400–(800) μs
Frequency	(200)–500 Hz
Intensity	0–100 mA
Stimulating sites	Cz (anode) 5 cm lateral to Cz (cathode)
Recording:	
Sensitivity	100 μV + 2 mV
Low bandpass	30–150 Hz
High bandpass	3 kHz
Recording sites	Thenar muscles Forearm flexor muscles Anterior tibial muscle

for the cathode and 5 cm lateral to Cz for the anode, being the active electrode. We used ECG electrodes for stimulation and fixed them with tape on the scalp.

In each patient we stimulated only the neurologically unaffected side. Our stimulating device is a constant current stimulator (Tönnies ES, Jaeger GmbH, Würzburg, Germany), which is controlled by a personal computer. Its maximum output is 100 mA. Our stimulating parameters are listed in Table 1.

Myogenic motor evoked potentials were recorded from thenar, forearm flexor muscles or anterior tibial muscle of the contralateral side in belly tendon fashion with needle electrodes. We use a Nicolet Pathfinder I (Nicolet Biomedical Instruments, Madison, Wisc., USA) for recording. The recording parameters are listed in Table 1.

We recorded myogenic motor evoked potentials under two types of general anaesthesia: the so-called Balanced Anaesthesia (BA) is the method usually employed in our institution. It consists of nitrous oxide and isoflurane or enflurane as a volatile anaesthetic. Analgesia is maintained by fentanyl.

We use Total Intravenous Anaesthesia (TIVA) during surgery where intraoperative monitoring of SEP is employed. It consists of propofol and alfentanil for analgesia.

Usually Norcuronium is given for relaxation if required. As it is given as a bolus, the level of relaxation is changing during surgery. Using Tracrium, which has to be applied by constant perfusion, it is possible to maintain relaxation at a constant level.

RESULTS

We started testing with our standard parameters (5 pulses, 500 Hz, 400 μs) and assessed threshold stimulation intensity. Then we tested the effect of different stimulating parameters on shape, amplitude and latency of mMEP.

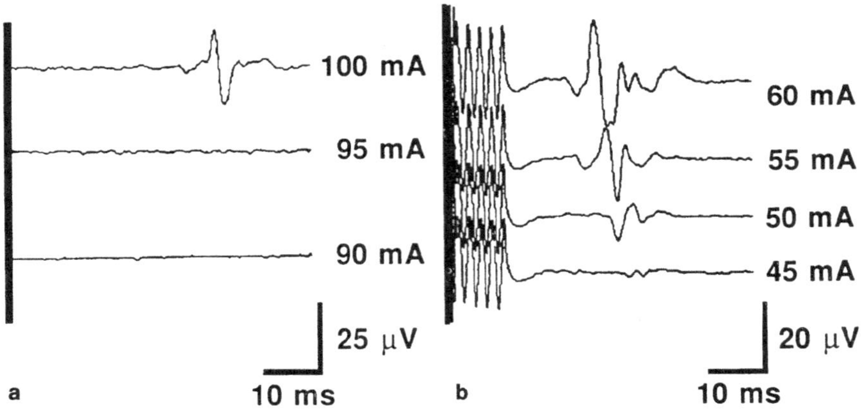

Figure 1 Different threshold intensities with different anaesthetic techniques. Records from the thenar muscles of two patients with different types of general anaesthesia. a. This patient received BA with nitrous oxide and isoflurane. MMEP after transcranial stimulation were only available with 100 mA. b. This patient received TIVA with propofol and alfentanil. A small triphasic potential was obtainable with 45 mA. In Figure 1a low bandpass was set to 150 Hz, so that no stimulation artefact could be obtained. The stimulating parameters for both records were: 5 pulses per stimuation, frequency 500 Hz, duration 400 μs

With our standard stimulating parameters threshold stimulation intensity for thenar varied between 45 to 65 mA under TIVA and 85 to 100 mA under BA (Figure 1).

Stimulation became effective with at least three pulses per stimulation. This stimulation resulted in a small biphasic potential. Stimulation with more pulses resulted in a more polyphasic potential of higher amplitude (Figure 2).

The most effective stimulating frequency varied in different individuals between 400 and 500 Hz. A decrease in frequency did not allow recording of mMEP. Our stimulating device did not permit stimulation at more than 500 Hz.

Alternating the duration of each pulse showed that the most effective stimulation was that with 400 and 600 μs. Stimulation with 200 and 800 μs led to a potential with lower amplitude.

We were also successful in recording mMEP from the anterior tibial muscle. We placed the stimulating electrodes on Cz for the anode and 5 cm behind Cz for the cathode. We used six pulses with 100 mA and a frequency of 500 Hz. With these parameters we recorded a small but clearly reproducible mMEP from the anterior tibial muscle (Figure 3).

We found differences between BA and TIVA; as mentioned above, with BA we needed a higher stimulating intensity to record myogenic motor evoked potentials than with total intravenous anaesthesia.

In one of our very first patients in this study we made an interesting observation in a patient suffering from an instability of the dens axis without any neurological deficit. The first attempt to stabilize the fracture failed, so

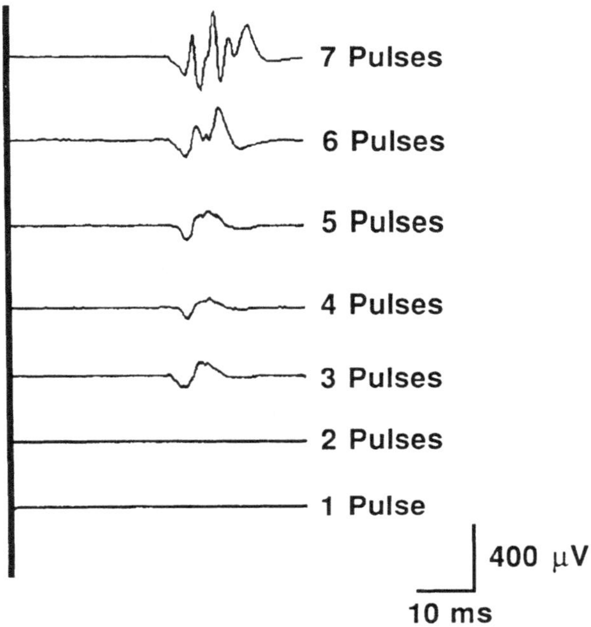

Figure 2 Effect of different numbers of pulses per stimulation in mMEP from the thenar muscles. Stimulation becomes effective with three pulses. The stimulating parameters are: intensity 75 mA, frequency 500 Hz, duration 400 μs. General anaesthesia was maintained with propofol (TIVA). Train-of-four showed two responses after four test stimuli

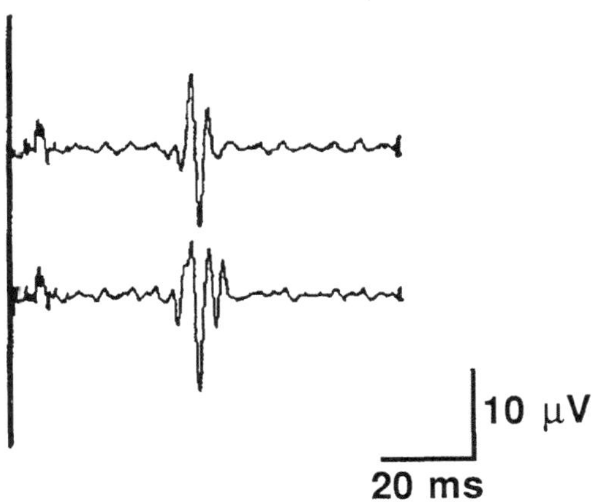

Figure 3 Records from the anterior tibial muscle. Stimulating electrodes were on Cz for the anode and 5 cm behind for the cathode. This patient received BA. Six pulses of 100 mA per stimulation were necessary to result in these potentials. Frequency 500 Hz, duration 400 μs

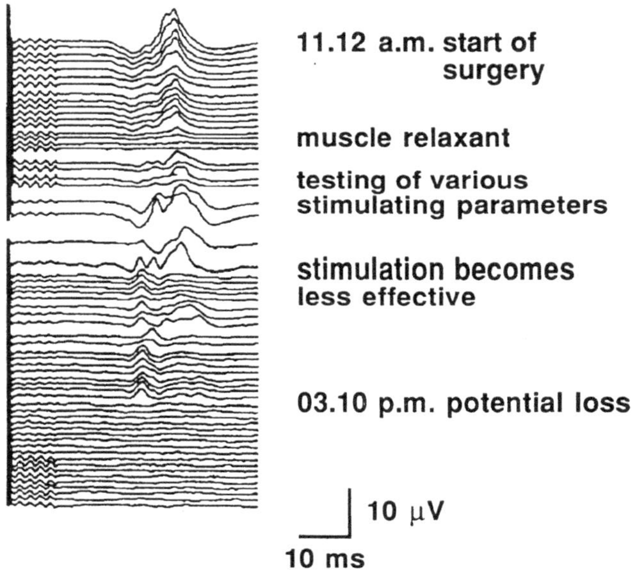

Figure 4 A 19-year-old male with instability C1/C2 without neurological deficit. Records from right forearm flexor muscles during anterior and posterior fusion of C1/C2. Loss of mMEP after 4 hours of BA without any postoperative neurological deficit

he underwent reoperation with anterior and posterior fusion of C1 and C2. During this operation BA was employed. At the beginning of surgery we recorded good potentials from the forearm flexor muscles. Application of muscle relaxants made the potentials vanish. After they recovered, we tested several stimulating parameters. After approximately four hours of anaesthesia the rate of effective stimulations decreased. Finally we lost our potentials. The patient awoke from anaesthesia without any neurological deficit. This phenomenon could be obtained in all our patients with BA who underwent surgery for such a long time. The reason for this is still unclear (Figure 4).

Myogenic motor evoked potentials after repetitive transcranial electric stimulation do not require averaging. Due to their large amplitude, they can be observed on the oscilloscope, so that real-time monitoring is possible. However they vary in amplitude and shape very clearly. Averaging of five to ten recordings leads to quite constant results, which markedly improves interpretation during intraoperative monitoring.

The results vary considerably in amplitude, which is due to variations of the site of the stimulating electrodes. The most appropriate positions led to high amplitudes of 200 to 400 mV.

Variation of amplitude or loss of mMEP can be due to muscle relaxants. In some cases we employed a very simple method to monitor the level of relaxation. We stimulated the median nerve at the wrist with 20 mA and a frequency of 0.4 Hz. Thus a change of amplitude due to relaxation could be clearly evaluated. In most cases the degree of muscle relaxation was controlled by the 'train-of-four' technique, and was maintained at two responses following four test stimuli.

DISCUSSION

There are various methods for recording motor evoked potentials. Stimulation can be delivered either transcranially or directly to the cortex. Magnetic and electrical stimulation are possible.

Recording sites can be either muscles, peripheral nerves or the spinal cord itself. We focused our interest on myogenic motor evoked potentials, i.e. compound muscle action potentials.

A new method for recording mMEP has to meet some general requirements. The stability of evoked potentials under general anaesthesia is very important. We want them to be inter- and intraindividually stable, i.e. the method has to be feasible in every patient and throughout the whole operation.

Motion artefacts must not occur, so that microsurgical procedures are not disturbed.

Intraoperative awareness has to be avoided under general anaesthesia. Recovery from anaesthesia should not be delayed. We want the patients to wake up as soon as possible after surgery, so that long stays on the ICU are not necessary.

Different types of anaesthesia may be used including general anaesthesia with either nitrous oxide and volatile anaesthetics or with intravenous anaesthetics. Local anaesthesia does not require any sophisticated method to assess myogenic motor evoked potentials[4,5], but is not employed in our clinic for brain and spinal surgery. We use two types of general anaesthesia in our clinic: BA and TIVA.

Myogenic motor evoked potentials after single pulse transcranial electric stimulation are known to be suppressed by general anaesthesia[1]. General anaesthesia seems to block motor conduction on two levels: at cortical level, where the generation of I-waves is strongly influenced and at spinal motoneuronal level, where not enough motoneurons are elicited[6].

Two possible techniques to solve this problem were worked out in our laboratory: The first is facilitation of mMEP by an additional peripheral stimulus (H-Reflex)[6]. This method allows us to record mMEP under general anaesthesia, but is impractical as a tool for intraoperative monitoring.

The second solution is repetitive high frequency stimulation of the cortico-motoneuronal tract. With this technique there seems to be a synaptical cumulation of EPSPs which bring the motoneuron to the firing status. Recording of mMEP after direct cortical stimulation was shown to be possible with this method under general anaesthesia[3]. In this study we tried to show another application of this method in the operating room. Repetitive high frequency transcranial electric stimulation for myogenic motor evoked potentials seems to be a promising method for intraoperative monitoring in surgery of the brain stem and the spinal cord.

A third solution is a type of anaesthesia that allows single pulse stimuli without motion artefacts. Some groups claim success with such a technique.

The advantages of transcranial evoked mMEP in contrast to other methods for motor evoked potentials are:

1. The sites of stimulation and recording are outside the field of surgery. Therefore the surgeon is not involved in the recording and surgery itself is not disturbed.
2. The stimulus is not focused on a small region. The dispersed stimulus seems to elicit a large number of neurons of the motor cortex. Thus signal conduction through a large part of the pyramidal tract can be monitored.

Our preliminary study brought us to two conclusions:

1. The best method in our hands to evoke mMEP under general anaesthesia is repetitive high frequency stimulation.
2. TIVA with propofol seems to be more appropriate than BA during prolonged surgery.

Further studies have to be employed to learn more about the characteristics of this new technique for recording myogenic motor evoked potentials. The degree of relaxation where mMEP are recordable has to be worked out. More detailed information about the influence of different anaesthetics has to be assessed.

References

1. Zentner J, Ebner A. Nitrous oxide suppresses the electromyographic response evoked by electrical stimulation of the motor cortex. Neurosurgery. 1989;24:60–62.
2. Zentner J, Kiss I, Ebner A. Influence of anesthetics – nitrous oxide in particular – on electromyographic response evoked by transcranial electrical stimulation of the cortex. Neurosurgery. 1989;24:253–256.
3. Taniguchi M, Cedzich C, Schramm J. A new electro-physiological examination technique of motor systems under general anaesthesia: Technical description. Neurosurgery. 1992 (submitted).
4. Penfield W, Holdrey E. Somatic motor and sensory representation in the cerebral cortex of man as studied by electrical stimulation. Brain. 1939;60:389–443.
5. Berger MS, Cohen WA, Ojemann GA. Correlation of motor cortex brain mapping data with magnetic resonance imaging. J Neurosurg. 1990;72:383–387.
6. Taniguchi M, Schramm J. Motor evoked potentials facilitated by an additional peripheral nerve stimulation. In: Levy WJ, Cracco RQ, Barker AT, Rothwell J, eds. Magnetic motor stimuation: Basic principles and clinical experiment (EEG Suppl. 43). Elsevier Science Publishers, 1991:202–211.

48
Electromyography of paravertebral muscles for evaluation of spinal disorders

S. SATO, K. IBARAKI, H. TAKARA, F. KANAYA and S. MORI

INTRODUCTION

The spinal nerves divide into two branches, which are the dorsal and ventral ramus. The former innervates the paravertebral muscles (PVMs), and the latter innervates the muscles of the extremities. There are two anatomical characteristics of PVMs, which are useful for the electromyographic diagnosis. (1) PVMs are segmentally innervated by a few adjacent dorsal rami. Therefore, the level of a spinal lesion could be determined by electromyography (EMG) of PVMs. (2) PVMs would be affected earlier than the muscles of the extremities, because PVMs have shorter motor branches from the spine than the muscles of the extremities. So early diagnosis of the spinal disorder could be done by EMG of PVMs[1].

One purpose of this study is to examine what EMG findings in PVMs are useful for clinical application. The other purpose is to demonstrate the usefulness of EMG in spinal cord lesions, which are hard to localize clinically or to demonstrate by imaging methods.

PATIENTS AND METHODS

Patients

Ten cases (five men, five women), 27–74 years of age, were examined. All patients suffered from myelopathy or weakness of the upper extremities caused by spinal disorder.

EMG methods

The skin was marked at even intervals before EMG. Metal wires were placed on the skin, and an A-P X-ray was taken to determine the spine level. The needle electrode was inserted 2 or 3 cm lateral to the spinous process on each mark. The tip of the electrode was inserted in the lamina, then withdrawn about 5 mm to examine the rotators and multifidi. The patient was asked to extend his neck or back while recording.

Evaluating methods of EMG findings (Study I)

EMG of PVMs was taken on four patients (three cases with cervical spondylotic myelopathy, one case with extramedullary spinal cord tumour), whose affected spine level was already determined by imaging methods, that is MR-imaging, myelography or computed tomography. EMG findings were compared with the degree of compression of the spinal cord.

Methods of clinical application

Study II

EMG of PVMs was applied to four patients with spinal disorders, whose lesions could not be completely evaluated by the imaging methods.

Study III

EMG findings of PVMs and the extremities were compared in two cases with intrinsic atrophy of the hand in order to differentiate incipient motor neuron disease (MND) from cervical spondylotic myelopathy.

RESULTS

Study I

High amplitude potentials (HAPs) were seen at the level of spinal cord compression or adjacent levels (Figure 1). Interference and polyphasic potential showed poor reproducibility. Fibrillation was less valuable, because pain during EMG prevented the relaxation of PVMs.

EMG of the upper extremities was also taken. The neurogenic pattern of the upper extremities was recorded from the same spinal segments as HAPs of PVMs. It was found that HAP was a valuable indicator.

Study II

EMG of PVMs proved to be helpful in evaluating the lesions of the spinal cord in four cases as follow:

Case I

A 55-year-old female with cervical myelopathy. Mild anterior compression on T1-weighted MR image and slightly high intensity area on T2-weighted MR image were seen. Gadolinium-contrast MR image (Gd-MR image) showed

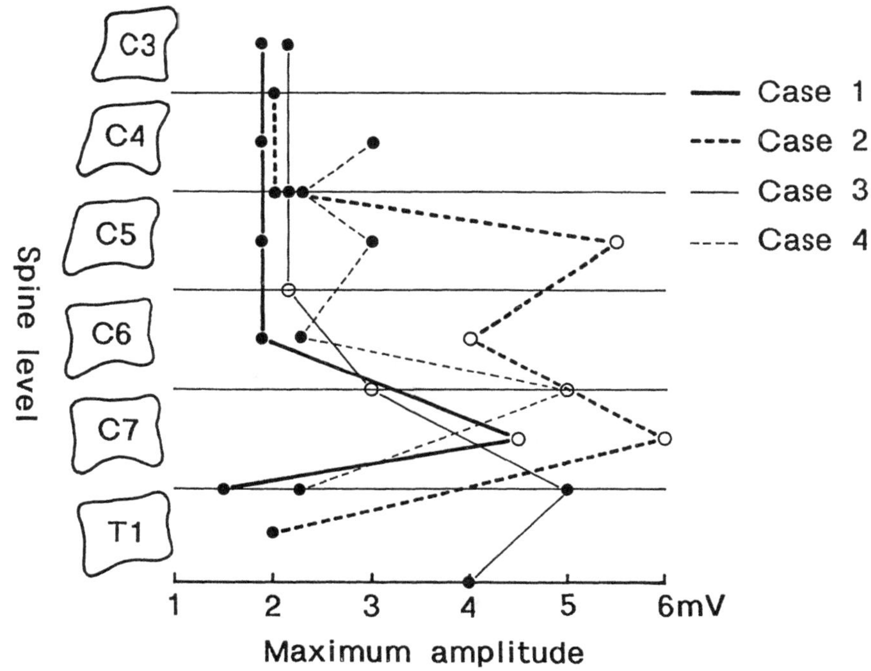

Figure 1 Maximum amplitudes on EMG of PVMs recorded in the different levels of the cervical spine. Open circle represents the compression site and closed circle represents the uninvolved site in images. In all cases HAPs were strongly correlated with the compression sites

intramedullary enhancement (Figure 2A). This finding of Gd-MR image was atypical with cervical spondylotic myelopathy. The lesion was considered as intramedullary change secondary to compression.

EMG of PVMs demonstrated that the lesion on Gd-MR image was responsible for the myelopathy (Figure 2B). An anterior decompression and fusion at C6–7 was performed and her symptoms disappeared.

Case 2

A 60-year-old female with tetraplegia. She underwent decompressive surgery with a diagnosis of cervical spondylotic myelopathy four years ago, but the tetraplegia has persisted. Gd-MR image showed intramedullary enhancement at the level of C5–6.

EMG of PVMs showed HAPs at the level of C4–5 to C6–7. This EMG finding proved that the enhanced area was functionally affected. A biopsy of the lesion revealed gliosis, which was thought to be irreversible.

Case 3

A 55-year-old male with the acute onset paraplegia. Skin blisters were found on the right back eight days after the onset (Figure 3A). The spine level of

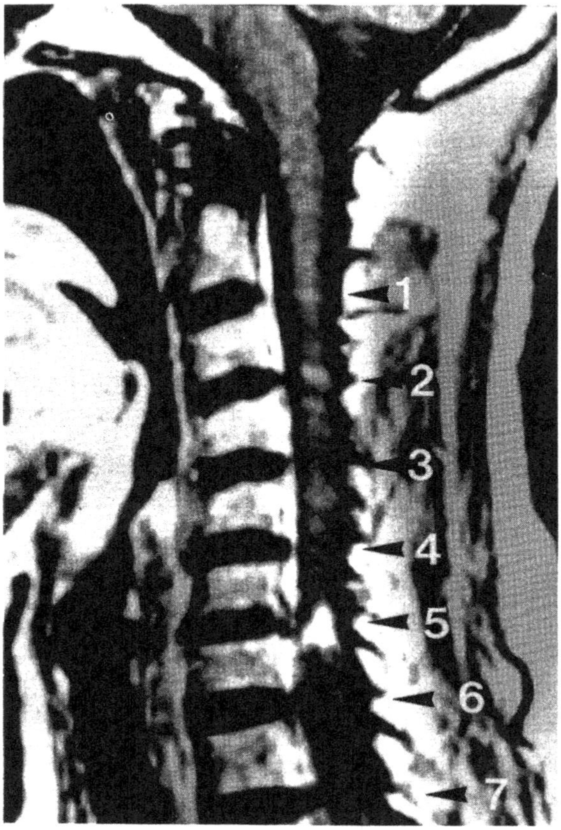

A

Figure 2 A 55-year-old female with cervical myelopathy. A: Intramedullary enhancement on Gd-MR image. B: HAPs were shown on EMG of PVMs recorded at the level of No. 5. This finding suggested that the enhanced area of the spinal cord was responsible for the myelopathy

HAPs was consistent with the dermatome of the skin blisters (Figure 3B). These findings strongly suggested that the myelopathy was caused by herpes zoster virus.

Case 4

A 56-year-old male suffering from left drop foot. Not only compression fracture of Th12 due to osteoporosis resulting from liver cirrhosis, but also lumbar spinal canal stenosis below L3 were listed as the lesions by the imaging methods. However, the responsible lesion for the left drop foot could not be localized.

EMG of the left PVMs revealed HAP of 5 mV only at the level of Th10, whereas the amplitude at other levels was less than 2 mV. These findings indicated Th12 fracture as the responsible lesion. An anterior decompression and fusion between Th11 and L1 resulted in immediate recovery of muscle power.

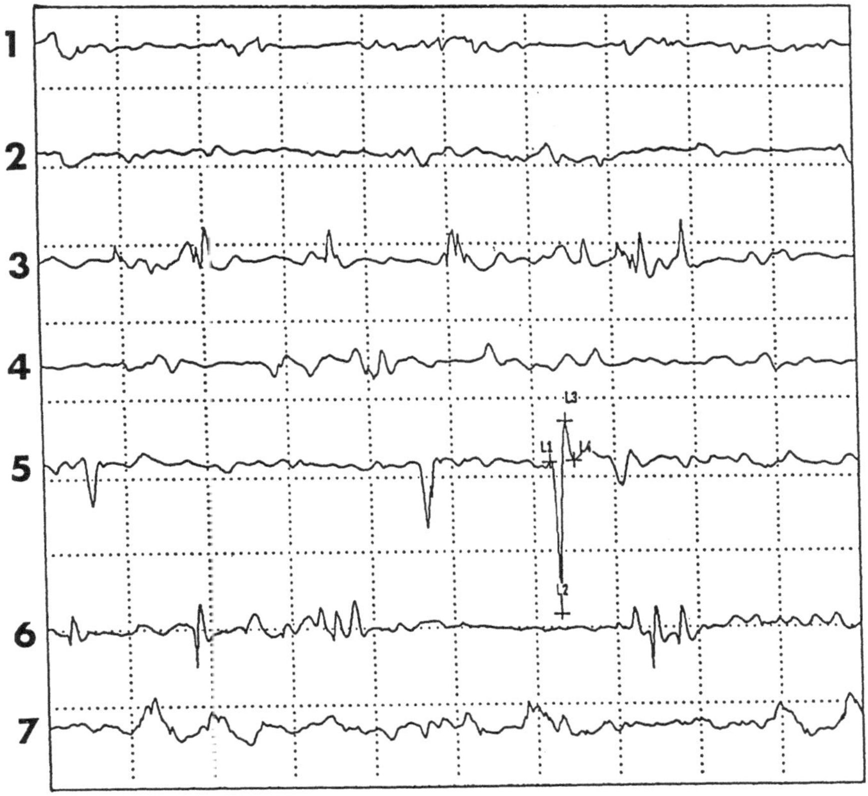

NO	AMPLIFIER SENS (V/DIV)	HI-F (Hz)	LO-F (Hz)	ACQUISITION ANLY (ms)
1	1m	5K	10	100
2	1m	5K	10	100
3	1m	5K	10	100
4	1m	5K	10	100
5	2m	5K	10	100
6	2m	5K	10	100
7	2m	5K	10	100

B

Figure 2 *Continued*

Study III

A 57-year-old male with intrinsic atrophy of the left hand and 56-year-old male with intrinsic atrophy of both hands.

In EMGs of these patients PVMs showed no HAPs at the spine level corresponding to the intrinsic muscles, whereas the muscles of the extremities revealed HAPs. These findings were different from cervical spondylotic myelopathy, which showed abnormal findings on both PVMs and the muscles of the extremities.

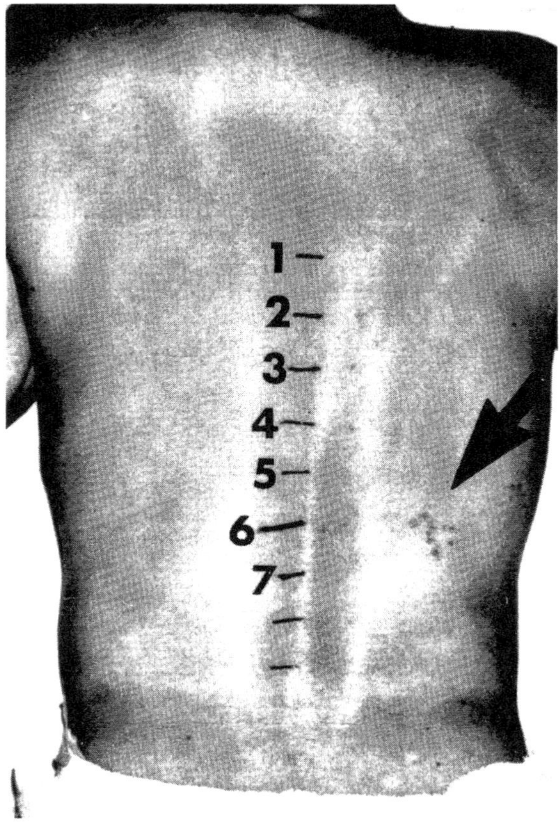

A

Figure 3 A 55-year-old male with the acute onset paraplegia. A: There were skin blisters of herpes zoster (arrow) on the right back. B: HAPs were recorded at the level of No. 2 in the right PVMs. It became clear that the lesion of the spinal cord was consistent with the dermatome of the skin blisters, and this condition was firmly diagnosed as herpetic myelopathy

These conditions were finally diagnosed as MND after the atrophy spread to other extremities.

DISCUSSION

EMG technique

For evaluating the affected spine level, it is important that the needle electrode should be inserted into the rotators or multifidi. When the electrode is inserted superficial to these muscles, EMG of the erector spinae is recorded. The posterior myotomes extend lower than the corresponding dermatomes and anterior myotomes. In the cervical and lumbar regions the posterior myotomes extend as many as six segments lower[2]. Therefore, a correct

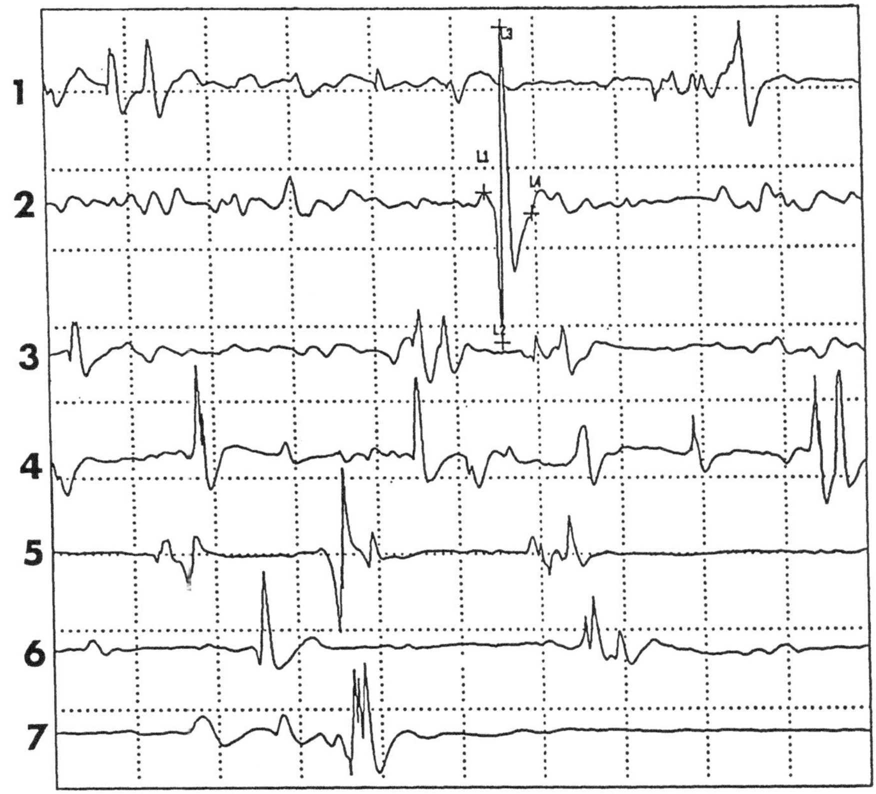

NO	AMPLIFIER SENS (V/DIV)	HI-F (Hz)	LO-F (Hz)	ACQUISITION ANLY (ms)
1	2m	5K	10	100
2	2m	5K	10	100
3	1m	5K	10	100
4	2m	5K	10	100
5	1m	5K	10	100
6	1m	5K	10	100
7	1m	5K	10	100

B

Figure 3 *Continued*

diagnosis of the affected level could not be made by EMG of the erector spinae.

Amplitude and duration of HAP on EMG of PVMs

The amplitude of normal motor unit potentials in the rotators or multifidi seems to be smaller than the amplitude of normal motor unit potentials in the extremities[3]. Since control subjects were not examined, no upper limit of normal amplitude could be established[4,5]. The potentials twice or three times higher than the potentials recorded at other sites were considered as HAPs, which ranged from 4 to 6 mV.

Similarly, motor unit potentials of PVMs are of short duration[2]. It may be the reason why the duration of HAPs of PVMs, which was about 6 ms in our cases, was shorter than the general definition of HAPs.

Study II

Intramedullary enhancement shown on Gd-MR images in Case 1 and Case 2 are rare, and the pathology of the intramedullary conditions is not known. EMG of PVMs could prove that these conditions caused the myelopathy.

In Case 3 the imaging methods showed no abnormal findings. Skin blisters were a clue to the diagnosis and EMG of PVMs revealed that the spinal cord lesion was located at the level corresponding to the blisters.

It is often difficult to localize the lesion, when the imaging methods show multiple compression like Case 4. The responsible lesion could be localized by EMG of PVMs.

Study III

MND occasionally shows intrinsic atrophy only in the hands. Patients with cervical spondylotic myelopathy may have intrinsic atrophy of the hand as the main finding. Therefore, in the incipient MND case with cervical spondylotic changes on X-ray films, it becomes very difficult to differentiate MND from cervical spondylotic myelopathy.

There was a discrepancy between EMG of PVMs and EMG of the extremities in our two cases with incipient MND. It could be assumed that only some of the anterior horn cells were selectively involved in these patients (Figure 4A). However, this discrepancy was not observed in cervical spondylotic myelopathy patients. The affected spinal segments on EMG of PVMs were consistent with those on EMG of the upper extremities in cervical spondylotic myelopathy patients in Study I, because the anterior horn cells of cervical spondylotic myelopathy patients were segmentally compressed (Figure 4B).

Therefore, an EMG pattern of 'PVMs-normal and extremities muscles-abnormal' strongly suggests incipient MND. It was thought that in consideration of these findings besides the neurological findings and data of other examinations one could make a more precise differentiation between the incipient MND and cervical spondylotic myelopathy with intrinsic atrophy alone.

CONCLUSION

1. High amplitude potentials were most useful among the EMG findings of PVMs for the clinical application to diagnosis of the spinal cord disorders.

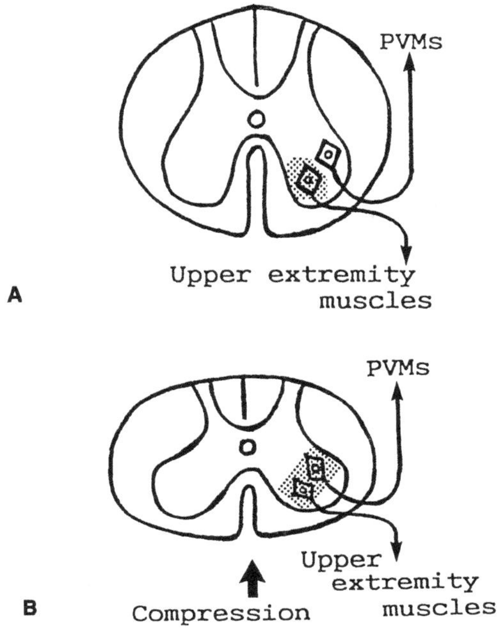

Figure 4 Schematic drawings of the involvement of the anterior horn cells in the incipient MND (A) and cervical spondylotic myelopathy (B). The shaded area represents the affected site. In a part of the incipient MND patients the anterior horn cells innervating the upper extremity muscles may be selectively involved, and the anterior horn cells innervating PVMs may be intact. However, in cervical spondylotic myelopathy patients both the anterior horn cells to the upper extremity muscles and the anterior horn cells to PVMs are segmentally compressed

2. Electromyography of the paravertebral muscles was helpful, not only in determining whether the lesion diagnosed by the imaging methods was responsible for myelopathy or not, but also in localizing the level of the spinal lesion.
3. Electromyography of the paravertebral muscles will help to differentiate the incipient motor neuron disease from cervical spondylotic myelopathy.

References

1. Johnson EW, Melvin JL. The value of electromyography in the management of lumbar radiculopathy. Arch Phys Med Rehabil. 1969;50:720.
2. Gough JG, Koepke GH. Electromyographic determination of motor root levels in erector spinae muscles. Arch Phys Med Rehabil. 1966;47:9–11.
3. Kimura J. Examination of non-limb muscles. Electrodiagnosis in diseases of nerve and muscle: Principles and practice. Kimura J, eds. Philadelphia: Davis; 1983:299.
4. Isch F. Les applications de l'électromyographie en neurologie. Rev Méd (Paris). 1962;3:1.
5. Richardson AT. A standard technique for clinical electrodiagnosis. Ann Phys Med. 1952;1:88.

49 Differential effect of isoflurane on motor evoked potentials elicited by transcortical electric or magnetic stimulation

T. SLOAN and D. ANGELL

INTRODUCTION

Intraoperative monitoring of spinal cord function has been developed in an attempt to reduce neural morbidity during surgery. In an attempt to monitor more specifically the motor aspects of the cord, motor evoked potentials have been developed. Motor evoked potentials can be produced by stimulation of the motor cortex by an electrical impulse applied to the scalp (EMEP) or via an intense, brief, magnetic field applied over the scalp (MMEP). Both techniques induce an electric current in the motor cortex that causes activation of the motor neurons and initiates a descending cascade of electrical signals that results in peripheral muscle activity. The evoked responses can be recorded as nerve action potentials in the spinal cord or peripheral motor nerves or as electromyographic (EMG) signals from muscle contractions.

The application of these two technologies to the monitoring of patients undergoing surgery is not fully understood due to a variety of factors that have yet to be fully characterized. Limited experience suggests that monitoring is feasible with EMEP[1] and MMEP[2] but a systematic study has not been conducted characterizing the many potentially confounding variables that may hinder the EMEP or MMEP interpretation in the operating room. These variables may account for the lack of successful monitoring in some series[2]. At an even simpler level, EMEP and MMEP have not been tested side-by-side to determine which technique has the higher probability of being useful during surgery since anaesthetic agents are known to alter the response. This study was designed to evaluate the effect of isoflurane on EMEP and MMEP to determine if differential effects occur.

METHODS AND MATERIALS

Eight adult female cynomologous (*Macaca fascicularis*) monkeys (weight 3.2–5.5 kg) were tested in this institutionally approved study. Food was withheld for six hours prior to receiving ketamine (15 $mg\,kg^{-1}$) with atropine (0.02 $mg\,kg^{-1}$)[1] intramuscularly in a squeeze cage. An intravenous line was then placed in a leg vein and ketamine (10–15 $mg\,kg^{-1}\,h^{-1}$) in balanced salt solution was infused at 2–4 $ml\,kg^{-1}\,h^{-1}$ sufficient to prevent reactivity to the endotracheal tube or movement to cutaneous stimulation. Lidocaine (approximately 0.2 ml of 2% solution) was placed on the vocal cords using direct laryngoscopy and the trachea was intubated approximately two minutes later using a 2.5–3.5 mm (inside diameter) endotracheal tube. After assuring proper tube placement, the tube was secured and the lungs were ventilated with 35% oxygen in air using a mechanical ventilator (Harvard Respirator 665, South Natick, MA). A tidal volume of 10 ml/kg was used with a ventilatory rate sufficient to produce an end tidal CO_2 ($ETCO_2$) of 35–40 mm Hg. The animal was placed in the right lateral decubitus position on a padded table with the head elevated on a soft pad and the left arm resting in a padded support to minimize movement artefacts. Blankets and hot water warming pads (Aquamatic K module, Bellville, OH) were used to maintain an oesophageal temperature of 36.5–37°C (Monotherm 6500, Precision Biomedical, Plano, TX). Electrocardiogram (Grass Instrument Co., Model 7D Polygraph, Quincy, MA), blood pressure (Omega 1400 NIBP, Tulsa, OK), oxygen saturation (Ohmeda Biox 3700, Boulder, CO), and end-tidal carbon dioxide (Instrumentation Laboratories IL 200, Lexington, MA) were continuously monitored during the study.

Magnetic cortical motor evoked potential (MMEP) stimulation was accomplished using a Cadwell MES 10 magnetic stimulator (Cadwell Laboratories, Kennewich, WA) using a tipped, 9 cm diameter coil placed over the animal's head so that the tipped edge of the coil was over the right parietal region of motor cortex appropriate for the left hand. The initial stimulation intensity used was 80%, which was in excess of the threshold intensity for stimulation of these animals (40–50%). Several MMEP stimuli were recorded at slightly different coil locations during each testing session to avoid position-related alterations in the data. These stimulations were separated by at least 60 seconds. The EMG responses to cortical stimulation were acquired using subdermal needle electrodes (Type E-2, Grass Instrument Co, Quincy, MA) placed in the left hand of the animal. The active electrode was placed subdermally over the thenar muscle group, the reference electrode laterally near the proximal interphalangeal joint and the ground electrode over the lateral epicondyle of the left humerus. Evoked response recording was done using an amplification of 500 (MMEP) with a 60 Hz notch filter, bandpass filtration of 1–3000 Hz and an analysis window of 50 milliseconds (ms) using a Biologic Navigator (Biologic Systems Incorporated, Mundelin, IL). The onset was measured as the time from cortical stimulation to the beginning of the EMG response and the amplitude was measured as the voltage from the most negative to the most positive component of the response. After baseline data were acquired, stimulation intensity was decreased by 5% increments until the response could no longer be produced. The lowest effective level of stimulation was recorded as threshold. If threshold was within 10% of the initial level of stimulation, additional recordings were taken as high as twice the threshold to be used as baselines. Stimulation never exceeded 100% of the stimulation output; if stimulation could not be obtained at this level then the animal was considered not to respond at this concentration of isoflurane.

Electrical cortical motor evoked potential (EMEP) stimulation was accomplished using a Digitimer D180A (Digitimer, Inc. Welwyn Garden, England) set at 50 microsecond pulse width. Using subdermal needles (Type E2, Grass Instrument Co, Quincy, MA) the cathode was placed midway between the vertex (Cz) and the left tragus and the anode was placed anterior to C_4. This configuration produced the shortest onset, maximal amplitude response in the thenar muscles of these animals. Acquisition electrodes and parameters were identical to MMEP. Several responses were initially conducted at twice threshold (45%) and onset and amplitude determined similar to the MMEP. Threshold was then determined in a similar fashion using 5% increments.

Approximately one hour after beginning the ketamine infusion, the initial recordings of EMEP and MMEP were made (no isoflurane). Isoflurane 0.26% was then introduced into the ventilatory circuit. After a twenty minute equilibration period the evoked responses were reacquired. Isoflurane concentration was increased in a stepwise fashion and evoked responses acquired after each re-equilibration. The process was continued until EMEP and MMEP had been tested at isoflurane concentrations of 0.75%. Isoflurane concentrations were calculated by mass spectrometry (Perkins Elmer, Pomona, CA). Following this, isoflurane and ketamine were discontinued and the animals allowed to recover. Data were analysed by later evaluation of the recorded electrical responses. All evoked responses were measured by an observer blinded to the isoflurane values.

RESULTS

EMEP and MMEP were easily acquired and all monkeys tolerated the testing protocol well. No monkeys clinically exhibited any evidence of neurological or behavioural disorders. No seizure activity was witnessed during the study. Figure 1 shows examples of the EMEP and MMEP acquired in the study.

Figure 2 shows a plot of the number of animals which responded at each of the tested levels of isoflurane. Animals consistently lost the MMEP at lower concentrations than the EMEP. If ED_{50} is defined as the concentration where 50% of the animals lost the response, the ED_{50} (MMEP) is 0.32% and that for EMEP is 0.42%.

Figure 3 shows the plot of average onset (milliseconds) versus isoflurane concentration. Figure 4 shows the plot of maximal amplitude (millivolts of the maximal response acquired at each isoflurane concentration.

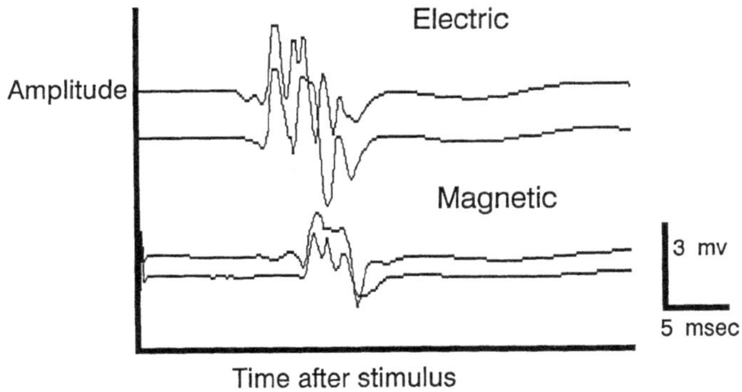

Figure 1 Shown are examples of the EMG responses from transcranial electrical and from transcranial magnetic stimulation. Shown to the right of the tracings are an indicator of amplitude in millivolts (vertical) and time post stimulation (horizontal axis)

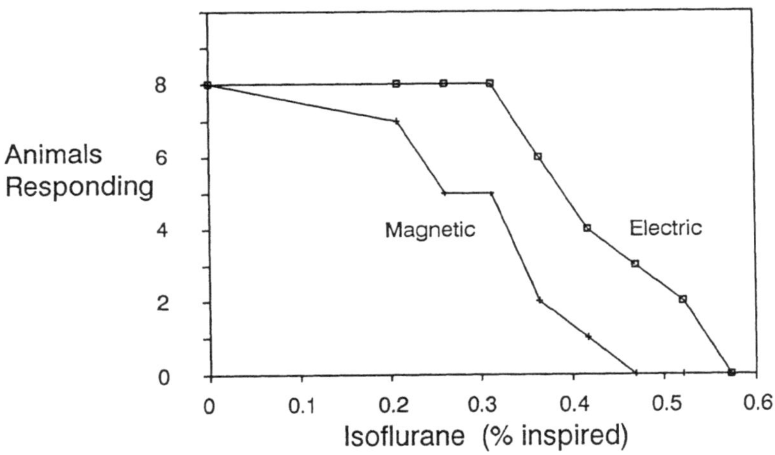

Figure 2 Shown is a plot of the number of animals in which a motor evoked response could be elicited by transcortical stimulation using the electrical or magnetic technique. The tested concentrations of isoflurane are shown along the horizontal axis

DISCUSSION

This study demonstrates that both EMEP and MMEP are sensitive to anaesthesia with isoflurane. In addition, MMEP appears to be more sensitive than EMEP as measured by stimulation threshold and the number of animals responding at any given concentration. This difference may well be the result of synaptic depression by isoflurane inhibiting the initial activation of the motor cortex by the magnetic technique but not the electric technique.

This differential sensitivity of the EMEP and MMEP responses has not been reported but is consistent with the known differences in neuroanatomic structure. A large amount of work has been done to refine the anatomy of the pathways of EMEP and MMEP in unanaesthetized animals and man.

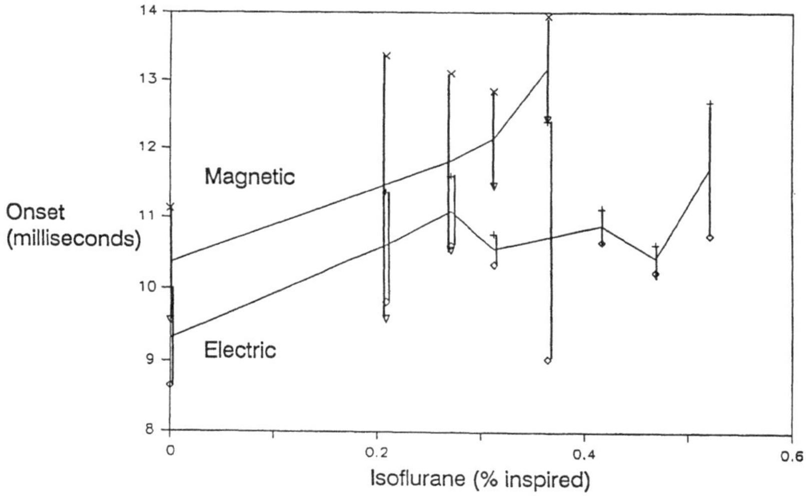

Figure 3 Shown are the mean ($\pm$ SEM) for onset (milliseconds) for the electrical and magnetic techniques at the tested concentrations of isoflurane

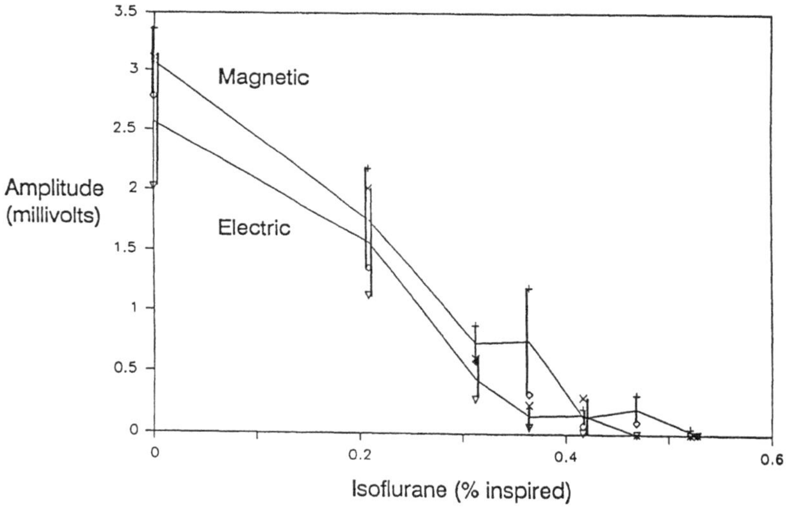

Figure 4 Shown are the mean ($\pm$SEM) for maximal motor evoked potential amplitudes for the electrical and magnetic techniques at the tested concentrations of isoflurane

These two stimulation techniques appear to cause a descending motor response by activation of the cortical motor neurons by different methods[3–6]. EMEP appears to produce sufficient intracortical current to cause direct cellular depolarization of axons of pyramidal cells and production of a D wave that travels the motor pathways to the anterior horn of the spinal cord where a peripheral motor response can be produced. In contrast, MMEP appears to cause depolarization of the cortical motor neurons by inducing a trans-synaptic current that results in weaker I waves. Since several of these I waves must summate in the anterior horn, the evoked muscle response of the MMEP is slightly delayed from that of the EMEP (i.e. slightly longer onset for MMEP). This difference between the techniques is the method of inducing the response (EMEP is via motor neurons and MMEP is via trans-synaptic activation) and the type of waves travelling down the spinal pathways.

In this study the changes in latency and amplitude were similar by both stimulation techniques (except in the range of isoflurane where the magnetic technique was inhibited and the electric technique remained). This suggests that the effect of isoflurane on onset and amplitude is by depressing the motor evoked responses in a common pathway such as depression of axonal conduction, anterior horn cell or neuromuscular junction inhibition.

The sensitivity of motor evoked potentials to inhalational agents has been shown in previous work. In general concentrations of isoflurane which produce little to no effect on sensory evoked responses produce marked alterations in motor evoked responses. The major effect of isoflurane appears to occur at low concentrations (0.2–0.5%)[7,8] similar to that seen in this study.

This study suggests the transcranial electric stimulation technique produces motor evoked EMG responses which are more resistant to the depressant effects of isoflurane than similar studies from transcortical magnetic stimulation. Further studies being conducted explore the effect of other anaesthetic agents.

Acknowledgement

This project was supported by a grant from the Morrison Trust, San Antonio, Texas.

References

1. Levy WJ. Clinical experience with motor and cerebellar evoked potential monitoring. Neurosurgery. 1987;20:169–182.
2. Shields CB, Paloheimo MPJ, Backman MH et al. Intraoperative use of transcranial magnetic motor evoked potentials. In: Chokroverty S, ed. Magnetic stimulation in clinical neurophysiology. Boston, MA: Butterworths; 1990:173–184.
3. Rossini PM, Caramia MD, Zarola F. Mechanisms of nervous propagation along central motor pathways: noninvasive evaluation in healthy subjects and in patients with neurological disease. Neurosurgery. 1987;20:183–191.
4. Thompson PD, Rothwell JC, Day BL et al. Mechanisms of electrical and magnetic stimulation of human motor cortex. In: Chokroverty S, ed. Magnetic stimulation in clinical neurophysiology. Boston, MA: Butterworlths; 1990:121–144.
5. York DH. Review of descending motor pathways involved with transcranial stimulation. Neurosurgery. 1987;20:70–73.
6. Amassian VE, Stewart M, Quirk GJ et al. Physiological basis of motor effects of a transient stimulus to cerebral cortex. Neurosurgery. 1987;20:74–93.
7. Haghighi SS, Green KD, Oro JJ et al. Depressive effect of isoflurane anesthesia on motor evoked potentials. Neurosurgery. 1990;26:993–998.
8. Haghighi SS, Madsen R, Green DK, Oro JJ, Kracke GR. Suppression of motor evoked potentials by inhalation anesthetics. J Neurosurg Anesth. 1990;2:73–78.

50
Temporal summation and motor evoked potential spinal cord monitoring in man

B. A. TAYLOR, M. E. FENNELLY, A. TAYLOR and J. FARRELL

INTRODUCTION

The spinal somatosensory evoked potential (SSEP) has been used as the routine method of intraoperative spinal cord monitoring during scoliosis surgery since described in 1982[1]. It has a good track record, with a negligible false-negative rate in over one thousand cases reported from The Royal National Orthopaedic Hospital[2]. There have, however, been sporadic reports of post-operative paraplegia despite an apparently unchanged intraoperative SSEP tracing, and this has prompted a desire for direct monitoring of motor pathways. This functional requirement is often confused with 'anterior cord monitoring' and the assumption made that one modality (MEP) will represent compromise of anterior cord blood supply and the other posterior supply (SSEP). A better reason for attempting to develop a motor evoked potential (MEP) intraoperative monitoring method is suggested by recent animal experiments that have demonstrated that spinally evoked neurogenic (NMEP) and compound muscle action potentials (CMAP) fall in amplitude many minutes in advance of the SSEP when the injury mechanism is reduced cord perfusion[3]. The reason for this has not been clearly identified, nor have the neurophysiological pathways responsible for conduction of the MEP yet been documented.

This paper describes experience in developing an intraoperative MEP cord monitoring technique designed to be both neurophysiologically and practically complementary to the SSEP already in use. It requires few extra electrodes and connections and provides rapid and stable feedback to the operating surgeon. Initial clinical experiences have been described previously[4] and the importance of temporal summation in MEP monitoring in man has been discussed[4]. During development of the technique several interesting insights into the neurophysiology of the human spinal cord have emerged and the central theme of this paper is the importance of a dual stimulus pattern to cause temporal summation, not only at the motoneuron but also at an apparent interneuron nucleus around the vertebral level of T8.

TECHNIQUE

To minimize connections the existing bipolar epidural catheter electrode usually used for recording the SSEP (Ortho Express, Amersham, England) was connected to the stimulator instead of the preamplifier, and the cathode made caudal. Thus the stimulus site was in the upper thoracic or lower cervical, dorsal epidural space. Recordings were made from lower limb muscle, most commonly tibialis anterior or gastrocnemius, via surface electrodes or a concentric needle electrode. CMAP recording was chosen for two reasons:

1. Response amplitude would be greater than for neurogenic potentials.
2. The likelihood of antidromic sensory conduction would be minimized.

Stimulation and recording were carried out with a Medelec MS91 using constant voltage stimulation and the BER filter setting (200–2000 Hz). Sweep duration was 50 ms and gain usually between 200 μV and 1 mV/cm. Responses were recorded as single sweeps without the need for averaging.

Anaesthetic technique was critical and some aspects are described in more detail in other papers in these Proceedings. In summary it was important to avoid nitrous oxide and any halogenated inhalational agent as these markedly decreased the response amplitude. Neuromuscular blockade was either avoided or monitored by observing the fatigue induced by a train of four stimuli at 2 Hz, and details of this technique are also elsewhere in these Proceedings. Satisfactory surgical anaesthesia was maintained with an infusion of propofol (Diprivan, ICI) to steady state, supplemented by intravenous fentanyl or alfentanil and ventilation was with air/oxygen mix. Blood pressure control if required was by sympathetic blockade.

STIMULATION TECHNIQUE

Initial attempts using *single* stimuli were very unrewarding. In the best circumstances when the stimulus was applied rostral to vertebral level T8 a response of 2–5 μV peak-to-peak could be obtained by averaging 50–100 sweeps. In contrast, caudal to T8 a single stimulus evoked a large response of several millivolts without averaging (Figure 1).

The addition of a second stimulus pulse, 1–2 ms after the first, dramatically augmented the response from levels rostral to T8 (Figure 1). It was now possible to evoke an apparently maximal lower limb CMAP by stimulating the upper thoracic cord in a way that was potentially useful for intraoperative cord monitoring (Figure 2).

The reason for this augmentation was sought. First the simple explanation of increased stimulus power was excluded by demonstrating a response of much larger amplitude to a dual stimulus of 0.2 ms than a single stimulus

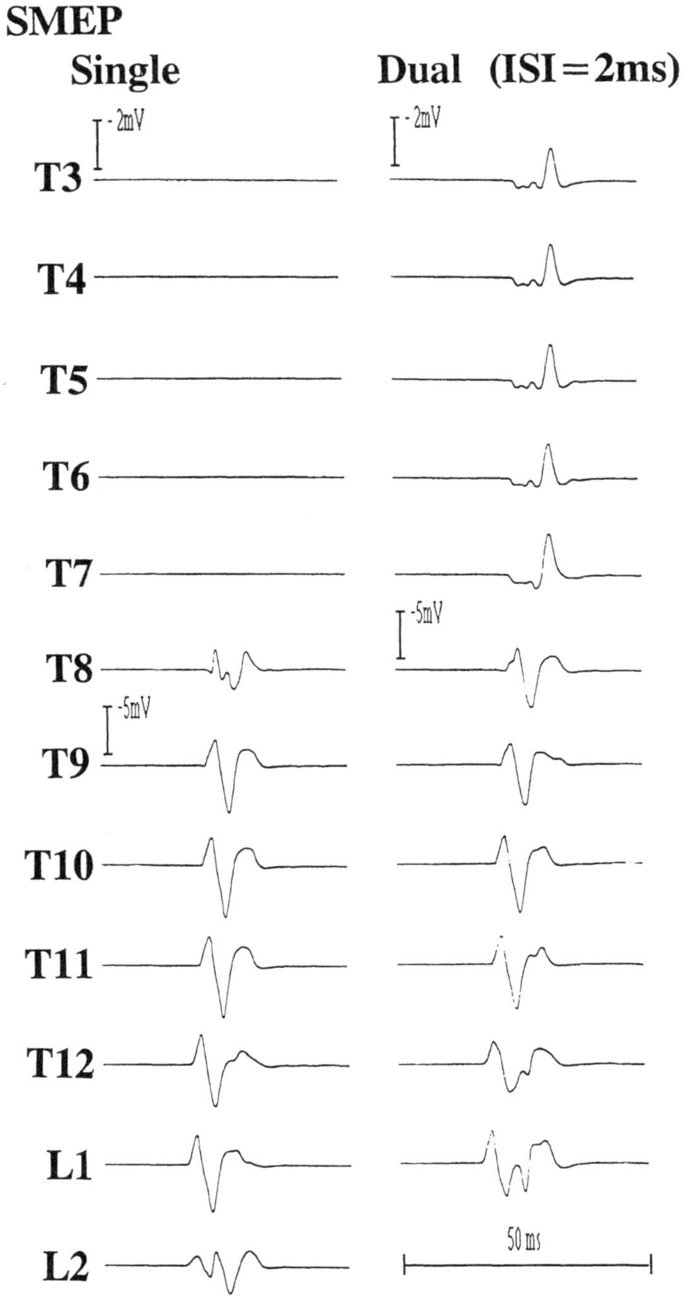

Figure 1 Spinal motor evoked potentials recorded with a concentric needle electrode from medial head of gastrocnemius in response to single or double pulse bipolar epidural stimulation of the spinal cord at each vertebral level from T3 to L2. Stimulus intensity 50 V, pulse width 0.5 ms. Each record is a single sweep, negative up. Intraoperative recordings during posterior spinal fusion for adolescent idiopathic scoliosis. No response to single stimuli above T8, maximal responses below T8. Maximal responses above T8 with dual stimulus with different form below T8

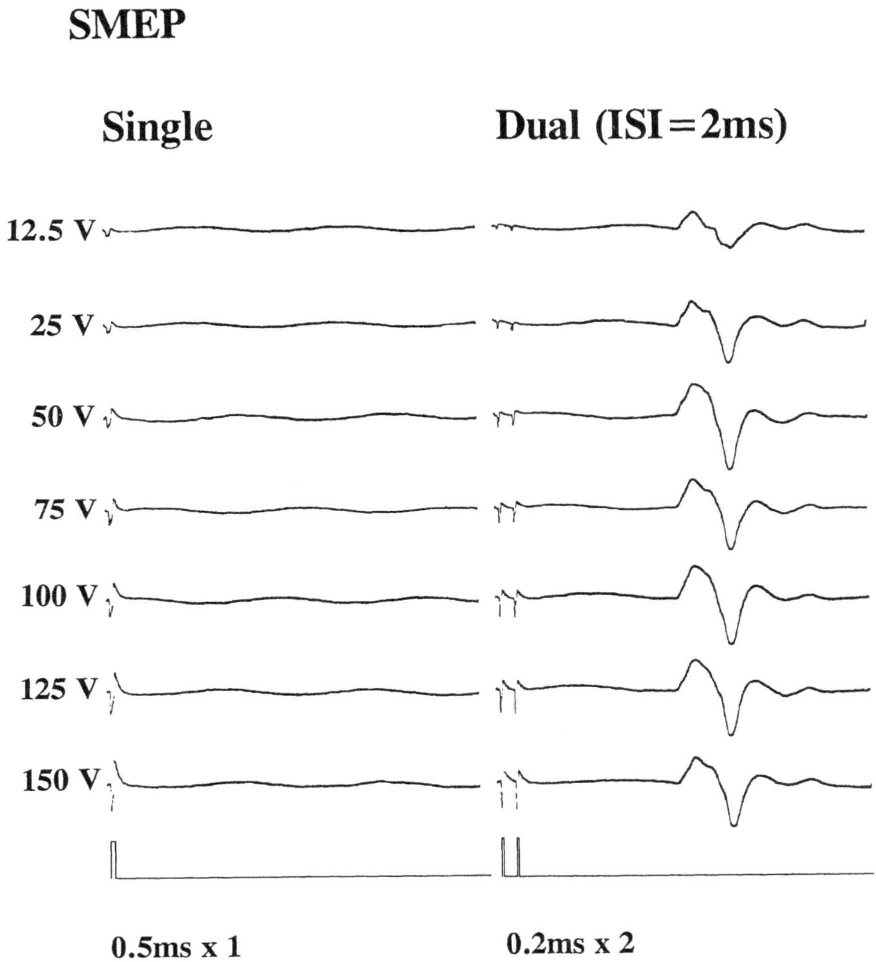

Figure 2 The ability to attain a maximal MEP reponse to epidural spinal cord stimulation with dual but not single stimuli. Stimulus site was T3, recording from gastrocnemius. Negative up, single sweep recordings. Increasing intensity of stimulation from 12.5 to 150 V with either single or dual stimuli. Intraoperative recordings during posterior spinal fusion for adolescent idiopathic scoliosis

of 0.5 ms at the same intensity (Figure 2). Secondly the latency of the response in relation to the interstimulus interval (ISI) was studied by maintaining the same stimulus intensity, duration and site whilst increasing the ISI from zero to 10 ms (Figure 3). This revealed an increase in latency of the whole response which was directly related to the increase in the ISI, suggesting that the generation of any response at all depended on both pulses being present (Figure 4a). Thirdly the amplitude of the response in relation to the interstimulus interval was measured (Figure 4b). This revealed that there was an optimal ISI of between 1 and 2 ms beyond which the amplitude of

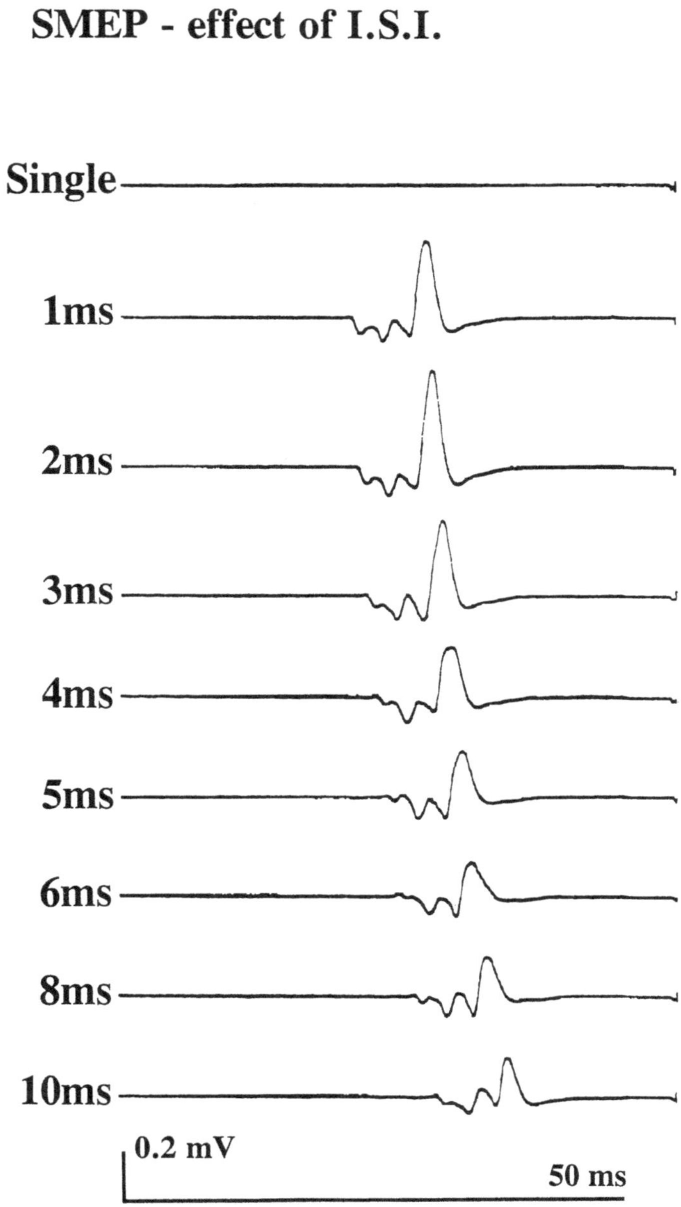

Figure 3 The effect on latency and amplitude of the SMEP stimulated at T3 and recorded from gastrocnemius muscle by varying the interstimulus interval from zero to 10 ms. The latency increases in proportion to the ISI. The amplitude reaches a maximum at an optimal ISI of 2 ms

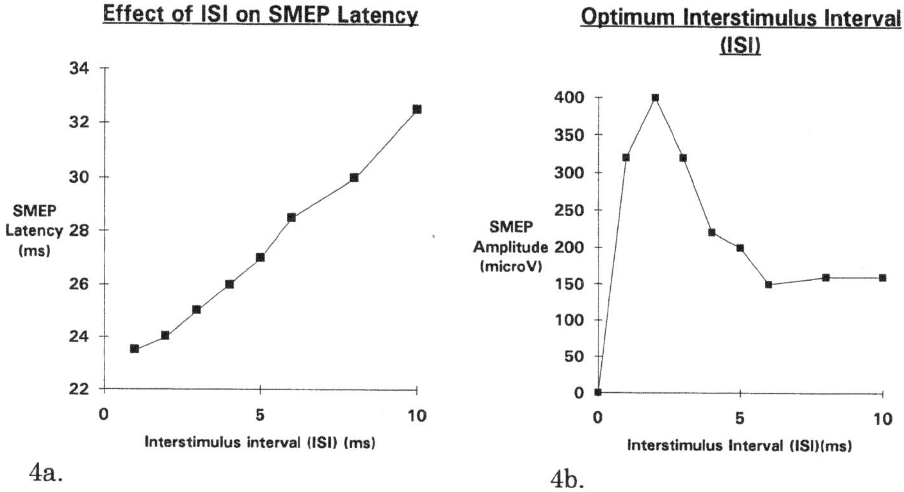

Figure 4 Graphs derived from Figure 3 to show (a) direct relationship between ISI and response latency, suggesting that both stimulus pulses are required for any propagation of the SMEP, and (b) optimal interstimulus interval of 2 ms. Decay part of the curve in 4b suggests that significant facilitation remains for more than 10 ms after cord stimulation

the response again fell away although the form of the response remained essentially the same.

DISCUSSION

These three sets of observations suggest that summation is occurring at some point in the pathway between the upper thoracic cord and the target muscle. This is more likely to be temporal than spatial summation as the characteristics of the two stimulus pulses in intensity, duration and site of application were identical. This would make it likely that both pulses were stimulating the same population of axons beneath the epidural electrode and the optimum interstimulus interval of 2 ms makes it unlikely that any of these axons were still refractory. Thus it is probable that this stimulus pattern generated two very similar volleys in all the ascending and descending pathways that were sufficiently close to the epidural electrode to be stimulated.

The site of the temporal summation was sought by applying the stimulus serially to each vertebral level of the spinal cord from T1 to the cauda equina at L2 (Figure 1). Initially the most obvious site for summation was the motoneuron, this being an inevitable synapse within the motor pathway. Surprisingly then, the effect of dual versus single stimulation on the cord at, and just rostral to, the vertebral levels where the motoneurons for the target muscles would be expected to lie (T11–12), was not nearly so dramatic as that observed after stimulation of the upper thoracic cord (Figure 1). Single stimulus pulses in the lower cord evoked very respectable CMAPs. Also as the interstimulus interval was increased it became clear that the form of the muscle action potential was becoming more complicated as the ISI became longer than the refractory period of the motor units, and a response to the second stimulus appeared in addition to the first. Similarly the initial latency of the response at these levels remained constant despite the increase in ISI.

The striking finding was of the apparent discontinuity in the characteristics of the spinal cord in response to epidural stimulation at vertebral level T8 (Figure 1). There appeared to be a constant form and amplitude of response rostral to this level and a change to a different, but again constant response between T9 and T11. The response characteristics changed again at L2, where stimulation would be expected to affect only the spinal roots of the cauda equina.

These observations suggest that the SMEP pathway from the upper thoracic spinal cord to muscles of the lower limb is interrupted not only by the neuromuscular junction and the synapse onto the motoneuron, but also by a further synapse onto an interneuron which lies around the level of the eighth thoracic vertebra in man, and that transmission across this level is not on a one-to-one basis but requires summation of at least two descending volleys.

The pathways in which these volleys travel have not been identified. The sharp discontinuity at T8 and the different response of the cord below this level to dual stimuli of varying interstimulus intervals makes antidromic conduction in sensory pathways (in particular the posterior columns) with dendritic spread to the motoneuron extremely unlikely. The possibility of gamma efferent stimulation in the upper cord with subsequent activation of the monosynaptic reflex arc was considered. To investigate this possibility the H reflex was elicited by stimulating the posterior tibial nerve at the popliteal fossa and recording muscle action potentials from the same electrode used to record the epidurally evoked response. The posterior tibial nerve stimulation site was 15 cm proximal to the recording site. The latency of the H reflex was 30 ms and the latency of the centrally evoked potential from T3 was 25 ms. Since the latency of the H-reflex was longer than the centrally evoked MEP, gamma efferent activation is effectively ruled out. Thus the pathway is probably a direct descending motor pathway which has an average conduction velocity of about 40 m/s.

SMEP INTRAOPERATIVE MONITORING

The SMEP is a technique that requires considerably more attention to detail than the SSEP. By utilizing dual stimulation, avoiding nitrous oxide and halogenated inhalational anaesthetic agents and closely monitoring neuromuscular blockade, stable intraoperative motor evoked potential spinal cord monitoring has been possible in over 40 cases, with technical failure in only one of the most recent twenty.

The presence of at least two synapses in the SMEP pathway in regions of the cord at risk during spinal deformity surgery is likely to be the reason why the SMEP is so sensitive to some anaesthetic agents. It is also possible that such synapses are responsible for the marked sensitivity of the MEP to ischaemia that has been shown in animal experiments. If synapses do exist around T8 in man their greater sensitivity to ischaemia may justify the greater difficulties of using the SMEP by providing a faster and more sensitive monitor of impending cord damage than can the axonally conducted SSEP. The success of dual stimulation in evoking motor potentials emphasizes the fact that this pathway must be recognized as neurophysiologically more complex than that of the SSEP, although complementary to it in terms of monitoring global spinal cord function.

It is interesting to speculate on the reason for an interneuron nucleus in the lower limb pathway. It is possible that this is analogous to the interneuron described by Jankowska in the cat[5,6], and that it may in some way be responsible for integrating gait patterns in man. It is also interesting to suggest that the preponderance of intraoperative spinal cord injury around the level of T8[7] may be due more to the greater sensitivity of the cord to ischaemia at this level than to a 'watershed' in arterial supply.

There is not yet enough experience to recommend criteria for interpreting the SMEP in terms of its significance for postoperative neurological status and this must await systematic observations during regular clinical use in association with the SSEP. It has, however, already proved useful in allowing intraoperative monitoring in patients with sensorimotor neuropathy (HSMN I, PMA, Charcot-Marie-Tooth disease)[8], in whom SSEP monitoring is generally impossible, and will on occasion allow monitoring in cases of partial spinal cord injury.

References

1. Jones SJ, Edgar MA, Ransford AO. Sensory nerve conduction in the human spinal cord: epidural recordings made during scoliosis surgery. Journal of Neurology Neurosurgery and Psychiatry. 1982;45:446–51.
2. Forbes HJ, Allen PW, Waller CS, Jones SJ, Edgar MA et al. Spinal cord monitoring in scoliosis surgery. Experience with 1168 cases. J Bone and Joint Surg. 1991;73-B(3):487–91.
3. Owen JH, Naito M, Bridwell KH. Relationship among level of distraction, evoked potentials, spinal cord ischaemia and integrity, and clinical status in animals. Spine. 1990;15(9):852–7.
4. Taylor BA, Fennelly M, Farrell J. Intraoperative spinal cord monitoring with combined motor and sensory spinal evoked potentials. Journal of Bone and Joint Surgery Orthopaedic Proceedings. 1992;74–B(Suppl.1):83–4.
5. Edgley SA, Jankowska E, Shefchyk S. Evidence that mid-lumbar neurons in reflex pathways from Group II afferents are involved in locomotion in the cat. J Physiol. 1988;403:57–71.
6. Jankowska E. A neuronal system of movement control via muscle spindle secondaries. Progress in Brain Research. 1989;90:299–303.
7. Ponte A. Postoperative paraplegia due to hypercorrection of scoliosis and drop of blood pressure. J Bone and Joint Surg. 1973;56-A(2):444.
8. Taylor BA, Fennelly ME. The spinal motor evoked potential allows intraoperative monitoring in sensorimotor neuropathy. Journal of Bone and Joint Surgery Orthopaedic Proceedings. 1992; (In press)
9. Taylor BA, Fennelly ME, Taylor A, Farrell J. Temporal summation – the key to motor evoked potential spinal cord monitoring in humans. Journal of Neurology Neurosurgery and Psychiatry. 1993;56:104–106.

51
Near maximal transcranial magnetic motor evoked potentials during surgery

B. A. TAYLOR, M. E. FENNELLY, T. R. MORLEY, M. HETREED and J. FARRELL

INTRODUCTION

There are two goals currently seen as desirable in the field of intraoperative spinal cord monitoring. The first is the introduction of methods that will test the integrity of motor pathways because of fears that somatosensory techniques may allow some spinal cord injuries, particularly those due to ischaemia, to occur unnoticed. The second is to evoke those responses using magnetic stimulation.

There are several advantages to the use of magnetic stimulation. It is non-invasive, and does not require electrical connection to the patient, obviating many of the problems of electrical isolation that can occur with electrical stimulators. It is also painless in use making additional pre- and postoperative examinations possible.

There are also, however, several problems. Firstly, stimulation of the motor cortex is necessary because of the inability, so far, to stimulate the spinal cord directly. This is more likely to be due to technical limitations of the stimulating coils than any fundamental physiological characteristic of the spinal cord, and will probably become possible later. Secondly, cortically evoked motor potentials are extremely sensitive to anaesthetic agents, and are usually abolished completely, especially from the lower limbs, upon induction of anaesthesia. This can render the technique useless for intraoperative monitoring. Thirdly, it is seldom possible to evoke a *maximal* compound muscle action potential (CMAP) from lower limb muscles, even in the fully conscious subject and with maximal stimulator output, without voluntary facilitation.

A stable, maximal response is important in any monitoring technique that depends on quantitative rather than qualitative criteria in its interpretation. Most clinicians involved in spinal cord monitoring accept that a drop of response amplitude to 50% or less of its initial, control value is likely to indicate impending cord damage, and use latency criteria as secondary indicators.

Previous experience with *spinally* evoked electrical MEPs during surgery has demonstrated the importance of a double pulse stimulation technique to produce stable, maximal CMAP responses from tibialis and gastrocnemius muscles[1,2], and has guided development of a suitable total intravenous anaesthetic (TIVA) protocol[3]. This paper describes initial experience with double pulsed, magnetic, cortical stimulation under general anaesthesia during scoliosis surgery.

ANAESTHETIC METHOD

Premedication

*Temazepam 200 mg

Induction

*Propofol (Diprivan, ICI) 2 mg/kg

*Fentanyl 10 μg/kg

*Vecuronium 0.1 mg/kg

Maintenance

*Propofol infusion to steady state

*Fentanyl bolus doses as required

*Air/oxygen ventilation

**Avoid nitrous oxide or isoflurane

Following induction of anaesthesia and after the initial posterior surgical exposure of the spine a 4F bipolar catheter electrode (Ortho Express, Amersham, England) was positioned under direct vision in the dorsal epidural space at T3. This was used to stimulate the cord for the spinal motor evoked potential (SMEP)[4] and identify a maximal SMEP response from the tibialis anterior surface electrodes (top trace, Figure 1). The effect of the vecuronium on the neuromuscular junction was then identified by applying a series of four stimuli at two per second to this epidural electrode and recording the first (T1) and fourth (T4) responses at the same tibialis anterior electrodes, according to the technique published elsewhere[5]. Cortical stimulation was employed once the T4/T1 ratio exceeded 70%, implying almost complete recovery from neuromuscular blockade.

RECORDING TECHNIQUE

Stimulation

*Magstim 'BISTIM' with double head coil
*Intensity setting 100%
*Interstimulus interval (ISI) 2 ms
*Contralateral cortex

Recording

*Medelec MS91
*Tibialis anterior surface EMG
*Single sweep (no averaging)
*Gain 0.2 mV/cm
*Timebase 5 ms/cm
*Filters 'BER' (200–2000 Hz)

The transcranial cortical magnetic MEPs were found to be extremely sensitive to coil positioning, so this was adjusted by hand whilst applying stimuli to find the optimal position. Several responses were then obtained at this site. The interval between successive stimuli was a minimum of 15 seconds.

RESULTS

The traces recorded from two patients are shown in Figures 1 and 2. In each figure the upper two traces are the T1 and T4 maximal responses to epidural stimulation with paired electrical pulses. The lower two traces are two successive T1 responses to paired cortical magnetic pulses. In Figure 1 the response to just a single cortical magnetic pulse is also shown for comparison.

For each patient the CMAPs evoked by epidural electrical and cortical magnetic stimulation are remarkably similar in form, although somewhat different between patients. For both patients it can be seen that there is an obvious stimulus artefact for the magnetic stimulation but minimal recorded artefact with the SMEP. The SMEP traces have been mounted so that the trace start time is the same as the magnetic ones, and latencies are therefore comparable. The T4/T1 ratios are 72% for Case 1 and 91% for Case 2.

The maximum peak-to-peak CMAP amplitude for magnetic stimulation is 0.88 mV for Case 1 and 0.28 mV for Case 2. The corresponding values for the SMEPs are 1.25 mV and 1.1 mV respectively. Thus magnetic stimulation achieved 70% of maximal SMEP amplitude for Case 1 and 25% for Case 2.

The latencies for the SMEPs are 22 ms and 20 ms, and for the magnetic MEPs 28 ms and 26.5 ms for Cases 1 and 2 respectively. This gives conduction times from central to spinal stimulation of 6.0 ms and 6.5 ms.

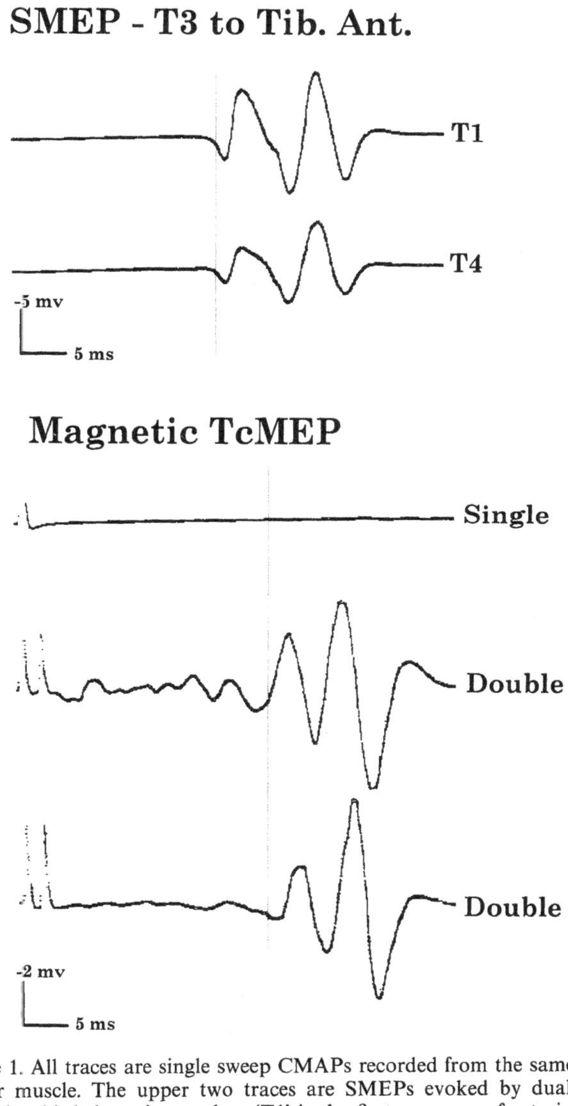

Figure 1 Case 1. All traces are single sweep CMAPs recorded from the same electrodes over tibialis anterior muscle. The upper two traces are SMEPs evoked by dual pulse electrical stimulation at the third thoracic vertebra. 'T1' is the first response of a train of four at 2 Hz and 'T4' is the fourth. The ratio of T4/T1 describes fade due to muscle relaxant in the MEP pathway. Magnetic stimulation to the head with single and dual pulses applied to the same site at 15 second intervals shows marked augmentation to the CMAP response by the dual stimulus, which then approaches the maximal amplitude of the spinally evoked response

DISCUSSION

This paper describes initial observations on the value of dual magnetic stimuli to evoke large amplitude lower limb CMAPs under anaesthesia during scoliosis surgery. The similarity in form of the SMEPs and magnetic

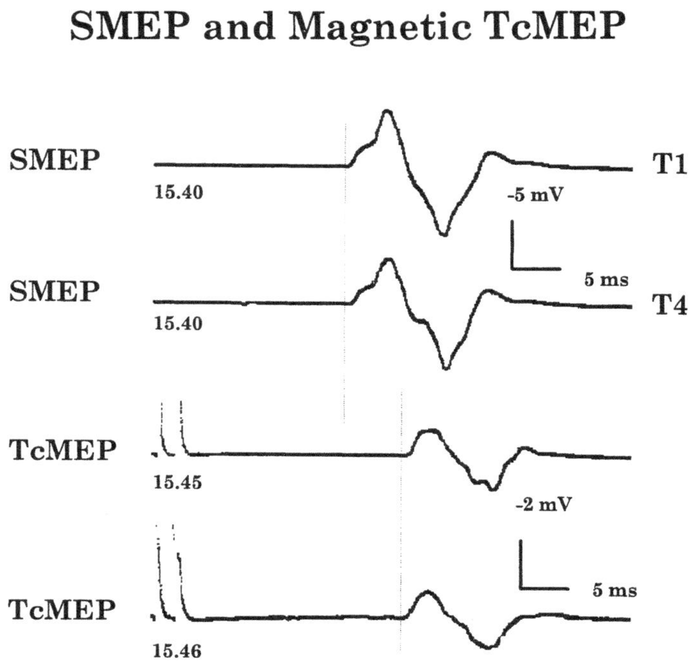

Figure 2 Case 2. Single sweep CMAP recordings with surface electrodes over tibialis anterior muscle. The upper two traces are the T1 and T4 (see text) responses to epidural electrical stimulation at the third thoracic vertebra. The lower two traces are consecutive responses to dual pulse, magnetic stimulation of the head

MEPs make it tempting to suggest that both modes of stimulation are activating similar descending pathways. The central motor conduction times of 6 ms and 6.5 ms are similar to those previously described[6].

In studies of the SMEP it has been possible to identify an optimal interstimulus interval of between 1 and 2 ms for the dual pulse stimuli[1,4]. In this paper it was not possible to repeat the systematic observations of the effect of changing ISI on CMAP amplitude because of the critical effect of coil positioning in relation to the cortex, and it may be that the 2 ms interval chosen is not the optimal one for cortical stimulation. Conversely, if the relationship of ISI to response amplitude is similar to that for the SMEP this would be further evidence that the two techniques share a common pathway. Other studies of transcranial electrical stimulation at high intensities[7] demonstrate corticospinal volleys that may be of relevance, with the large, short latency D-wave being followed by multiple (usually dual) I-waves at intervals of between 1 and 2 ms. It is possible that the dual SMEP stimuli mimic these I-waves reproducing a more physiological stimulus of the lower spinal cord. Since the dual magnetic stimulus was most effective at the maximum output setting, and positionally critical, it too may be stimulating

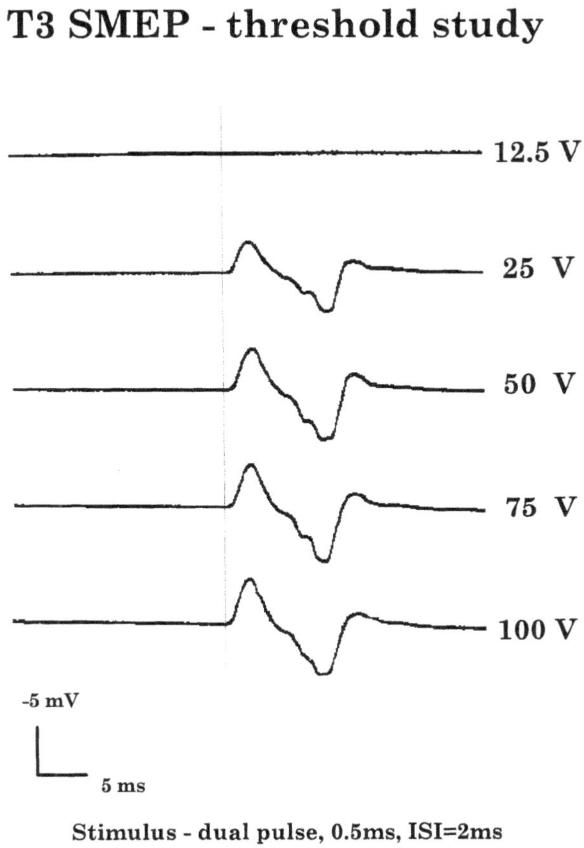

Figure 3 Case 2. SMEP threshold study. Epidural dual pulse electrical stimulation at T3. CMAP recording from tibialis anterior muscle. Stepwise increase in stimulus intensity from 12.5 to 100 volts shows a maximal response at 75 volts which was then used in subsequent SMEP recordings (upper traces in Figure 2)

the corticospinal pathways at a sub-cortical level. It is not clear whether the multiple I-waves recorded by Burke et al. represent repeated volleys in the same axons or volleys in different pathways, but it is thought that they are of cortical origin and they may therefore represent a more integrated and purposeful motor control signal than the single stimuli usually used in spinal cord monitoring techniques.

Intraoperative magnetic motor evoked potential spinal cord monitoring is not yet in routine use in scoliosis surgery because of the problems of stimulating a sufficiently large and stable response even in the absence of spinal cord injury. Anaesthesia of sufficient depth to allow surgery to continue significantly reduces response amplitudes, relegating magnetic cortical MEPs evoked with conventional, single pulses, to a qualitative rather than quantitative role. In many studies it is clear that anaesthesia has been purposefully lightened to allow such MEP recordings, which endows the technique with many of the potential problems of the wake-up test. Double pulse magnetic stimulation appears to augment the MEP sufficiently to allow monitoring without lightening anaesthesia. This may be because it excites the cortico-spinal pathways at a sub-cortical level, mimicking the SMEP. Further investigation will be directed towards identifying the optimal interstimulus characteristics and examining the descending spinal volleys evoked by this mode of stimulation.

References

1. Taylor BA, Fennelly ME, Taylor A, Farrell J. The method, rationale and results of spinal motor evoked potential (SMEP) spinal cord monitoring in scoliosis surgery. J Bone Joint Surg. 1992;74–B Suppl III:301.
2. Taylor BA, Fennelly ME, Taylor A, Farrell J. Temporal summation – the key to motor evoked potential spinal cord monitoring in humans. J Neurol Neurosurg Psychiat. 1993;56:104–106.
3. Fennelly ME, Taylor BA. A method to assess the effect of neuromuscular blocking agents on the spinal motor evoked potential during spinal surgery. In Jones SJ, et al., eds. Handbook of Spinal Cord Monitoring. Dordrecht: Kluwer;1993:277–80.
4. Taylor BA, Fennelly ME, Taylor A and Farrell J. Temporal summation and motor evoked potential spinal cord monitoring in man. In Jones SJ, et al., eds. Handbook of Spinal Cord Monitoring. Dordrecht: Kluwer;1993:368–75.
5. Fennelly ME, Taylor BA, Hetreed M. Anaesthesia and the motor evoked potential. In Jones SJ, et al., eds. Handbook of Spinal Cord Monitoring. Dordrecht: Kluwer;1993:272–6.
6. Gandevia SC. Human motor cortical output studied with transcranial stimulation. Curr Opin Neurol Neurosurg. 1989;2:782–7.
7. Burke D, Hicks RG, Stephen JPH. Corticospinal volleys evoked by anodal and cathodal stimulation of the human motor cortex. J Physiol. 1990;425:283–99.

52
Rapid changes in killed-end corticospinal MEP in patients with acute spinal cord injury

T. YAMAMOTO, Y. KATAYAMA, T. TSUBOKAWA, T. FUKAYA, T. HIRAYAMA and S. MEIJIMA

INTRODUCTION

During recording of spinal cord evoked potentials, a monophasic positive potential is observed when a nerve impulse approaches but does not pass beyond the recording site. This type of potential has been termed a killed-end evoked potential (KP). While this potential is useful for localizing the injured site clinically, no other clinical value has yet been reported for it. The KP is usually higher in amplitude than the positive wave of normal evoked potentials. The higher amplitude has been attributed to background injury currents resulting from a low transmembrane resistance at the damage site of axons. The amplitude of the KP may therefore represent a clinically available indicator of injury currents, which have been implicated in secondary processes of axonal degeneration.

We reported previously that the KP of the spinally evoked spinal cord evoked potential and corticospinal motor evoked potential (corticospinal MEP) can be readily recorded from the spinal epidural space. Unlike the spinally evoked spinal cord evoked potential, the corticospinal MEP directly reflects conduction disturbances occurring along corticospinal tract axons. The KP of the corticospinal MEP may thus provide information that is clinically important in the evaluation and treatment of spinal cord injury.

In the present study, we recorded the KP of the corticospinal MEP in patients with spinal cord injury, and investigated the time courses of the changes in KP. In addition, we undertook experimental studies on the effect on the KP of an artificially applied bucking voltage to counteract the injury currents.

METHODS

Clinical recording and monitoring of KP

KPs of the corticospinal MEP were recorded from seven patients with acute spinal cord injury. All patients gave informed consent for the recording of the corticospinal MEP and monitoring of the KP. The corticospinal MEP was recorded monopolarly from a flexible platinum electrode (PISCES, Medtronic Inc.) located in the epidural space, with a reference electrode placed in the paravertebral muscles. The motor cortex was stimulated epidurally with a RESUME electrode (Medtronic Inc.). At the time of spinal surgery (laminectomy and local cooling), the recording electrodes were inserted into the epidural space, one placed rostral to the site of injury and one placed at the site where the monophasic positive wave (KP) was recorded with the largest amplitude. The electrodes remained *in situ* for 4 to 6 days. Stimuli were applied with monophasic square wave pulses of 0.5 ms duration triggered with ECGS, and three stimulations were applied between each heart beat. Signals from the electrodes were fed into an amplifier with a bandpass range of 5 Hz–5 kHz and were averaged for 32–46 sweeps using a signal processor.

Changes of KP induced by experimental DC application

Nine adult cats weighing 2.5–3.5 kg were used. The animals were first anaesthetized with ketamine (10 mg/kg, i.m.), and anaesthesia maintained subsequently through periodic administration of pentobarbital sodium (10–30 mg/kg, i.v.). A pair of silver ball electrodes, separated by 2.5 mm, were placed on the dura overlying the motor cortex to induce motor cortex stimulation, and also placed on the dura overlying the cervical (C_{6-7}) spinal cord for inducing spinal cord stimulation. In order to deliver a direct current (DC) to the injured level of the spinal cord (Th_{12}), silver ball electrodes were placed on the dura overlying the spinal cord at the level of Th_{10} and L_2. Spinal cord injury was made at the Th_{12} level employing acute clip compression, and the KP was recorded from the silver ball electrode placed at the injured level (Th_{12}) of the spinal cord. The motor cortex and spinal cord were stimulated bipolarly, and the intensity of stimulation was adjusted to supramaximal levels for the component of interest.

RESULTS

Clinical recording and monitoring of KP

The corticospinal MEP was recorded from points rostral to caudal to the site of the presumed level of spinal cord injury. In the region rostral to the lesion, a normal corticospinal MEP was obtained, an initial biphasic positive negative wave (D wave) followed by several additional biphasic positive negative waves (I-waves) was observed. Although the I-waves were sometimes absent during recordings under general anaesthesia, the D-wave was always recorded with a stable amplitude. As the recording site was moved in the caudal direction, the latency of the D-wave became progressively prolonged and the amplitude of the D-wave gradually decreased. At the presumed level of the lesion, the D-wave changed from a biphasic positive negative wave to a high amplitude monophasic positive wave. The amplitude decreased progressively as the recording site was moved to more caudal sites, whereas the latencies remained constant (Figure 1). Recordings within an area 2–5 cm caudal from the presumed level of the spinal cord injury, showed no response.

Two recording electrodes were left in the epidural space: one was placed at the site where the monophasic positive wave was recorded with the largest amplitude, and the other was placed rostral to the site of injury. Although the corticospinal MEP recorded from the rostral electrode exhibited no obvious change, the amplitude of the KP of the corticospinal MEP rapidly decreased during the initial 24 hours after injury. The amplitude decreased by over 50% in all cases, and decreased by over 70% in three of seven cases. The KP of the corticospinal MEP further decreased gradually until monitoring was terminated (Figure 2).

Effects of DC on the killed-end evoked activity

DC application to the injury site, in which the electrodes were placed epidurally with the cathode proximal and the anode distal to the injury site, markedly reduced the amplitude of the KP. These effects of DC application on the KP were commonly observed in the case of both the corticospinal MEP and spinally evoked spinal cord potential. This effect appeared as soon as the commencement of the DC application. The amplitude of the KP decreased as the DC was increased. A DC of 500 μA gave rise to a 40% reduction in the amplitude of the KP (Figure 3). In contrast, DC application with the anode proximal and the cathode distal to the injury site increased the amplitude of the KP (Figure 4). These results suggest that appropriate application of a DC to the spinal cord can reduce the amplitude of the KP. Also, the characteristic amplitude of the KP recordings may be explained in relation to the injury currents which can be changed with DC current application.

DISCUSSION

In experimental studies, it has been reported that a large monophasic positive potential (KP) is seen at the site of spinal cord lesions when the fibre tracts within the spinal cord are activated. Further, the KP of the spinally evoked spinal cord evoked potential has been recorded intraoperatively in patients with spinal cord lesions. We have also found that the KP of the corticospinal MEP recorded in patients exhibits identical features to those reported for spinally evoked spinal cord potentials. Injury to the axons provides a low-

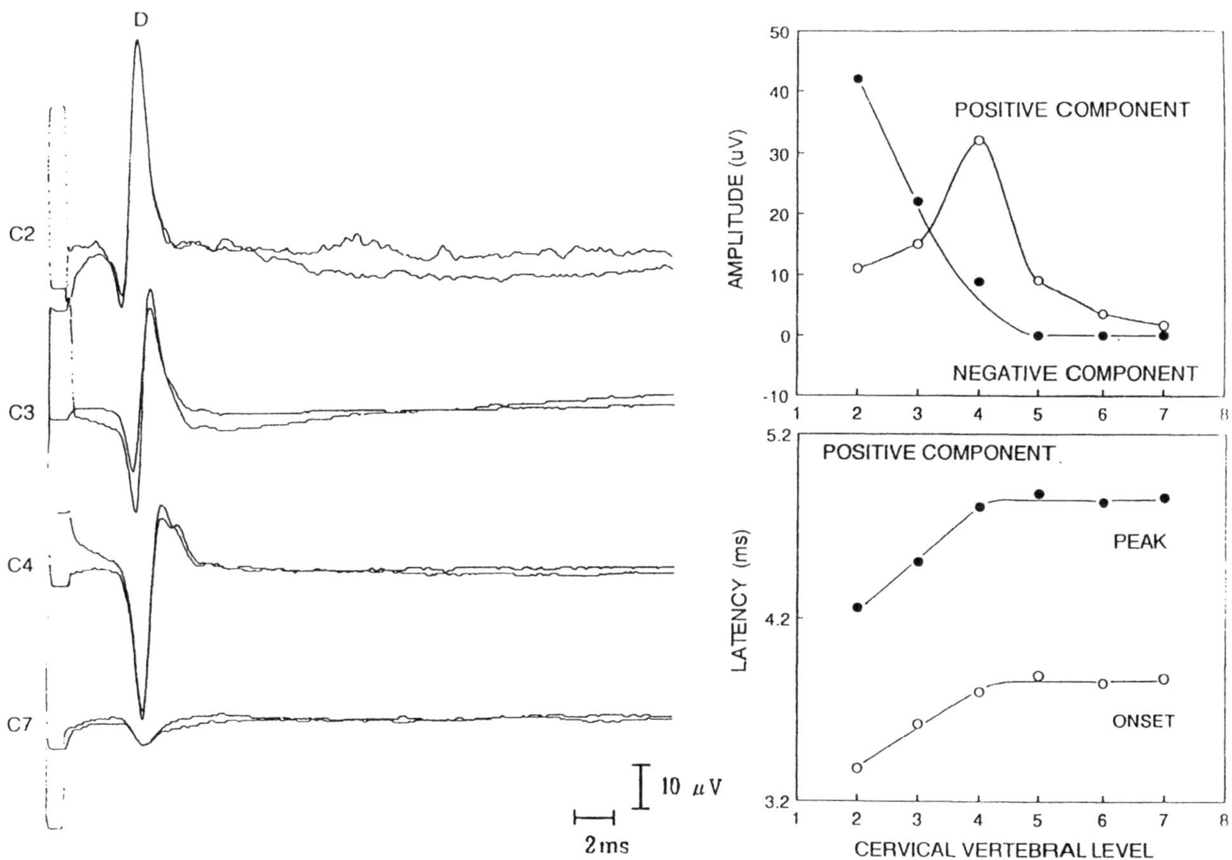

Figure 1 Representative example of killed-end evoked potentials of the corticospinal MEP in a patient with cervical spinal cord injury while the electrode was moved from rostral to caudal to the site. At the level of C_4, the presumed level of the lesion, monophasic positive wave was maximally observed and gradually decreased as the electrode moved to more caudal direction. Negative component rapidly decreased and disappeared around the level of C_4. The onset as peak latencies of the positive component increased until the electrode was moved to the level of C_4, but remained constant thereafter. Upwards corresponds to negativity

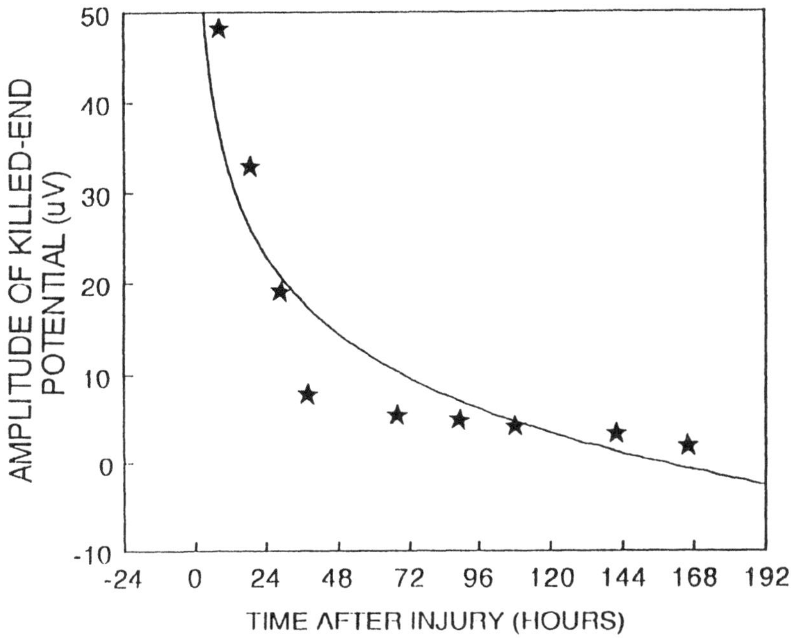

Figure 2 Time course of KP of corticospinal MEP recorded in a patient with cervical spinal cord injury. Note that the KP decreased rapidly during the initial 48 hours after injury

resistance pathway for current flows at the damaged site. Such injury currents result in distally negative voltage gradients, the classical demarcation potential. The higher amplitude of the KP as compared to the normal evoked potential has been explained on the basis of this demarcation potential. When the impulse approaches the site of injury, the current flows from the approaching volley are opposite in direction to the current flows for the demarcation potential. Thus, the volume-recorded positivity of the approaching volley appears more positive. This feature of the KP can be accounted for by the presence of inward injury currents occurring at the injured end of the axons, and the amplitude of the KP reflects the density of the injury currents.

We monitored the KP of the corticospinal MEP during 4 to 7 days after spinal cord injury in humans, and found that the KP underwent a rapid decrease in amplitude during the initial 24 hours after the injury. The time course of the KP observed in this study resembled that of the KP recorded in experimental spinal cord injury of cats. In the acute stage of spinal cord injury, a strong inward injury current may be caused mainly by Na^+ and Ca^{2+}. Influx of Ca^{2+} is thought to disrupt the neurofilaments and to induce axonal degeneration. Balentine et al.[1] suggested that calcium is a prime mediator of secondary injury in spinal cord injury. They reported that evaluation of the total calcium in the injured spinal cord is progressive from 45 min until 8 hours, and the data correlate well with the fact that it requires

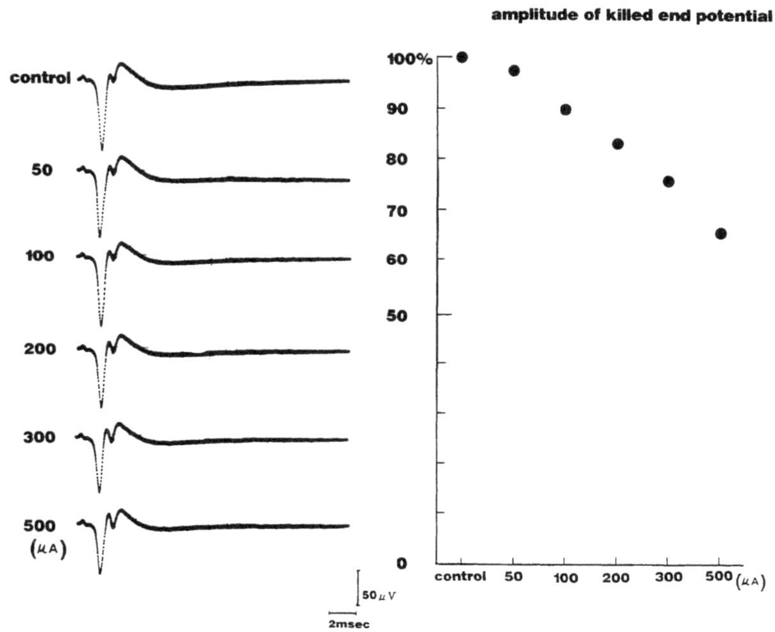

Figure 3 Comparison between the amplitude of the killed-end evoked potential and the intensity of direct current (DC) application

8 to 24 hours for maximal spinal cord necrosis to develop following experimental impact injury.

The occurrence of the KP reflects a block of conduction of the nerve fibres and does not indicate that the nerve fibres are already killed. Thus, we do not have to adopt a pessimistic outlook on the occurrence of a KP. It has been reported that long-term weak DC application can produce an improvement in neurological function and anatomical findings. It has also been found that cultured dorsal root ganglia or developing amphibian neuroblast axons tend to grow towards the negative pole of a field applied across their chamber. These results demonstrate the importance of the regeneration effect caused by very weak DC application. We propose another possible effect of the DC application; namely, that the DC counteracts injury currents in the acute stage of spinal cord injury. In order to establish a therapy along such lines, the amount of DC necessary to reduce the injury current and the appropriate electrode placement need to be determined. Characterization of the time course of changes in the KP of the corticospinal MEP in individual patients may yield useful information for developing such a therapeutic approach to spinal cord injury.

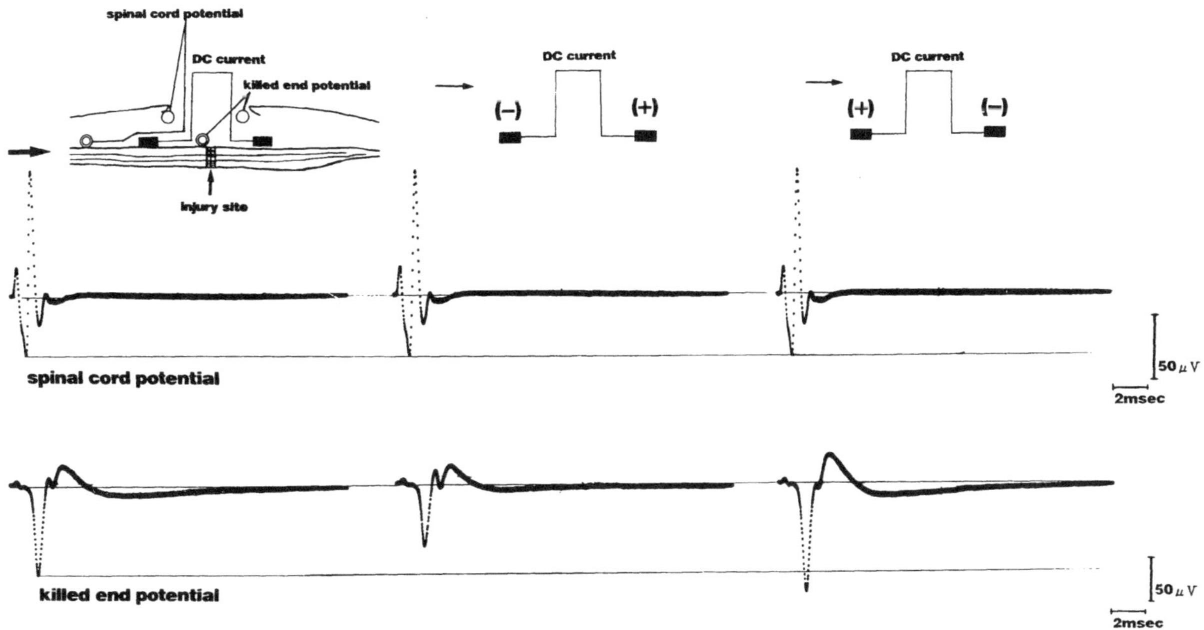

Figure 4 Upper recordings present the spinally evoked spinal cord potential recorded at Th_8 in response to C_6 stimulation, and the lower recordings represent the killed-end evoked potential recorded at the injury site (Th_{12}) in response to C_6 stimulation. DC application with the anode or cathode proximal to the injured site showed a quite different effect on the amplitude of the killed-end evoked potential

References

1. Balentine JD, Horgann EL, Banik NL et al. Calcium and the pathogenesis of spinal cord injury. In: Dacy RG Jr et al., eds. Trauma of the central nervous system. New York: Raven Press; 1985:285–295.

Animal studies

53 Electrophysiological phenomena recorded from spinal cord slice preparation in the adult rat

H. BABA, Y. SATO, K. SHIMOJI, M. YOSHIMURA and H. HIGASHI

INTRODUCTION

The dorsal root potential (DRP-V) is believed to have the same origin as the slow positive potential of the segmental spinal cord potential. Furthermore it is also thought to be a reflection of primary afferent depolarization (PAD) which is an index of presynaptic inhibition[1]. It is believed that primary afferent terminals are depolarized by GABAergic neurons in the substantia gelatinosa (lamina-II). However, intracellular recording from substantia gelatinosa (SG) neurons has been impossible *in vivo* and very difficult *in vitro*, because of the high vulnerability of SG neurons to anoxic and mechanical damage in adult rat together with their small size. An *in vitro* spinal preparation has been developed in the newborn rat. However, it is known that the mode of synaptic transmissions in the adult rat is not the same as in the newborn rat[2,3]. We therefore developed an *in vitro* adult rat spinal cord slice preparation to achieve stable intracellular recordings from SG neurons simultaneously with electrotonic potentials from the dorsal root for studying the relationship between them.

MATERIALS AND METHODS

Preparation of the slice

Adult Wistar rats were anaesthetized with urethane (1.5–2.0 g/kg ip). A lumbo-sacral laminectomy was performed and 1.5–2 cm length of spinal cord with attached dorsal roots was excised. Then, the spinal cord was placed in pre-oxygenated and chilled (4–6°C) Krebs solution. After removal of the dura mater, pia-arachnoid membrane and all ventral and dorsal roots with the exception of one or two on one side, the spinal cord together with the attached dorsal roots was placed in a shallow groove formed in an agar

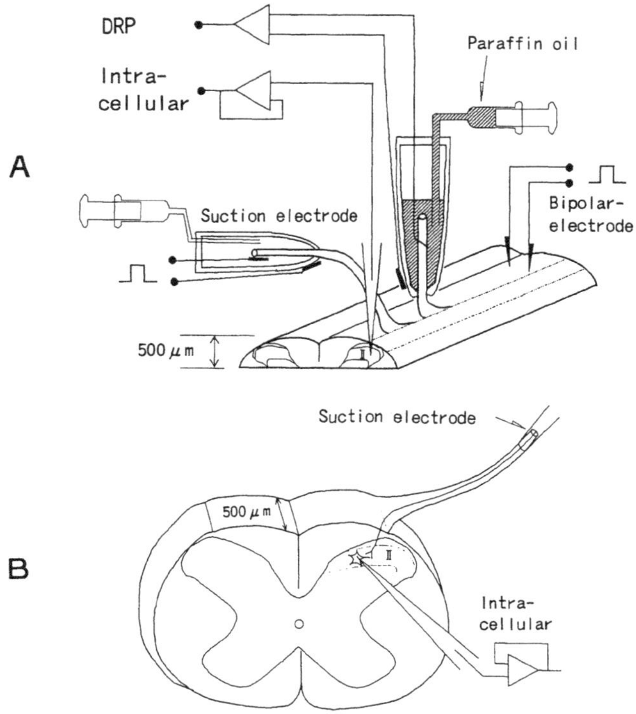

Figure 1 A, B Recording of dorsal root potential and intracellular potential in horizontal (A) and transverse slice (B) preparations. A: In the horizontal slice, recordings of both DRPs and intracellular potentials were made. The dorsal root and dorsal column were stimulated by a suction electrode and a bipolar electrode, respectively. B: In the transverse slice, the recording microelectrode was inserted into the SG under visual control. II: lamina-II, substantia gelatinosa

block. The spinal cord was immersed in Krebs solution and a horizontal or transverse slice with a thickness of 500 μm was made using a razor blade of a microslicer (Figure 1). The slice was mounted on a nylon mesh in a recording chamber and fixed with silver wires and titanium electron microscope grid. The slice was completely submerged and superfused continuously with Krebs solution (20–30 ml/min) equilibrated with 95% O_2–5% CO_2 at 36 ± 1°C. The composition of the Krebs solution was as follows (in mM): 117 NaCl, 3.6 KCl, 2.5 $CaCl_2$, 1.2 $MgCl_2$, 1.2 NaH_2PO_4, 25 $NaHCO_3$ and 11 glucose.

Dorsal root potential and intracellular recording

In the horizontal slice, DRPs were recorded with suction electrodes filled with paraffin oil (Figure 1A). In the transverse slice, with transillumination under a binocular microscope, the SG was identifiable as a translucent band across the dorsal horn. The recording microelectrode was then inserted into the SG under visual control (Figure 1B). In the horizontal slice, the SG

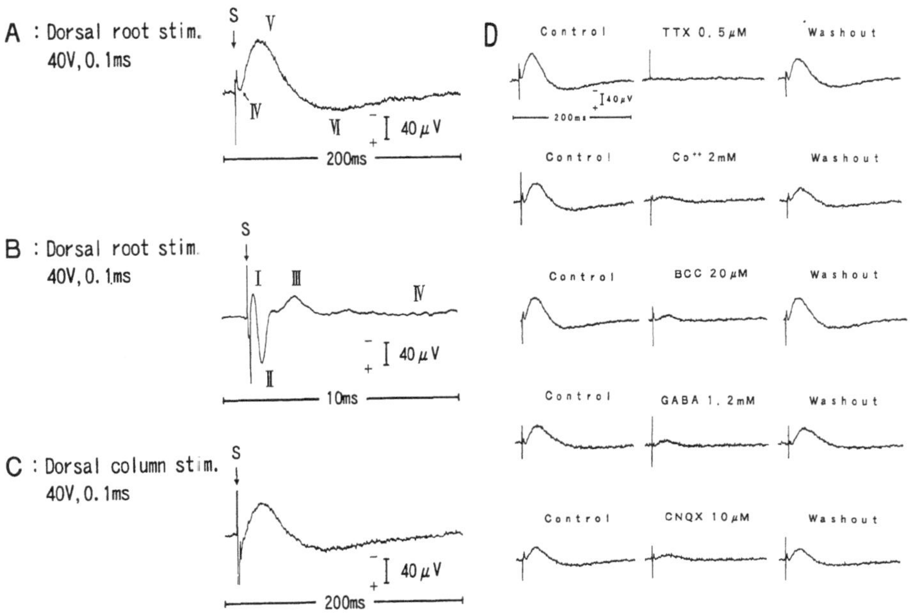

Figure 2 A, B, C, D DRPs evoked by dorsal root and dorsal column stimulation in the horizontal slice (refer to Figure 1A). A: Slow components of DRPs (DRP-IV, V, VI) evoked by dorsal root stimulation. B: Fast components of DRPs (DRP-I, II, III) evoked by dorsal root stimulation. C: DRP evoked by dorsal column stimulation. D: The effects of TTX, bicuculline, Co^{++}, GABA and CNQX on DRPs. DRP-V and VI were profoundly suppressed by TTX (0.5 μM), bicuculline (20 μM), Co^{++} (2 mM), GABA (1.2 mM) and CNQX (10 μM). Stimulus intensity = 40V, 0.1 ms. Signals were amplified by an AC amplifier. Time constant = 0.1 s

neurons were identified by the distance from the dorsal surface and electrophysiological criteria that we have previously analysed in transverse slices[4]. Microelectrodes were filled with 4M potassium acetate (resistances, 120–200 mΩ). Orthodromic stimulation of the dorsal root, which had a length of 5–10 mm, was achieved with a suction electrode. The dorsal column was also stimulated using a bipolar electrode.

Drug application

Drugs were applied by changing the perfusion solution without altering the perfusion rate and the temperature. The time required for the drug solution to flow from the tap to the recording chamber was about 5 seconds.

RESULTS

As shown in Figure 2, we could record DRPs consisting of five or six components in response to stimulation of both the dorsal root and dorsal

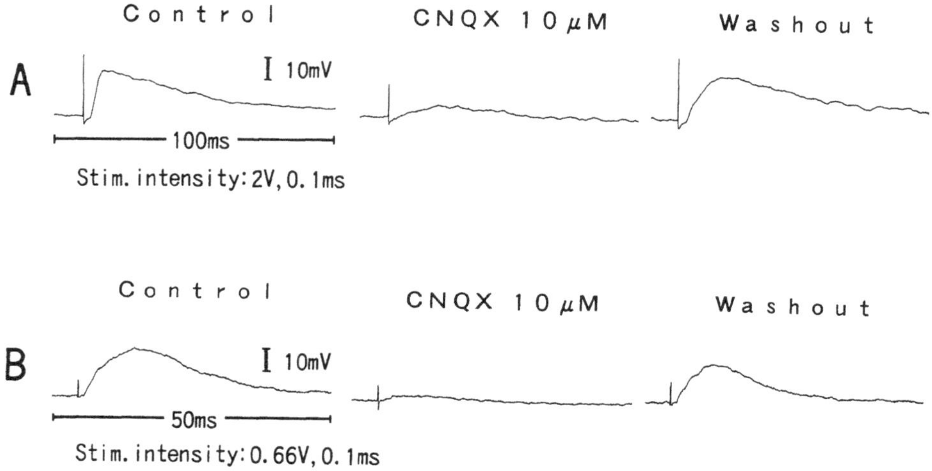

Figure 3 A, B EPSPs of SG neurons evoked by dorsal root (A) and dorsal column (B) stimulation. EPSPs were profoundly but not completely suppressed by CNQX (10 μM) in both A and B

column in the horizontal slice. DRPs evoked by dorsal column stimulation have a similar waveform to that by dorsal root stimulation (Figure 2C). DRP-V and VI derived from the stimulation of both dorsal root and dorsal column were profoundly suppressed by application of tetrodotoxin (TTX 0.5 μM), bicuculline (20 μM), $CoCl_2$ (Co^{++} 2 mM), GABA (1.2 mM) and 6-cyano-7-nitroquinoxaline-2, 3-dione (CNQX 10 μM) (Figure 2D). But DRPs were not augmented by midazolam (10^{-5}, 10^{-4}, 10^{-3} M) and thiamylal (10^{-5}, 10^{-4}). Only a comparably higher concentration of 10^{-3} thiamylal augmented DRPs (data not shown).

Figure 3 shows presumably monosynaptic EPSPs recorded from SG neurons. EPSPs were evoked by stimulation of both the dorsal root and dorsal column. They were profoundly but not completely suppressed by CNQX, an antagonist of non-NMDA type glutamate receptors, while they were completely suppressed by simultaneous application of CNQX (10 μM) and D(-)-amino-5-phosphonovaleric acid (APV 50 μM), an antagonist of NMDA type glutamate receptors (data not shown).

DISCUSSION

The central role of the substantia gelatinosa in the integration of nociceptive information has been well established[5]. It has been believed that primary afferent terminals are depolarized (PAD) by GABAergic neurons in the SG, and that PAD is the origin of the DRP-V and the P2 wave of segmental and descending SCP. However, a high vulnerability of SG neurons in the adult rat to anoxic and mechanical damage, together with their small size, has made it essentially impossible to record intracellular potential in SG neurons *in vivo* and even difficult to achieve this *in vitro*. An *in vitro* spinal cord preparation has been developed in the newborn rat. However, there are significant changes in the extent of dendritic arborization and presumably rearrangements in synaptic circuitry in the SG over the first few post-natal weeks[2,3]. The use of an adult spinal cord slice ensures that the functional circuitry of the SG reflects the mature state and not a developmental intermediate.

We recorded DRPs and EPSPs from SG neurons evoked by stimulation of the dorsal root and the dorsal column. The results suggest that both glutamate receptors and GABA receptors participate in the generation of DRPs.

We also found that the monosynaptic EPSPs in SG neurons were mainly mediated by glutamate binding to the non-NMDA receptors. Our hypothesis about the generation of DRPs evoked by dorsal root and dorsal column stimulation is as follows. The GABAergic neurons in the SG receive monosynaptic primary afferent input mediated by glutamate and then form axo-axonic synapses with primary afferent terminals. The impulses generated by dorsal column stimulation are conveyed to GABAergic neurons through the dorsal column tract antidromically and through collateral pathways. However, there is morphological, electrophysiological and neurochemical heterogeneity in the SG. Therefore further immunohistological studies are needed to clarify whether the neurons recorded in our preparations are GABAergic and form axo-axonic synapses with primary afferent terminals.

Further, it is possible that the DRPs recorded in our slice are generated in sensory areas (especially nociceptive areas), since the ventral horn and most of the nucleus proprius of the dorsal horn were resected in our horizontal slices.

CONCLUSION

1. We demonstrated in the present study that recording of dorsal root potential and intracellular potential in adult rat spinal cord slice preparation is possible.
2. The method seems useful for studying the mechanisms of synaptic transmissions of DRPs and the effects of drugs on the SG neurons.

References

1. Shimoji K, Ito Y, Ohama K, Sawa T, Ikezono E. Presynaptic inhibition in man during anesthesia and sleep. Anesthesiology. 1975;43:388–391.
2. Beal JA, Bicknell HR Jr. Development and maturation of neurons in the substantia gelatinosa (SG) of the rat spinal cord. In: Rowe M, Willis WD Jr. eds. Development, Organization and processing in somatosensory pathways. New York: Liss; 1985:23–30.
3. Falls W, Gobel S. Golgi and EM studies of the formation of dendritic and axonal arbors: the interneurons of the substantia gelatinosa of Rolando in newborn kittens. J Comp Neurol. 1979;187:1–18.
4. Yoshimura M, Jessell TM. Membrane properties of rat substantia gelatinosa neurons in vitro. J Neurophysiology. 1989;62:109–118.
5. Cervero F, Iggo A. The substantia gelatinosa of the spinal cord. J Neurol. 1980;103:717–772.

54
Transcranial magnetic motor evoked potentials recorded from the spinal cord in cats: normal waves and spinal cord injuries

Y. HIRAIZUMI, E. FUJIMAKI, E. E. TRANSFELDT and N. KAWAHARA

INTRODUCTION

Intraoperative sensory evoked potential monitoring has been used to evaluate the functional integrity of the ascending sensory pathway and does not necessarily reflect descending motor function. The wake-up test has been reported to have false-negatives and requires the patients' cooperation and frequently a prolonged wake-up time. Motor evoked potential (MEP) monitoring has been recommended for operations in which the motor system is threatened. MEP elicited by transcortical electrical stimulation was first reported by Levy in 1984[1,2]. However, it requires high voltage for stimulation transcranially over the motor cortex. In recent years several studies of magnetic-MEPs have been reported; however, this method uses compound muscle action potentials. It involves numerous components such as anterior horn cells, synapses, peripheral nervous system, and neuro-muscular junction. It is very sensitive to anaesthesia and muscle relaxants. In this study, we were able to measure transcranial magnetic-MEPs from the normal and injured spinal cords in animals.

I. PHYSIOLOGICAL DATA FOR NORMAL ANIMALS

Materials and methods

Fifteen adult mongrel cats, weighing 3.5–4.5 kg, were anaesthetized with pentobarbiturate (Nembutal) intraperitoneally, and were oxygenated with room air after intubation. Muscle relaxation was achieved by administration with pancronium bromide (Myoblock) intraperitonally. A Magstim 200 by

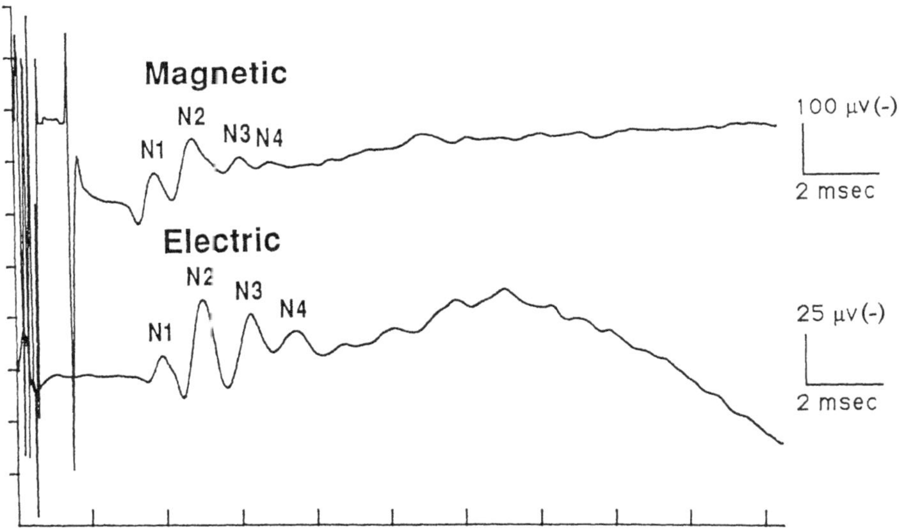

Figure 1 Waveforms of magnetic-MEPs and electric-MEPs in normal animals. Reproducible transcranial magnetic-MEPs were obtained by transcranial magnetic stimulation and spinal cord recording. Magnetic-MEP waveforms consisted of approximately two to five large negative peaks (N1–N4) and were followed by smaller polyphasic signals. N3 and N4 were relatively smaller than N1 and N2. Electric-MEPs also showed two to five large negative peaks and were followed by smaller polyphasic waves. N3 and N4 were relatively smaller than N1 and N2 as well. Each negative peak (N1–N4) of magnetic-MEPs appeared to coincide with each negative peak of electric-MEPs. Amplitudes of magnetic-MEPs were much larger

Novametrix was used as a magnetic stimulator and magnetic stimuli triggered and recorded by a Pathfinder II by Nicolet Ltd. Stimulation sites used were the international 10–20 system. The magnetic coil ring was fixed with a stereotaxic holder to obtain constancy of direction and site of stimulation. Electric stimulation of the motor cortex was performed transcranially by a pair of needle electrodes placed at the motor cortex (positive) and Cz (negative). Monopolar epidural electrodes as recording electrodes were placed at the epidural space of T7and L3 and reference needle electrodes were placed at the same levels of paravertebral muscles. Each waveform was averaged ten times.

Results

Reproducible transcranial magnetic MEPs were obtained by transcranial magnetic stimulation and spinal cord recording. Magnetic-MEP waveforms consisted of approximately two to five large negative peaks (N1–N4) and were followed by smaller polyphasic signals. N3 and N4 were relatively smaller than N1 and N2 (Figure 1). On the other hand, electric-MEPs also showed two to five large negative peaks and were followed by smaller polyphasic waves. N3 and N4 were relatively smaller than N1 and N2 as well. Each negative peak (N1–N4) of magnetic-MEPs appeared to coincide with each negative peak of electric-MEPs. Nerve conduction velocities of magnetic-MEPs were 82.1 ± 6.6 m/s in N1, 72.1 ± 14.6 m/s in N2, 57.2 ± 10.5 m/s in N3 and 43.6 ± 6.6 m/s in N4, respectively. Amplitudes of magnetic-MEPs were 54.7 μV in N1, 66.4 μV inN2, 20.7 μV in N3 and 13.6 μV in N4, respectively in the L3 recording level. Magnetic-MEP amplitudes were much larger.

Waveforms and latencies of magnetic-MEPs were affected by several factors. First, the centre of the coil ring was placed at the vertex and the inclination of the ring was varied. Latencies of each negative peak were the shortest and N1 and N2 amplitudes became larger when the inclination of the coil ring was 39°–45° to the occipito-mental line. Second, the coil ring was placed horizontally and was placed at various sites according to the international 10–20 system. Recording electrodes were placed at the right side of the spinal cord. Latencies of N1 and N2 were the shortest and amplitudes were the largest when the centre of the ring was placed between C4 and Cz, or C4 and Fp2. N3 was more distinct when the ring was placed a little anteriorly than C4 or Cz. When the stimulation intensity of magnetic stimulation increased, latencies became shorter and amplitudes became larger. Threshold intensity was about 50% of the maximum output. Maximal responses were usually obtained at 90% to 100% of the maximum output. To confirm that these magnetic-MEPs represent genuine descending motor pathways, a collision technique was applied. A descending volley was achieved by magnetic motor cortex stimulation and an ascending volley by simultaneous electrical bilateral sciatic nerve stimulation. Stimulation intervals between transcranial magnetic stimulation and electrical sciatic nerve stimulation were varied by 10 to −7 ms. The magnetic-MEP waveforms and latencies of each negative peak were not affected by colliding with each other. This indicated magnetic-MEPs represented no antidromic sensory neuron volleys (Figure 2). The electrical-MEPs, likewise, showed the same results as the magnetic-MEPs.

II. WAVEFORM CHANGES AFTER SPINAL CORD INJURY

Materials and methods

Thirteen adult cats were used. For the analysis of injury potentials after spinal cord injury, spinal cord lesions were produced at the L2 spinal cord level by partial sectioning of the spinal cord and by spinal cord compression posteriorly with Anderson's compression device[3]. Magnetic-MEPs and electric-MEPs were recorded at the injury level (L2), 1 cm above (L1), and 1 cm below the injured level (L3), respectively. In addition, ascending spinal cord evoked potentials (ascending-SEPs), descending-SEPs, and somatosensory evoked potentials recorded from the cortes (C-SSEPs) and from the spinal cord (S-SSEP) were simultaneously recorded for comparison.

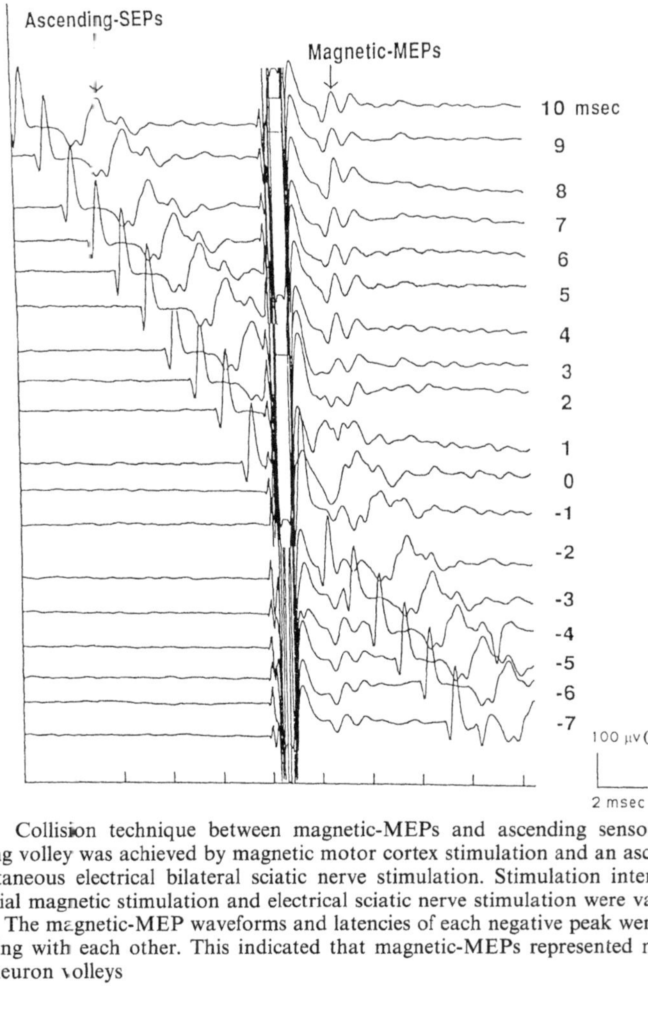

Figure 2 Collision technique between magnetic-MEPs and ascending sensory volleys. A descending volley was achieved by magnetic motor cortex stimulation and an ascending volley by simultaneous electrical bilateral sciatic nerve stimulation. Stimulation intervals between transcranial magnetic stimulation and electrical sciatic nerve stimulation were varied by 10 to −7 msec. The magnetic-MEP waveforms and latencies of each negative peak were not affected by colliding with each other. This indicated that magnetic-MEPs represented no antidromic sensory neuron volleys

Results

Lesion studies

The spinal cord lesion was produced by sectioning the posterior column, posterior half, posterior 2/3 and then the total spinal cord. Following the sectioning of the posterior columns, no change in these magnetic-MEPs occurred. However, after the posterior hemisection, especially after the posterior 2/3 section, N3 disappeared, though N1 and N2 still remained. After the total spinal cord section, N1 and N2 finally disappeared (Figure 3).

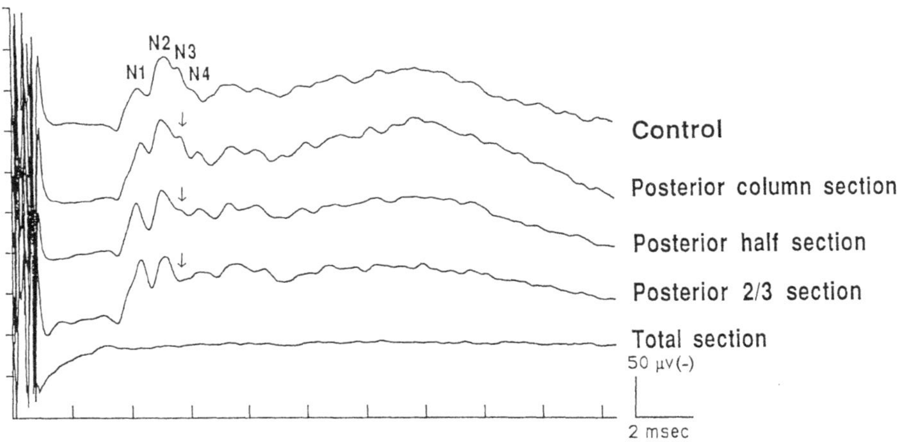

Figure 3 Lesion studies. The spinal cord lesion was produced by sectioning the posterior column, posterior half, posterior 2/3 and then total spinal cord. Following the sectioning of the posterior columns, no change in these magnetic-MEPs occurred. However, after the posterior hemisection, especially after the posterior 2/3 section, N3 disappeared, though N1 and N2 still remained. After the total spinal cord section, N1 and N2 finally disappeared

Compression studies

Spinal cord injury was produced by placing a 70 g stainless steel rod on the posterior aspect of the spinal cord. The weight was increased by 10 g every 15 minutes by adding weights to a tray on top of the rod. The weight was thus increased until waveforms finally disappeared. Following posterior spinal cord compression, amplitudes of N1, N2 and N3, especially N2 and N3 recorded from at and below the lesion level decreased and peak latencies began to increase after 70 g of 1 min compression. After 90 g, 1 min compression, positive-going potentials appeared and N1 finally disappeared below the lesion level. N1 amplitudes recorded above the lesion level decreased after 70 g, 1 min compression, whereas amplitudes then increased after 80 g, 1 min compression (Figure 4). Those changes were as sensitive as other simultaneously recorded monitoring methods such as electric-MEPs ascending SEPs, descending SEPs.

DISCUSSION

Use of magnetic stimulation and monitoring from the spinal cord in place of other conventional methods

Spinal cord recording of magnetic-MEPs, especially in small animals, has long been considered to be difficult because of large background artefact due to the proximity of the stimulating and recording sites. However, this recording was possible if its recovery from overload signals were improved by successful shielding, stabilization and correct placement of the coil-ring holder and appropriate averaging. Thus, magnetic-MEPs recorded from the

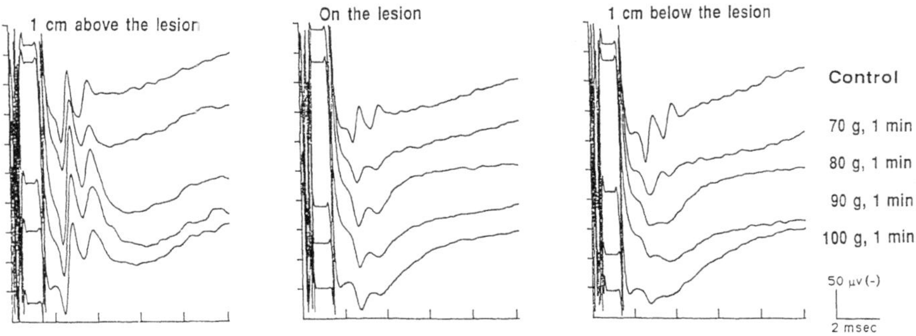

Figure 4 Compression studies. Spinal cord injury was produced by placing a 70 g stainless steel rod on the posterior aspect of the spinal cord. The weight was increased by 10 g every 15 minutes by adding weights to a tray on top of the rod. Amplitudes of N1, N2 and N3, especially N2 and N3 recorded from at and below the lesion level decreased and peak latencies began to increase after 70 g of 1 min compression. After 90 g, 1 min compression, positive-going potentials appeared and N1 finally disappeared below the lesion level. N1 amplitudes recorded above the lesion level decreased after 70 g, 1 min compression, whereas amplitudes then increased after 80 g, 1 min compression

spinal cord could be observed as reproducible waveforms. Stimulation rate was relatively slow. At present, one stimulus was obtained every 3 seconds at maximal output.

There are a number of clinical reports showing the failure to reveal motor deficits produced during operations. It has also been found in cats that experimentally induced cord ischaemia can result in damage to descending systems which may not be detected by monitoring SEPs[4]. In our series of spinal cord posterior compression, all negative peaks, especially N2 or peaks with longer latencies, were sensitive to incomplete paralysis. After 70 g, 1 min compression, waveform changes recovered when animals were left for several hours. This incomplete paralysis appeared to reflect spinal cord ischaemia. Whereas, compression with more than 90 g showed disappearance of all negative peaks and positive-going large potentials (injury potentials) appeared below the lesion site, which meant an irreversible lesion was produced. Those potentials never recovered to normal even after three hours of observation. On the other hand, the amplitude augumentation occurred at the recording site proximal to the injury. This augmentation phenomenon has been originally reported in SEP monitoring by several authors[5]. These alterations of the magnetic-MEPs were as sensitive as SEPs recorded simultaneously.

Activation of motor cortex and its descending tracts

Currently, it is uncertain which neural elements are excited by scalp stimuli: dendrites, presynaptic terminals, cell bodies, efferent axons, or some combination are all possibilities. From single motor unit recordings from small hand muscles, it is thought that the site of stimulation with weak magnetic stimuli is probably the presynaptic terminals, with stronger stimuli exciting cell bodies or axons, or both. Levy et al.[1] advocated that the transcranial electrical-MEPs should be a pure motor signal. All of the sensory systems have synapses at some subcortical location and, because the signals from stimulated axons cannot cross a synapse retrogradely, there should be no sensory component[4]. They used two methods to activate motor tracts: direct stimulation of the dorsolateral part of the spinal cord[6] and transcranial (palate to scalp) stimulation of brainstem[2]. Both techniques produced descending volleys in the spinal cord and, in some cases, visible muscle contraction in the hands and feet. However, there was no evidence that the signals recorded from the spinal cord did not involve activities of sensory fibres. To identify whether these magnetic-MEPs recorded from spinal cord represent pure descending motor pathways and rule out involvement of sensory volleys, we applied a collision technique. Our collision technique indicated that magnetic-MEPs did not involve any descending antidromic sensory volleys. Our spinal cord partial sectioning study also indicated that no detectable portion of the MEPs was travelling through the dorsal column[1]. It is highly unlikely that this method of stimulation activates sensory pathways.

It needs to be considered that magnetic-MEPs have been recorded from peripheral nerves or muscles, which are highly dependent on spinal cord gray matter, peripheral nerves, muscle function, anaesthetics and muscle relaxant. Calculation of the MEP conduction time from the scalp to muscle requires subtraction of the peripheral nerve conduction time from the total, whereas magnetic-MEPs recorded from the spinal cord reflect only spinal cord and upper tract function. The ability to record transcranial magnetic-MEPs from muscles during near-maximal blockade of neuromuscular function with neuromuscular blockers and anaesthetics may seem improbable. Amplitude variation of MEPs by transcranial magnetic stimulation and recording from lower limb muscles will limit clinical use if SEP-style criteria for transmission failure are employed. Direct MEP recording from the spinal cord seems to be more useful for monitoring of spinal cord motor function.

Potential distributions of magnetic-MEPs and their tracts in the spinal cord

In Levy's early experimental studies it was thought that MEP reflected the corticospinal tract activity in the cat[1]. However, there is a controversy about the origin and the conduction pathway of MEPs. In our spinal cord sectioning study, no detectable portion of the MEPs was travelling through the posterior columns but N3 was localized to the dorsolateral column of the spinal cord, which corresponded to the corticospinal tract (pyramidal tract). This result was supported by analysis of individual nerve conduction velocities. Nerve conduction velocity of the corticospinal tract is reported to be around 50–70 m/s[2]. N3 showed conduction velocities appropriate to the corticospinal tract. The persistence of both N1 and N2 after transection of the dorsal half of the spinal cord indicates that a significant part of the MEPs is conducted through non-pyramidal tracts which exist in the ventral funiculus of the spinal cord. These results were also supported by analysis of nerve conduction velocities. Buchanan[7] has shown in the cat that the vestibulospinal tract (VST) and the reticulospinal tract (ReST) descend the ventral funiculus of the spinal cord. These vestibulospinal fibres, probably in conjunction with reticulospinal fibres, form a fast-conducting pathway (70–100 m/s)[8]. The largest fibres and, therefore, the fastest part of the spikes (the field potential distribution of N1 and N2), has been regarded as part of the pyramidal response[8]. However, it was shown in this paper that there were large-fibre tracts with exactly these ranges of conduction velocities in the ventral funiculus of the spinal cord. They corresponded to the distribution of the descending extrapyramidal tracts. Thus, it is more appropriate to call these descending motor tracts the cortico-motoneuron pathway or descending motor pathway than to refer to the corticospinal tract alone[1]. N4 and other later peaks were so small and difficult to reproduce that it was impossible to identify their origin.

Waveforms and latencies of magnetic-MEPs

Reproducible transcranial magnetic-MEPs were obtained in this study. Magnetic-MEPs consisted of approximately two to five large negative peaks, and waveforms were similar to electric-MEPs. There were no significant differences in latencies and in nerve conduction velocities of each corresponding negative peaks. Since magnetic-MEPs are produced by locally induced current by alteration of magnetic field, it seems there is basically no difference between magnetically and electrically induced MEPs. Though 5–10% of electric current reaches the brain[1], magnetic stimulation is not attenuated by bone structure; thus the amplitude of each negative peak was much larger in magnetic-MEPs. As the motor cortex of lower-limbs is located deeper in the cortex than that of upper-limbs, magnetic-MEPs are affected by the position or direction of the coil ring. Slight movements of the position of the coil often produced substantial alterations in the size of the recorded potentials. Amplitudes decreased when averaged with hand-holding. A stable ring-holder was essential to record constant reproducible waveforms. The strength of the magnetic field measured in the centre of the coil is about 1.5 Tesla at maximum intensity. There have been reports that a maximal response could be obtained when the rim of the coil-ring was positioned over the motor cortex. Thus, the largest waveforms were usually obtained when the centre of the coil was in the region between C4 and Cz, or C4 and Fp2. When the coil ring was inclined to 30–45° to the occipito-mental line, signals were optimal, latencies were the shortest and amplitudes were the largest. This suggested magnetic flux could reach deeper and elicited more neural elements of descending motor neurons or interneurons in deeper layers when the inclination of the ring was greater. As stimulation intensity was raised, the latency of the volley decreases. This could be due either to recruitment of faster-conducting axons or to spread of stimulation current to more distal parts of the cortical axons. If the decrease in latency was due to recruitment of faster conducting axons, then the conduction velocities measured between two recording sites on the spinal cord should change as stimulation strength is raised. However, conduction velocities were independent of stimulus intensity, although latencies of the volleys decreased as stimulation intensity was raised. This could be due to spread of stimulation current to more distal parts of the cortical axons. Higher intensity for stimulation should be used to stimulate motor cortex supplying the lower limbs which lies more deeply in the cortex than that supplying the upper limbs.

CONCLUSION

This monitoring method reflects descending motor tract function and is as sensitive as other monitoring methods such as SEPs or electric-MEPs. Magnetic-MEP monitoring, as an indicator for motor functional integrity, has important intraoperative applications for operations in which the motor system is threatened. Basic analysis of magnetic-MEPs recorded from the spinal cord will provide the necessary physiological data for this promising technique for clinical applications such as intraoperative spinal cord monitoring.

References

1. Levy WJ, McCaffrey M, York DH, Tanzer F. Motor evoked potentials from transcranial stimulation of the motor cortex in cats. Neurosurg. 1984;15:214–227.
2. Levy WJ, York DH, McCaffreyM, Tanzer F. Motor evoked potentials from transcranial stimulation of the motor cortex in humans. Neurosurg. 1984;15:287–302.
3. Anderson DK, Prockop LD, Means ED. Cerebrospinal fluid lactate and electrolyte levels following experimental spinal cord injury. J Neurosurg. 1976;44:715–722.
4. Levy WJ. Clinical experience with motor and cerebellar evoked potential monitoring. Neurosurg. 1987;20:169–182.
5. Lee C. The change of the spinal evoked potential caused by constant-speed subacute compression of the spinal cord. J Jpn Orthop Assoc. 1985;59:949–960.
6. Levy WJ, York DH. Evoked potentials from the motor tracts in humans. Neurosurg. 1983;12:422–429.
7. Buchanan AR. The course of the secondary vestibular fibers in the cat. J Comp Neurol. 1937;67:183–204.
8. Brookhart JM, Morris RE. Antidromic potential recordings from the bulbar pyramid of the cat. J Neurophysiol. 1948;11:387–398.

55
In vitro animal study of the spinal cord evoked potentials

T. IWAHARA, Y. ATSUTA, M. IMAI, Q. ZHAO, Y. HARADA and Y. TAKEMITSU

INTRODUCTION

Recently developed *in vitro* techniques have a number of advantages for the analysis of neuronal organization. In this method, the condition of the material is strictly controlled so that the effect of any single factor can be investigated. In the present study the *in vitro* spinal cord preparation was used and the effects of temperature, anoxia, potassium concentration and pH on the spinal cord potentials were studied.

MATERIALS AND METHODS

A total of 32 adult rats weighing 150–250 g were anaesthetized with halothane. Following the laminectomies, the spinal cord from the cervical to lumbar enlargement was excised. The isolated spinal cord was immersed in a bathing chamber filled with artificial, oxygenated cerebrospinal fluid (aCSF) maintained at 32°C (composition of normal aCSF in mM concentrations: NaCl 128, KCl 3, NaH_2PO_4 0.5, $CaCl_2$ 1.5, $MgSO_4$ 1, $NaHCO_3$ 21, glucose 30, pH7.4). The spinal cord evoked potential elicited by stimulation of the dorsal spinal cord surface was recorded using insulated bipolar platinum electrodes with exposed tips. For the stimulation and recording, an electro-diagnostic system (Dantec Counterpoint) was used and the band-pass filter was set at 5 Hz–5 kHz. The stimulation was delivered at 2 Hz with 0.05 ms square pulses. To study the effect of temperature on the evoked potential, the temperature of the aCSF was varied between 15 and 36°C. To study the effect of anoxia, the aCSF saturated with oxygen was replaced with that saturated with nitrogen. In one part of the anoxia study, the bath was separated into the stimulating site and recording site, and anoxia was induced only in the recording site. In addition, the effect of potassium concentration was studied by replacing the normal aCSF with high and low potassium

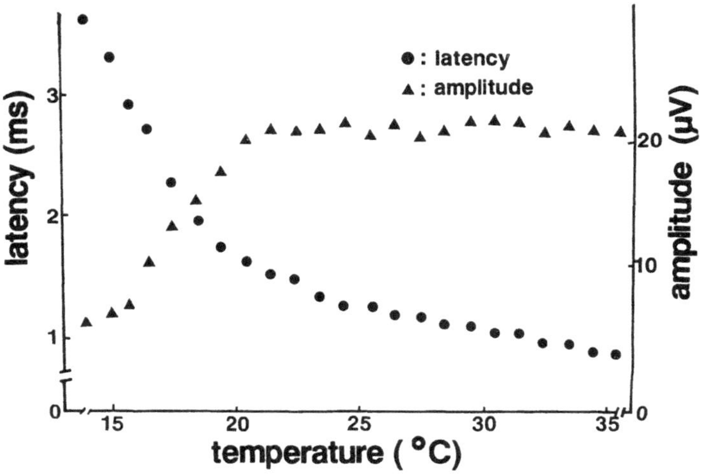

Figure 1 Effects of temperature on the amplitude (filled triangles) and latency (filled circles) of N1 potential. For explanation, see the text

concentration. Finally the effect of lowered pH was studied by injecting lactate into the bathing aCSF.

RESULTS

The data of the present study were based on 32 animals. The evoked potential consisted of early and late negative potentials (N1,N2) which were identical with those of *in vivo* experiments. The N2 potential had a lower threshold to electrical stimulation than the N1. It was confirmed that the evoked potential could be recorded for over 8 hours without a major change.

Effects of the temperature

Figure 1 shows a representative case of the results. Both the amplitude and peak latency of the Nl1 potential were affected by the temperature. The amplitude did not show a significant change between 20 and 36°C, but decreased progressively below 20°C. In contrast, the peak latency showed a gradual delay when the temperature was lowered to 20°C and the delay was accelerated below 20°C. The changes of the amplitude and latency were completely reversible so that the original potential was recorded when the temperature was returned to the original level.

Effects of anoxia

A change in evoked potential was observed after oxygen deprivation (Figure 2). In this case, the amplitude of the N2 potential was significantly decreased within 3 minutes, and following this change, the amplitude of the N1 potential decreased within 10 minutes. The time course of the amplitude decrease was

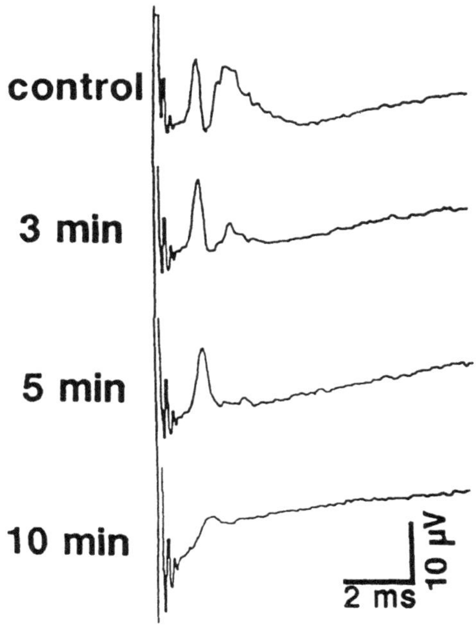

Figure 2 Effects of anoxia on the evoked potential (upward negative). The time after oxygen deprivation is shown on the left

different among animals, but consistently N2 amplitude decreased earlier than N1. Little recovery of the potential was observed by reoxygenation.

Effects of potassium concentration

The control evoked potential was recorded in the aCSF with the potassium concentration at 3 mM. Immediately after changing the aCSF to potassium-free aCSF, the amplitude of the potential increased. In contrast, when the aCSF with 53 mM potassium was used, the potential was markedly suppressed (Figure 3). These changes were reversible in both cases by returning to the original aCSF with 3 mM potassium.

Effects of lowered pH by lactate

Lactate was injected into the bathing CSF to lower the pH. When the pH was lowered from 7.4 to 7.2, the amplitude of the N1 potential was decreased by about 20%. By further lowering the pH to 6.8, the N1 and N2 potentials were markedly suppressed. The potential did not return to the original amplitude even after restoring the pH to 7.4.

DISCUSSION

In vitro techniques for analysis of various parts of the neuronal organization have been developed. In this method, the neurophysiological analysis of the

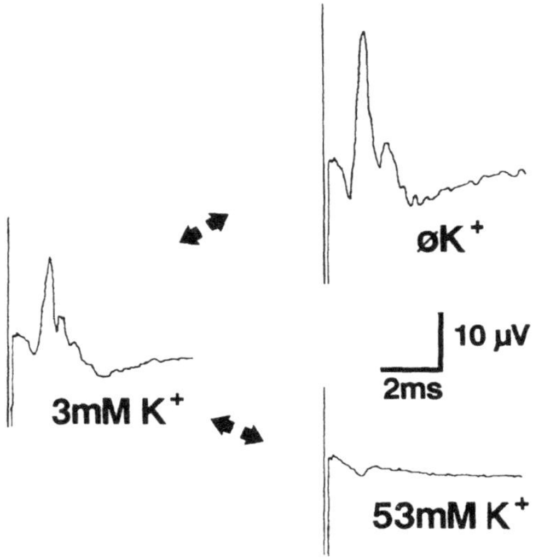

Figure 3 Effects of the potassium concentration on the evoked potential (upward negative)

target tissue can be easily conducted in a bathing chamber. Furthermore, using this preparation, environmental and physiological factors such as temperature, pH, ion concentrations, etc. and any pharmacological agents can be strictly controlled which makes it possible to analyse the pure effects of any single factor. We previously used a neonatal rat spinal cord–hindlimbs preparation to analyse motor function *in vitro*[1,2]. In the present study, adult rat spinal cord was employed for the analysis of the spinal cord potential evoked by spinal cord stimulation. This potential known to be generated mostly from tract fibres was identical to that of *in vivo* preparations. And this preparation lasted long enough for the experimental procedures to be continued. One great advantage of this *in vitro* model was that the experimental conditions were not affected by physiological changes of the animal due to anaesthesia, blood pressure, body temperature and metabolism. The present study analysed temperature, oxygen deprivation, potassium concentration and pH. The effect of these factors would have been difficult to analyse if an *in vivo* preparation had been used.

In regard to temperature, the result of the study showed that the latency of the N1 is more sensitive to the temperature decrease than is the amplitude. This information suggested to us that when only the latency of the potential is changed during spinal cord monitoring, temperature change may be a possible cause. Pure anoxia could be easily induced in the *in vitro* preparation by replacing the oxygen with nitrogen. This is similar to *in vivo* asphyxia and ischaemia experiments. However, asphyxia and ischaemia change not only the tissue oxygen pressure but also the carbon dioxide pressure, pH and local metabolism simultaneously. The result of the study showed that the N2 potential is more sensitive to anoxia than the N1. This information indicates that the N2 potential needs to be given considerable attention during spinal cord monitoring. Another advantage of the *in vitro* preparation was that partial anoxia, such as anoxia of only the stimulating site or recording site, could be produced. We have observed a different pattern of potential changes in these cases (not shown). Further investigation of anoxia effects will contribute to the understanding of the complicated pathophysiology of the spinal cord.

It is known that the potassium ion plays an important role in the regulation of neuronal activity. Extracellular potassium concentration is changed at the site of spinal cord injury[3,4]. It is also affected by the serum potassium level. Extreme changes of the potassium concentration were tested in the present study, and it was noted that the amplitudes of N1 and N2 were strongly affected by environmental potassium concentration. This phenomenon is thought to be caused by the change of axonal membrane excitability.

The results of the study demonstrated that this *in vitro* model is a strong tool in investigating physiological and pharmacological factors. This *in vitro* model will also be available for the study of quantitative spinal cord compression.

CONCLUSIONS

1. An *in vitro* experimental model for the study of spinal cord evoked potential was developed.
2. Using this model, the pure effects of temperature, anoxia, potassium concentration and pH could be studied.

Acknowledgement

A part of this study was supported by the Ministry of Education Grant 02454340.

References

1. Atsuta Y, Garcia-Rill E, Skinner RD. Characteristics of electrically induced locomotion in the rat *in vitro* brainstem–spinal cord preparation. J Neurophysiol. 1990;64:727–735.
2. Iwahara T, Atsuta Y, Garcia-Rill E, Skinner RD. Locomotion induced by spinal cord stimulation in the neonate rat *in vitro*. Somatosensory and Motor Res. 1991;8:281–287.
3. Eidelberg E, Sullivan J, Brigham A. Immediate consequences of spinal cord injury: Possible role of potassium in axonal conduction block. Surg Neurol 1975;3:317–321.
4. Lewin MG, Hansebout RR, Pappius HM. Chemical characteristics of traumatic spinal cord edema in cats. J Neurosurg 1974;40:65–75.

56
Comparison of spinal evoked potentials by transcranial magnetic and electrical stimulation – a canine study

H. KAJIHARA, Y. YAMAUCHI and M. NOZAWA

INTRODUCTION

Since Baker et al.[1] developed a method to record the electromyogram evoked by transcranial magnetic stimulation in 1985, there have been many reports on the usefulness of various potentials evoked by transcranial magnetic stimulation. However, the localization of the conduction pathway has not necessarily been clarified. Further, recording of the spinal evoked potential after transcranial magnetic stimulation has been considered to be relatively difficult.

We have performed an experimental study of efferent spinal potentials evoked by transcranial magnetic stimulation in dogs as a means to monitor spinal function. In this experiment, a comparative study of the efferent spinal potentials evoked by transcranial magnetic and electric stimulation was made in dogs, using our newly designed mapping spinal electrode.

MATERIALS AND METHODS

Under intravenous anaesthesia and after administration of a muscle relaxant, a laminectomy was performed.

For experiment 1, bipolar wire electrodes were attached to the dorsal midline epidural space at the levels of T7, T11 and L2. The potentials evoked by transcranial magnetic stimulation were recorded using a magnetic stimulator (DANTEC) and a recorder (Model 7S12, NEC SAN'EI Inst. Ltd.). The magnetic stimulation coil was 13.8 cm in outside diameter and 2.7 cm in inside diameter. The coil was attached at the mid-point of the line that connected the auricles, and 10 measurements were averaged in each session.

For experiment 2, an 8-channel mapping spinal electrode was attached to

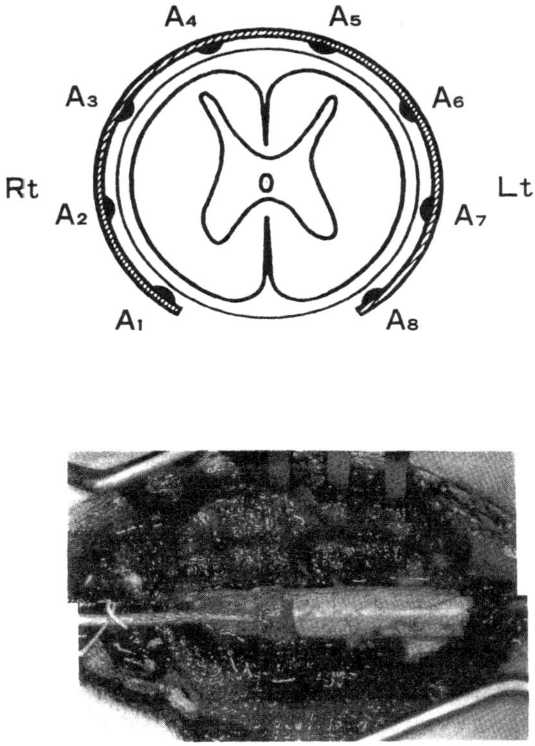

Figure 1 Mapping electrode and block diagram of experiment 2. An 8-channel mapping spinal electrode was attached to the dura at the level of T11 and potentials evoked by transcranial magnetic and electrical stimulation were recorded

the dura at the level of T11 in six young mongrel dogs and the potentials evoked by transcranial magnetic and electrical stimulation were compared. The electrical stimuli were delivered with needle electrodes, and the cathode was placed in the upper hard palate according to Levy[2]. The electrical stimulation and evoked potential recording were made with Neuropack 8 (Nihon Koden Kogyo Co.). The electrical stimuli were 0.2 ms square waves delivered at 20 Hz, and 100 measurements were averaged in each session. The mapping spinal electrode unit was a silicone device in which silver ball electrodes were arranged in two rows longitudinally and eight rows transversely. The unit was attached to the dura in such a way that the electrodes surrounded the dura in a ring shape, and eight bipolar electrodes were used for recording (Figure 1).

RESULTS

Efferent spinal potentials evoked by magnetic stimulation were recorded from all the bipolar electrodes attached to the epidural space. The potentials

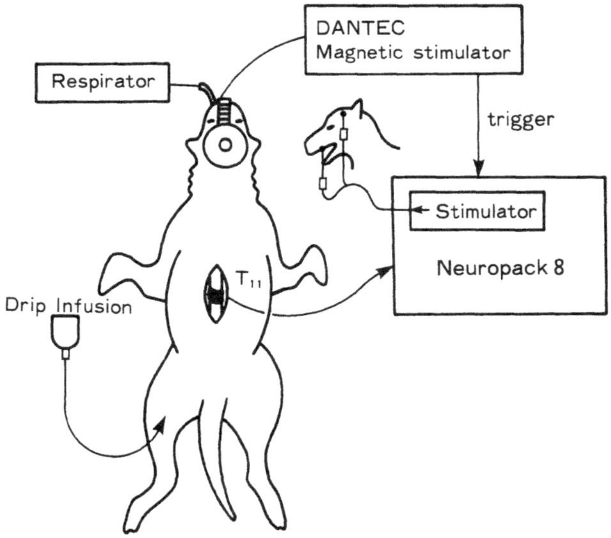

Figure 1 *Continued*

consisted of a first component of high amplitude and of relatively short latency and a second polyphasic component. With prolongation of the anaesthesia by intravenous administration of 2.0 mg/kg Nembutal, the amplitude of the polyphasic wave was reduced, and it disappeared from the recording. The first component, on the other hand, showed no change at all in either amplitude or latency. The potential began to appear at about 40% of the maximal stimulation intensity of 1.5 Tesla and its amplitude gradually increased but its latency showed no change. The rise of the evoked potential tended to be affected by artefacts with increasing intensity of stimulation (Figure 2).

With transection of the dorsal column under microscope between cord levels T13 and L1, the first component showed little change in either latency or amplitude.

The first component of the potential evoked by magnetic stimulation was recorded on all eight channels. No wave showed any change in latency, nor any notable change in amplitude.

Unlike after magnetic stimulation, the evoked potential after electrical stimulation tended to be polyphasic consisting of a fast component of high amplitude followed by polyphasic waves. When electrical stimulation was of

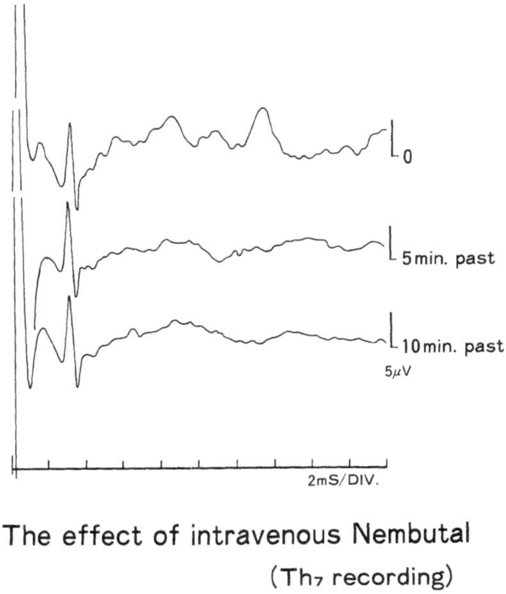

Figure 2 Potentials evoked by transcranial magnetic and electrical stimulation; recordings from bipolar wire electrode

low intensity, the evoked potentials showed a single component of high amplitude, but some were bifurcated. With mild electrical stimulation, some potentials were lateralized depending on the position of the stimulating electrode.

The latency of the first component became shorter with increasing intensity of electrical stimulation, and many waves divided into two or three peaks (Figure 3).

The difference in wave pattern due to laterality of the stimuli became unnoticeable with increasing current of stimulation, but there was a tendency for ventrally recorded waves to have a larger amplitude than the dorsally recorded ones.

The first component of the potential evoked by magnetic stimulation showed no change in latency, no matter whether it was recorded ventrally or dorsally, and the average latency for six dogs was 4.42 ms. The first component of the potential evoked by electrical stimulation was bifurcated in some cases, but the waves recorded dorsally consisted of a single peak. The average latency of these waves was 4.41 ms, about the same as the average latency of the first component of potentials evoked by magnetic stimulation.

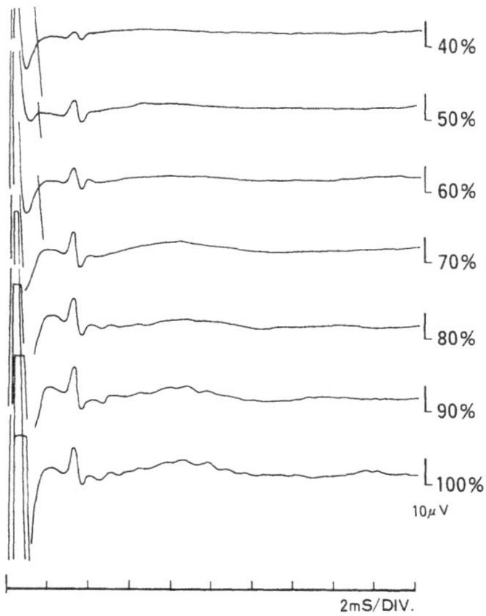

Changes due to magnetic stimulator output
(Max. 1.5 tesla, Th_7 recording)

Figure 2 *Continued*

DISCUSSION

Efferent spinal potentials evoked by transcranial magnetic stimulation could be recorded in dogs. Generally, it is considered to be difficult to record spinal evoked potential through this method and this is mainly attributed to artefacts at the time of recording. However it seemed possible to record the evoked potentials by appropriate stimulation conditions and a suitable recording method.

The magnetic stimulation evoked efferent spinal potentials consisting of a first component with high amplitude and relatively short latency and a polyphasic component with longer latency. With prolongation of the intravenous anaesthesia, this first component showed no change at all in either amplitude or latency. Also, with transection of the dorsal column, the first component was stable. Our previous experiment by lateral compression of the spinal cord showed that this first component tended to correlate with motor function[3]. From the above findings, it was suggested that the first component evoked by magnetic stimulation would mainly reflect the function of the ventral portion of the spinal cord.

Whereas the electrical stimulus evoked varied spike waves depending on the strength of stimulation, the latency of the first component to a mild electrical stimulus was similar to that produced by the magnetic stimulation. Further, the potentials evoked by magnetic stimulation showed little change from the dorsal through the ventral portion, whereas the electrical stimulation

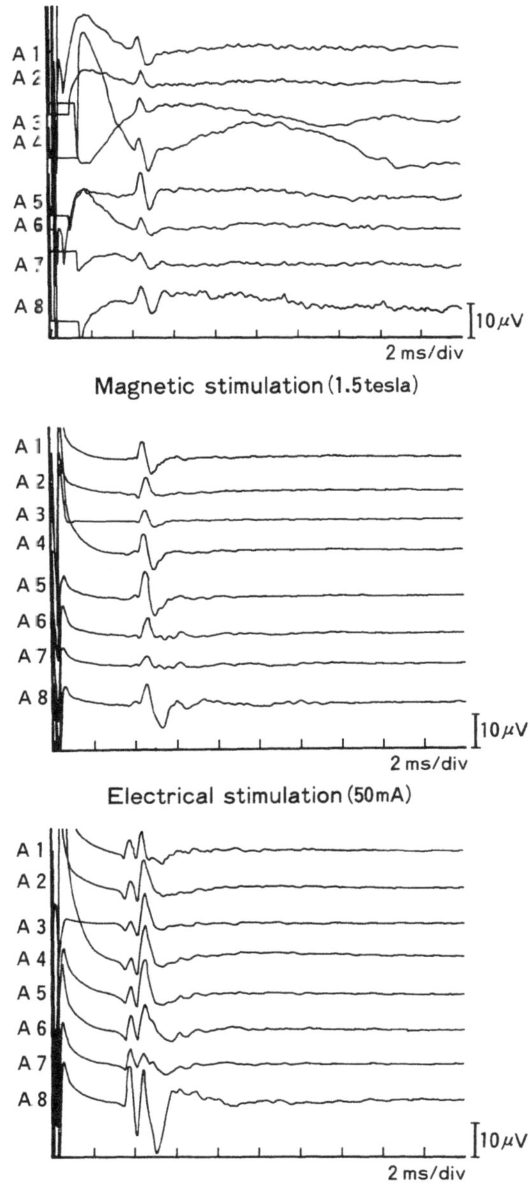

Figure 3 Potentials evoked by transcranial magnetic and electrical stimulation; recordings from mapping spinal electrode

evoked various changes. From this, we believe that the area of evoked activity from the magnetic stimulation was wider than that by electrical stimulation.

SUMMARY

1. Efferent spinal potentials evoked by transcranial magnetic stimulation were recorded in dogs.
2. It appeared that these potentials mainly reflected the function of the anterior portion of the spinal cord similar to the potentials evoked by electrical stimulation.

References

1. Baker AT, Jalinous R, Freeston IL. Noninvasive magnetic stimulation of the human motor cortex. Lancet. 1985;1: 1106–1107.
2. Levy WJ, York DM. Evoked potentials from the motor tracts in humans. Neurosurg. 1983;12:422–429.
3. Kajihara H, Yamauchi Y, Nozawa M et al. Spinal cord evoked potential by cranial magnetic stimulation in dogs. J Jpn Orthop Assoc. 1990;64:S1121 (in Japanese).

57 Comparison between descending segmental evoked spinal cord potentials and conductive evoked spinal cord potentials under acute spinal cord compression

H. KOMORI, K. SHINOMIYA, M. YOKOYAMA, T. MATSUOKA and K. FURUYA

INTRODUCTION

The conventional method of using Conductive Evoked Spinal Cord Potentials (hereinafter abbreviated C-ESCP) to monitor the spinal cord function during surgery on the cervical or thoraco-lumbar region is the recording of ascending spinal cord potentials, and it is not possible to detect dysfunctions of motor pathways or grey matter of the spinal cord at an early stage. To improve this status, we have reported since 1987 on action potentials recorded in the median nerve after a stimulus to the upper cervical spinal cord as a new monitoring method (Descending Segmental Evoked Spinal Cord Potentials; hereinafter abbreviated DS-ESCP), especially on the origin of the potentials and the conductive pathways[1,2]. In the present study, we have compared the changes in the potentials and pathological findings in cases of acute posterior compression of the spinal cord in order to check this method for its reliability.

METHOD

Intratracheal intubation was established in adult cats anaesthetized by Ketamine (30 mg/kg). The animals were immobilized with pancuronium bromide (0.1 mg/kg/h) to induce controlled ventilation. Stimuli were given to the epidural space at C1/2 on the dorsal side using catheter-type electrodes with a distance of 10 mm between two electrodes. DS-ESCP was recorded at the median nerve at the elbow using bipolar hook-type electrodes, while the descending C-ESCP was recorded at the thoracic epidural space using catheter-type electrodes. DISA 1500 system and EVOMATIC 8000 system (DANTEC Denmark) were employed and the bandwidth of the recording system (10–1000 Hz) was kept constant during recording. Stimulation was applied with an intensity of 1–30 mA, a frequency of 5–52 Hz and a duration of 0.2 ms.

Posterior compression of the spinal cord was made in a persistent manner by placing a cylindrical weight with a diameter of 3.5 mm and a weight of 6.6 g upon the spinal cord at C5/6 level after C4–6 enblock laminectomy and incision of the dura mater. The degree and duration of the compression were made to vary, and potentials were recorded during compression and for 1 hour after release. The cats were sacrificed by large doses of thiobarbiturate to collect pathological specimens.

RESULTS

Conductive pathways

DS-ESCP recorded from the median nerve at the elbow level of adult cats was found to be composed of two negative spikes, with inter-peak latency of approximately 1 ms. The first peak showed hardly any change in amplitude and latency according to the frequency of stimulation, whereas the second peak showed reduction in amplitude with more frequent stimulation.

To clarify the conductive pathway, the nerve roots were severed one by one from the T1 posterior root rostrally. The first peak showed 58% reduction in amplitude after additionally cutting the C8 posterior root, and almost disappeared after cutting the C7 posterior root. At this point, the amplitude of the second peak was reduced by approximately 20% and the peak latency was delayed by 0.2 ms.

After transection of the posterior roots from C4 to T1, the anterior roots were severed one by one from the T1 anterior root rostrally for observation of changes in the remaining peak. This amplitude decreased to 17% with severing of the C8 anterior root and almost disappeared with additional C7 anterior root severance.

In an experiment with partial transection of the spinal cord, the first peak decreased in amplitude to 10%, while the second peak was reduced by only 55% when the dorsal column had been mainly transected at the C3 level. When the dorsal half of the spinal cord had been transected, the first peak almost disappeared, while the second remained at 13% in amplitude. When the entire spinal cord had been transected, both peaks disappeared.

Acute posterior compression of the spinal cord

Compression by 120 mmHg

When the duration of the compression was less than 30 minutes, both potentials of DS-ESCP were decreased in amplitude immediately after the compression. On the other hand, C-ESCP showed no change of its amplitude, but latency remained delayed during the whole period of the compression. After releasing the compression, DS-ESCP showed prompt improvement of its amplitude and could be restored to approximately 80% of the control at 1 hour after release. C-ESCP also showed a delay of latency, but no further delay was found after release of the compression. There were no abnormal pathological findings.

When the duration of the compression exceeded 30 minutes, the improvement of amplitude of DS-ESCP was poorer than that of the cases with shorter compression. On the other hand, C-ESCP showed hardly any changes in amplitude, but further delay of latency was found after release of the compression (Figure 1). In this circumstance, notable oedema was found on the dorsal side of the spinal cord at the compressed region, and haemorrhages were noted around this area. In terms of pathological findings, no change was seen in the white matter of the spinal cord, but minor haemorrhage in the grey matter was occasionally noted.

Compression by 160 mmHg

Even when the duration of compression by 160 mmHg was less than 15 minutes, the improvement of amplitude of DS-ESCP after release was poorer than that seen with compression by 120 mmHg. N2 of C-ESCP showed a small amplitude reduction only during compression. Delay of latency of C-ESCP, however, was found not only during compression but after release (Figure 2). In terms of pathological findings, deformation of the central canal was notable and relatively large haemorrhagic specks were found in the grey matter (Figure 3).

When the duration of the compression was made longer, the distal latency of C-ESCP was delayed up to approximately 120% of the control at 1 hour after release.

DISCUSSION

Conductive pathways

The first peak of DS-ESCP recorded under anaesthesia by ketamine and pancuronium bromide is a potential that is mainly conducted antidromically through the dorsal column in the spinal cord and via the C7 and C8 posterior roots. Therefore, the first peak can serve as a guideline for dysfunction of sensory tracts. On the other hand, the second peak reflects potentials not only conducted through the dorsal half of the spinal cord, but also conducted through the ventral half, and approximately 80% of this potential involves a synapse at the C8 spinal cord segment and proceeds via the C8 anterior root. Therefore, this potential does not serve as a monitor for the corticospinal tract in the pure sense, but it is likely to reflect the functions of the grey matter of the spinal cord.

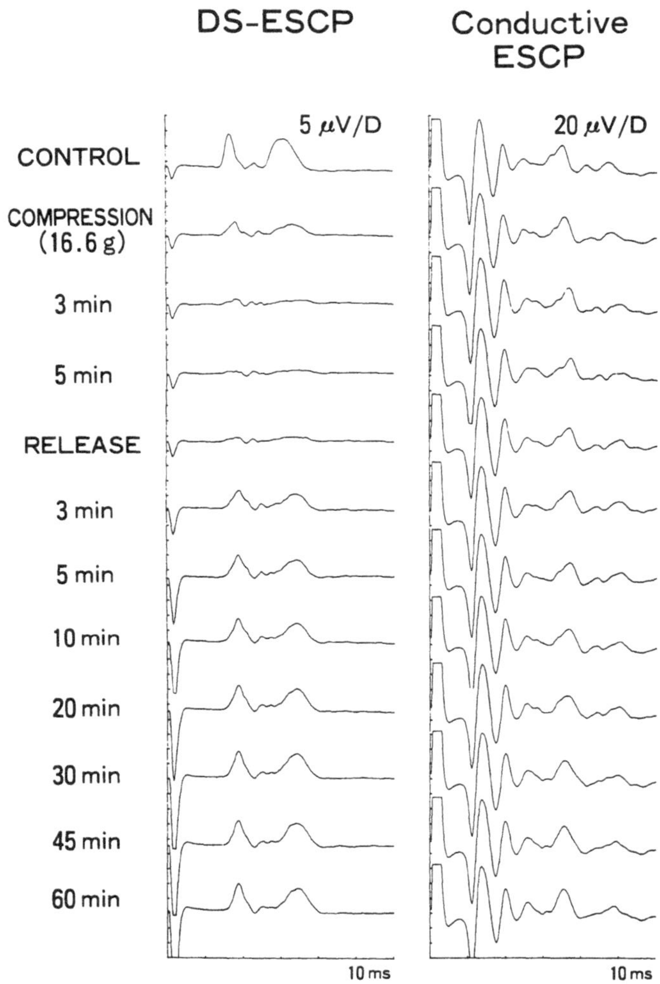

Figure 1 Posterior compression by 120 mmHg (16.6 g) for 5 minutes. Both potentials of DS-ESCP were decreased in amplitude immediately after the compression. After releasing the compression, DS-ESCP showed prompt improvement of its amplitude. On the other hand, C-ESCP showed no change of its amplitude, but latency remained delayed during the whole period of the compression

Acute posterior compression of the spinal cord

It can be realized from the results of the experiment with spinal cord compression that the spinal cord disturbance caused by compression varies from reversible to irreversible in nature, depending on the degree and duration. The changes in the potentials when irreversible changes were induced in the grey matter can be characterized by the poor improvement of the amplitude in DS-ESCP and the further delay of latency in C-ESCP

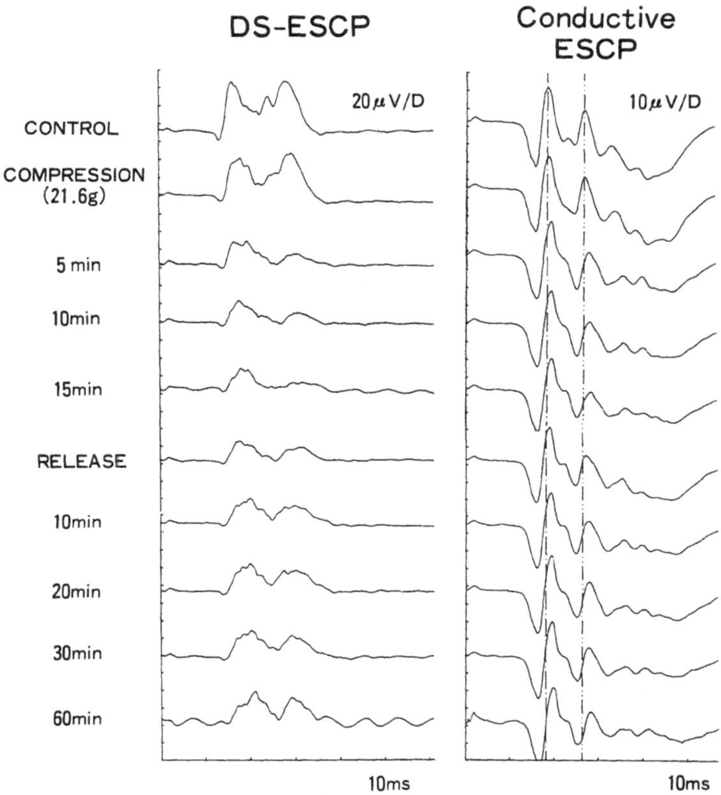

Figure 2 Posterior compression by 160 mmHg (21.6 g) for 15 minutes. The improvement of amplitude of DS-ESCP after release was poorer than that seen with compression by 120 mmHg. Delay of latency of C-ESCP was found not only during compression but also after release

after releasing compression. Consequently, when spinal cord monitoring is performed using only C-ESCP, it is likely that judgment of spinal cord dysfunction at an early stage would be difficult even when compression lasts so long as to cause irreversible pathological changes in the grey matter. On the other hand, if DS-ESCP is combined, the amplitude will be markedly reduced immediately after application of compression, which can serve as a warning to the surgeon.

Problems in clinical applications

The advantage of our method is that the functions of both upper and lower extremities can be monitored at the same time with a single site of stimulation. Besides, dysfunctions of grey matter can be sensitively examined by the second peak of DS-ESCP. However, if dysfunctions of the spinal cord are already present, it will be highly likely that DS-ESCP can hardly be recorded[3]. Besides, it is necessary to strengthen the stimulus more than 15 mA in animal

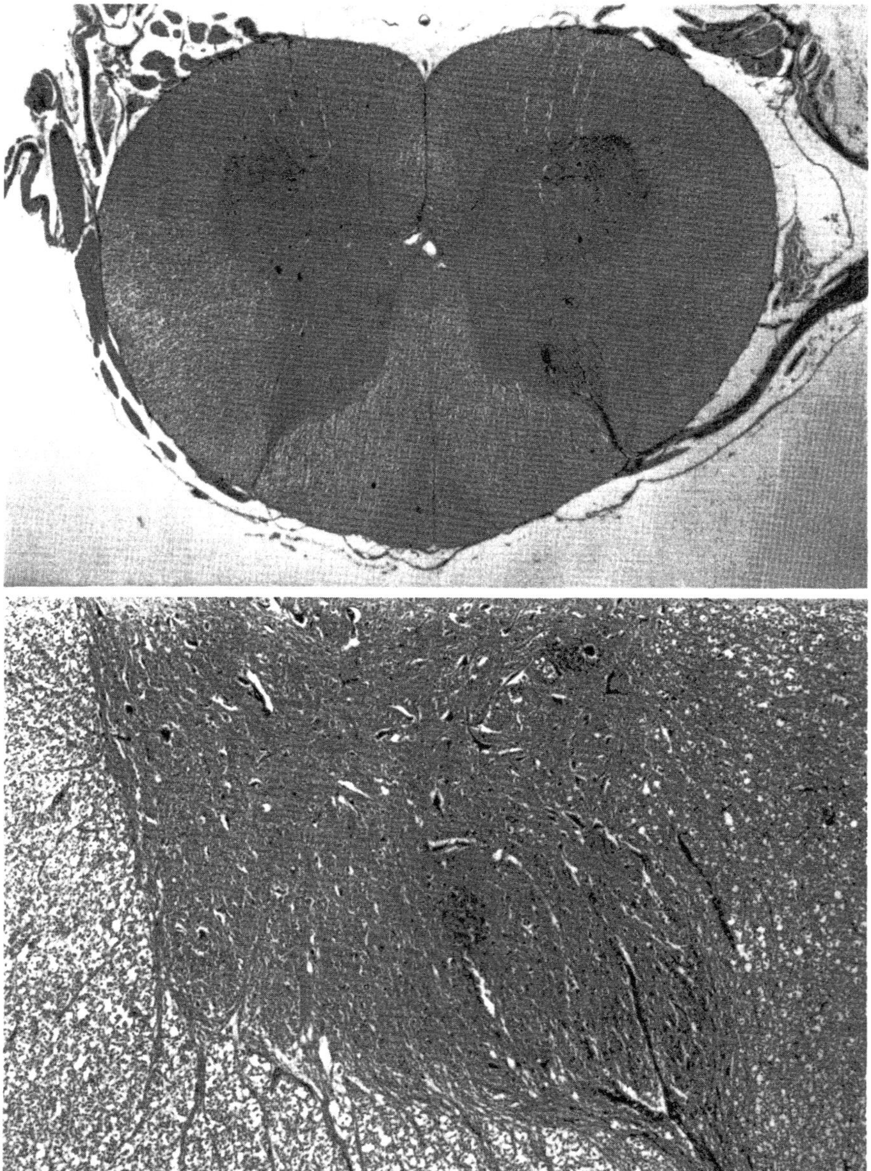

Figure 3 Histological findings after compression by 160 mmHg for 15 minutes. a: × 3.3 H&E. b: posterior horn × 16 H&E

experiments in order to stimulate the whole spinal cord[2], and action potentials via a synapse can hardly be recorded under anaesthesia commonly used in recent years[3–6]. Consequently, it will be an important issue in the future for an anaesthetic method to be established that will permit recording of trans-synaptic potentials, and for a stimulation method utilizing the characteristics of synapses such as post-tetanic potentiation to be designed.

SUMMARY

1. The first peak of DS-ESCP recorded under anaesthesia by ketamine and pancuronium bromide was a potential conducted through the dorsal column and via the posterior roots. The second peak was found to be a potential that descended through the whole spinal cord and mainly proceeded via the C8 anterior root.
2. Sensitive responses to compression of the spinal cord and its release were noted in DS-ESCP. C-ESCP showed a persistent delay of latency in the cases of long-lasting or strong compression.
3. In terms of pathological findings, changes such as haemorrhages in the grey matter of the spinal cord were noted in the cases of weak compression over a long period and strong compression over a short period.
4. Combined use of DS-ESCP is desirable for spinal cord monitoring, as this is likely to reflect the functions of grey matter of the spinal cord.

References

1. Yokoyama M, Shinomiya K, Sato H, Komori H. The experimental study of nerve action potentials recording from the median nerve by stimulating the cervical spinal cord. J Electrodiagnosis Spinal Cord. 1987;10:27–30.
2. Yokoyama M, Shinomiya K, Sato H, Komori H, Furuya K. Monitoring motor function of the spinal cord: the Descending Segmental Evoked Spinal Cord Potential. In: Shimoji K et al., Spinal Cord Monitoring and Electrodiagnosis. Berlin Heidelberg: Springer-Verlag; 1991: 27–35.
3. Shinomiya K, Yokomaya M, Komori H, Matsuoka T, Mutoh N. Clinical application of evoked spinal cord potentials for spinal surgery. Orthopaedic Surgery and Traumatology. 1991;34:31–37.
4. Matsumoto M, Kawamura H. The effects of Halothane and Enflurane on SEP (Spinal Evoked Potentials) and VRP (Ventral Root Potentials). J Electrodiagnosis Spinal Cord. 1987;10:39–42.
5. Matsumoto M, Namiki A, Kawamura H. The effects of intravenous anesthetics on the VRP (Ventral Root Potentials). J Electrodiagnosis Spinal Cord. 1988;11:25–28.
6. Matsuoka T, Shinomiya K, Komori H, Furuya K. An experimental study of Descending Segmental Spinal Cord Potentials recorded through Sciatic and Peroneal nerves. J Electrodiagnosis Spinal Cord. 1991;13:36–39.

58
Refractory period of motor evoked potentials and spinal cord evoked potentials in the cat

K. T. TAMURA, T. O. OHIRA and S. T. TOYA

INTRODUCTION

Reports have appeared in the literature describing the refractory period in peripheral nerves, and its clinical usefulness is already established. There are however few such investigations in the central nervous system. The D-wave in motor evoked potentials (MEPs) and the N1-wave in spinal cord evoked potentials (SCEPs) have been reported as non-synaptic potentials in previous studies. Accordingly a refractory period as in peripheral nerves may be suggested in MEPs and SCEPs. Both potentials are important in the diagnosis of pyramidal tract damage. The aim of the present study was to demonstrate the existence of the refractory period in MEPs and SCEPs, and to determine its physiological characteristics compared with peripheral nerve in an animal study. Additionally, in an intracerebral haemorrhage model we tried to demonstrate the possibility of the application of the refractory period in MEPs as an index of pyramidal dysfunction.

MATERIALS AND METHODS

Normal refractory period

Animal preparation

These studies were performed on 17 adult cats in SCEPs, and on 35 adult cats in MEPs. The weight of the cats was from 2.5–3.5 kg. The cats were premedicated with atropine (0.02 mg/kg), and after intramuscular injection of ketamine (35 mg/kg), tracheostomy was performed and oxygenation and normocapnia were maintained with controlled ventilation. Anaesthesia was maintained with an intramuscular injection of ketamine and intraperitoneal injection of 1% α-chloralose. An infusion catheter was placed into the femoral vein and a femoral arterial line was established to allow continuous monitoring of blood pressure and gas analysis. Neuromuscular block was established with intravenous administration of pancuronium bromide. Rectal temperature was maintained within a range of 37.5–38.0°C. Cats were fixed in the sphinx position.

Recording of SCEPs

Laminectomy was performed at the vertebral level of C1 and T10. Dual-electrode epidural catheters were inserted into the epidural space at the vertebral level of C3 and T8. To record the ascending SCEP, the T8 electrodes were used for stimulation and the others for recording. Stimulation and recording electrodes were exchanged to record the descending SCEP.

Recording of MEPs

A small right frontal craniectomy was performed just above the anterior sigmoid gyrus, where bipolar 2 mm silver ball electrodes were placed for stimulation. After C1 laminectomy, an epidural catheter electrode (anodal) was inserted in the epidural space at the C2–C3 vertebral level. A cathodal needle electrode was inserted in the paraspinal muscle at the same level as the anodal electrode.

Stimulation and recording protocols

A constant-current unit (VC-11 and SEN-3301: Nihon Kohden) was used to deliver rectangular pulses, of 100 μs duration. Stimulation at supraminimal intensity output was used (about 2–3 mA). An evoked potential averager (QC-111J: Nihon Kohden) was used to record both SCEPs and MEPs and the data were recorded on a personal computer (PC-9800VM: NEC). Four to eight signals were averaged digitally. The analysis time was 10 ms and longer, according to the stimulus interval. The high and low cut-off filters were set at 10 kHz and 300 Hz respectively. The amplification gain ranged from 10 to 50 μV/division.

Construction of recovery curves

The paired shock technique was used to construct a recovery curve, using the digital subtraction method. The amplitude of the wave (N1-wave in SCEPs, D-wave in MEPs) was determined, as elicited by test stimuli (amplitude 1: A1) and conditioning stimuli (amplitude 2: A2). A2 was expressed as a percentage of A1 (R) and plotted as a function of the stimulus interval, which ranged from 0.4 ms to 10 ms in SCEPs, 0.4 ms to 500 ms in MEPs. When the wave elicited by conditioning stimuli was covered by the test stimuli wave, the subtraction method was used. The frequency of each paired stimulation was 2 Hz.

Intracerebral haemorrhage model

Adult cats were prepared in the same way as for MEPs recording (see above). First the control MEP was recorded and its recovery curve was calculated. A microballoon was then stereotactically inserted just lateral to the anterior limb of right internal capsule; A13, H2, L10, by Snider atlas[1], on the same side as the stimulated cortex.

After one hour, the balloon was inflated with saline (total 0.2 ml) over 30 minutes. Sixty minutes after full inflation, the balloon was rapidly deflated. The volume, 0.2 ml, in the cat is equal to about one-third of the average volume of the intracerebral haemorrhage in the human.

During this procedure, the D-wave from both testing and conditioning stimuli were measured every 15 minutes during inflation and every 30 minutes for an hour after deflation. Stimulus intervals we analysed were 1, 2, 3, 4 and 5 ms.

Those cases with macroscopic haematoma were excluded and finally 20 cats were used in this study.

RESULTS

Normal refractory period

SCEPs

Figure 1 presents the recovery curve of N1-wave for ascending and descending SCEPs. These curves were similar to that of peripheral nerve. The absolute refractory period was about 0.4 ms in both. The two curves exhibited a difference in relative refractory periods.

MEPs

Figure 2 presents the recovery curve of D-wave for MEPs. As is the case for the peripheral nerve, it shows an absolute refractory period (0.4 ms) and hypernormal phase (6–10 ms).

However in the relative refractory period (0.5–5 ms) the recovery curve showed a notch, 'step formation', around 2 to 3 ms. And in the longer stimulus intervals than hypernormal phase, it was reduced again: the 'suppression phase'. Neither phenomenon has been observed for peripheral nerve.

Intracerebral haemorrhage model

No significant change was seen during this study in the amplitude of the D-wave by test stimuli (Figure 3). However when the stimulus intervals were 3 ms and 1 ms, the excitability of D-wave by conditioning stimuli was changed significantly, which resulted in change of the recovery curve (Figure 4). On inflation of the balloon, the excitability of the D-wave by conditioning stimuli decreased at 1 ms ($0.001 < P < 0.01$) and 3 ms ($P < 0.001$). Upon deflation,

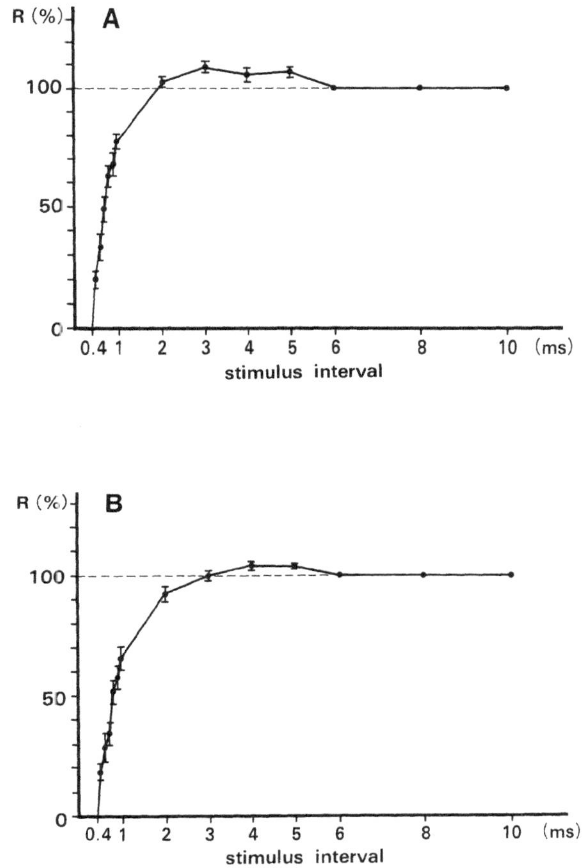

Figure 1 A: Recovery curve of the N1-wave in ascending SCEPs. B: Recovery curve in descending SCEPs

the excitability was increased at 3 ms ($P < 0.001$). The range of stimulus intervals in our analysis, (1–5 ms) was matched to the relative refractory period of MEPs which had been shown in the first part of the study.

DISCUSSION

There have been many reports about the peripheral nerve refractory period. Lehmann[2] reported its clinical usefulness in some neuropathies. However there are only a few reports about the refractory period of the central nervous system, especially in somatosensory evoked potentials. Regarding the pyramidal tract, Inghilleri[3] reported the recovery cycle in peripheral muscle potentials by transcranial stimulation of the motor cortex in the human. Direct recording from the spinal cord however has not yet been discussed.

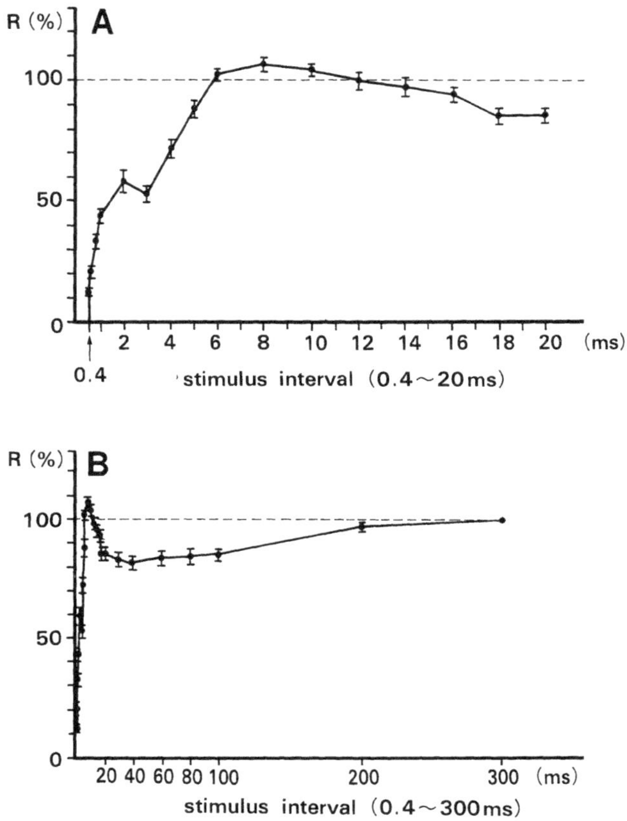

Figure 2 Recovery curve of the D-wave in MEPs. The upper curve is at stimulus interval of 0 to 20 ms, the lower curve of 0 to 300 ms

We tried to find a new direct parameter concerning the pyramidal tract in the central nervous system which could be used during surgery.

The N1-wave in SCEPs[4] and D-wave in MEPs[5] are thought to be non-synaptic potentials and indicate pyramidal tract activity. Our study demonstrates the existence of a refractory period in SCEPs and MEPs. In SCEPs its recovery curve was similar to that of peripheral nerve but the recovery curves of ascending and descending SCEPs were not the same for the relative refractory period. Perhaps it indicates that both conduction tracts are different.

In one patient with spinal cord disease, we measured the recovery curve during surgery. The results were similar.

On the other hand the recovery curve of MEPs was different from SCEPs in two respects. While it essentially consisted of an absolute and a relative refractory period and a hypernormal phase, the curve was more complex in the relative refractory period than SCEPs. There was a 'step formation'

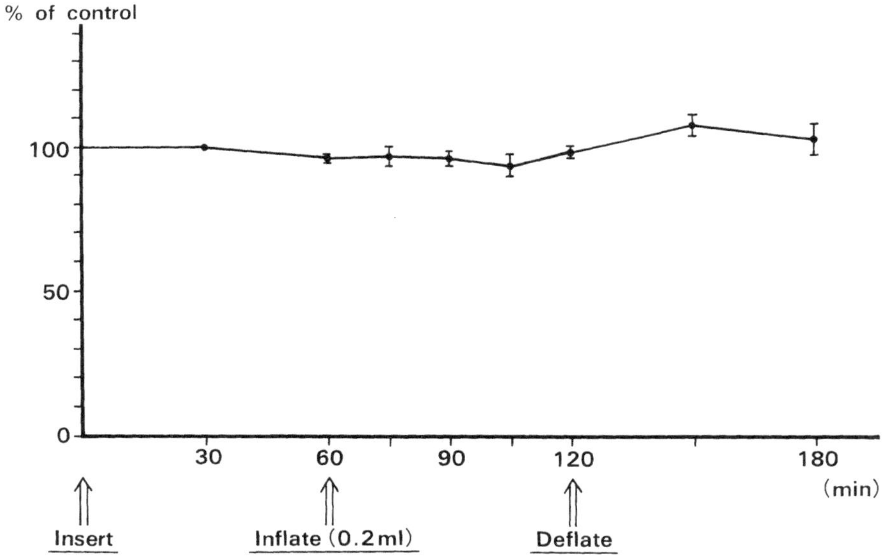

Figure 3 The change in the amplitude of the D-wave by single stimulation during Part 2 study. The amplitude of D-wave before insertion is 100%

between 2–3 ms. The pyramidal tract (from cortex to spinal cord) is known to consist electrophysiologically of many kinds of fibres, thus perhaps resulting in this 'step'. Furthermore our study showed one more period longer than the hypernormal phase, which was termed the 'suppression phase'. This phenomenon is difficult to explain by individual cellular activity because of the time. We assume it is the effect of other suppressive pathways driven by interneuron networks.

In the intracerebral haemorrhage model our preliminary data suggested that, in the relative refractory period, the change of amplitude in the D-wave was more sensitive with conditioning stimuli than with testing stimuli. Currently it is common to use the amplitude of the D-wave by single stimulation when making evaluation of the pyramidal tract by MEPs. One must measure the normal amplitude of the D-wave for control but it is technically difficult in the human when using direct cortical stimulation. In addition the amplitude of the D-wave is easily changed by the conditions of stimulation, especially during cranial surgery. However, measuring the refractory period has several advantages. First, one needs only damaged-side stimulation. Second, because depending on the ratio of amplitude to paired shocks, more reliable data can be acquired than by the single stimulation method using the absolute amplitude during surgery.

We predict it will be possible to determine the constant value of the refractory period in SCEPs and MEPs in the human. We have not yet examined the usefulness of SCEPs in spinal cord injury but as is the case for peripheral nerve and MEPs, the change of refractory period of N1 will

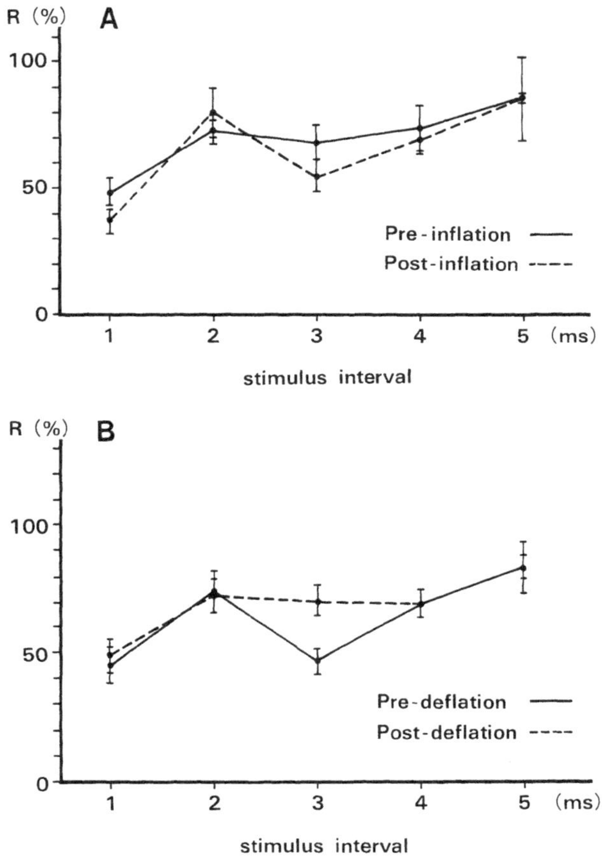

Figure 4 The change of recovery curve at the stimulus interval of 1 to 5 ms during Part 2 study. A: The recovery curves in pre-inflation and 15 minutes after inflation. B: The recovery curves in pre-deflation and 30 minutes after deflation

be more sensitive than that of absolute amplitude of N1 for evaluating spinal function. We are not continuing to examine this point.

References

1. Snider RS, Niemer WT. A stereotaxic atlas of the cat brain. University of Chicago Press; 1964.
2. Lehmann HJ, Tackmann W. Neurographic analysis of trains of frequent electric stimuli in the diagnosis of peripheral nerve disease. Europ Neurol. 1974;12:293–308.
3. Inghilleri M, Berardelli A, Cruccu G, Priori A, Manfredi M. Motor potentials evoked paired cortical stimuli. Electroenceph clin Neurophysiol. 1990;70:382–389.
4. Imai T. Human electrospinogram evoked by direct stimulation on the spinal cord through epidural space. J Jap Orthop Ass. 1976;50:1037–1056.
5. Patton HD, Amassian VE. Single and multiple-unit analysis of cortical stage of pyramidal tract activation. J Neurophysiol. 1954;17:345–357.

59
The effect of anaesthetic agents on descending spinal cord evoked potential and the compound muscle action potentials elicited by stimulation at the motor cortex and the spinal cord

Y. TODA, T. TAMAKI, B. TAYLOR and R. OGAWA

INTRODUCTION

Intraoperative spinal cord monitoring using motor evoked potentials is a desirable goal, but the sensitivity of most techniques limits their value during anaesthesia. This paper examines the effects of various anaesthetic agents on three different MEP recording techniques in 41 adult cats.

METHOD

The three motor evoked potential techniques are examined (Figure 1). The MEP is evoked by monopolar stimulation of the motor cortex and bipolar epidural recording at Th5. The gastrocnemius compound muscle action potential evoked by bipolar epidural stimulation at Th5 is called CMAP stimulated at spinal cord. The CMAP stimulated at the cortex combines these two methods by stimulating the cortex and recording compound muscle action potentials from gastrocnemius.

The anaesthetic agents studied were as follows:

Group 1 – inhalational agents enflurane, halothane and isoflurane were administered individually at 1 M.A.C. and 2 M.A.C. (minimum alveolar concentration) concentrations with pure oxygen.

Group 2 – 67% nitrous oxide was given with oxygen.

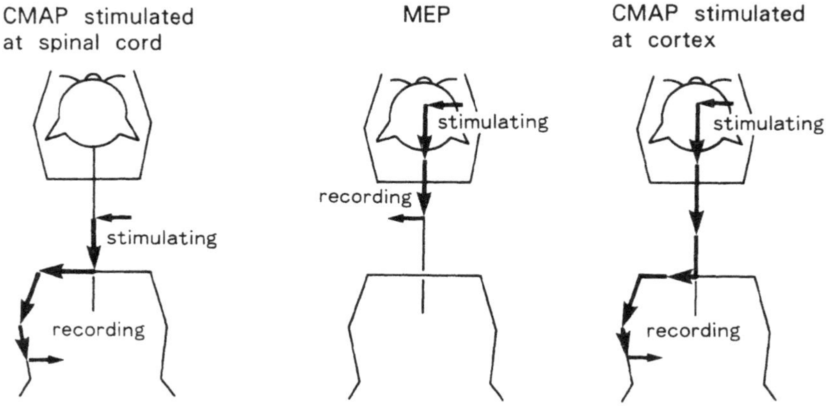

Figure 1 Diagrammatic representation of three motor evoked potential pathways assessed

Group 3 – 67% nitrous oxide was added with inhalational agents of group 1.

Group 4 – modified neuroleptanalgesia regime of diazepam 0.4 mg/kg, pentazocine 1.0 mg/kg and 67% nitrous oxide.

For stimulation, constant current stimuli of 0.2 ms duration at three per second were used in all studies. For epidural cord stimulation the intensity was set at 5 mA for transcranial cortical stimulation at 60 mA.

High cut filters were set to 3000 Hz for all recordings and low cut filters to 20 Hz for compound muscle action potentials and 150 Hz for the motor evoked potentials; 200 sweeps were averaged for each recording.

RESULTS

Control waveforms were recorded for each evoked potential following intramuscular pentazocine and hydroxizine. The MEP consisted of a short latency spike or D-wave and polyphasic I-wave.

The compound muscle action potentials evoked by stimulation of the cortex or spinal cord were simpler in form.

High frequency stimulation at 50 Hz or administration of muscle relaxant abolished the I-wave and the compound muscle action potential.

The effects of the inhalational agents at 1 M.A.C. for 30 minutes were observed (Figure 2). In the MEP the D-wave remained stable in all cases. The I wave amplitude was reduced to approximately 50% of the control value but then stabilized. The compound muscle action potential stimulated from the spinal cord was greatly reduced and the compound muscle action potential stimulated from the cortex was virtually abolished. A marked reduction in amplitude of both modalities of the compound muscle action potential occurred within 10 minutes of introducing the inhalational agents. The spinally evoked response was reduced to less than 30% of its initial value and the cortical response was abolished.

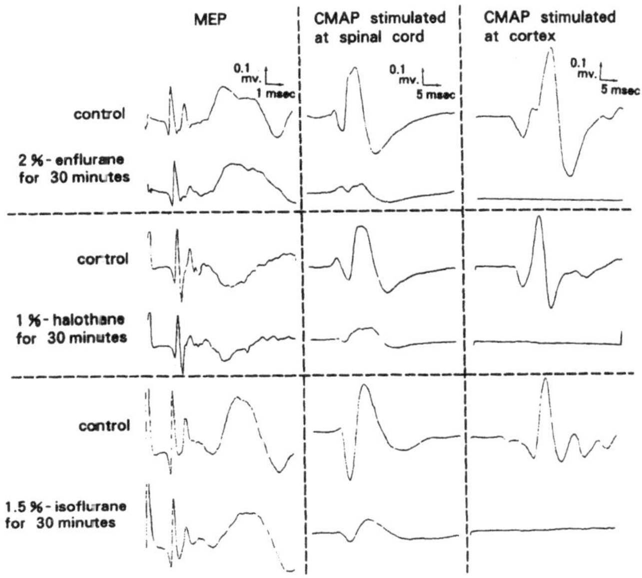

Figure 2 Traces of motor evoked potentials and compound muscle action potentials showing the effect of administration of inhalational anaesthetics at 1 M.A.C. for 30 minutes

All these effects of inhalational agents were found to be dose dependent by comparison of the effects of the inhalational agents at 1 M.A.C. and 2 M.A.C. Once again in the MEP the D-wave is least affected and the compound muscle action potential after cortical stimulation is most affected. The maximum effect took between 10 and 20 minutes to occur.

67% nitrous oxide showed similar effects on the recorded potentials over a similar time of administration. It was however still possible to record a compound muscle action potential from cortical stimulation of about 20% of the initial amplitude, unlike the total loss that occurred with halothane.

When 67% nitrous oxide was added to each inhalational agent the reduction in amplitude of all responses other than the D-wave was emphasized (Figure 3).

In the graph the upper line is D-wave amplitude and this was significantly affected. The second line is the I-wave amplitude which was reduced to approximately 30% of the control value. The third and fourth line are compound muscle action potentials after spinal cord stimulation which were abolished within 10–15 minutes.

The effects of an intravenous bolus injection of diazepam 0.4 mg/kg and pentazocine 1 mg/kg with 67% nitrous oxide were observed (Figure 4). These intravenous agents made little difference to the response amplitudes when compared to nitrous oxide alone, and each potential, including the compound muscle action potentials from cortical stimulation continued to be recordable.

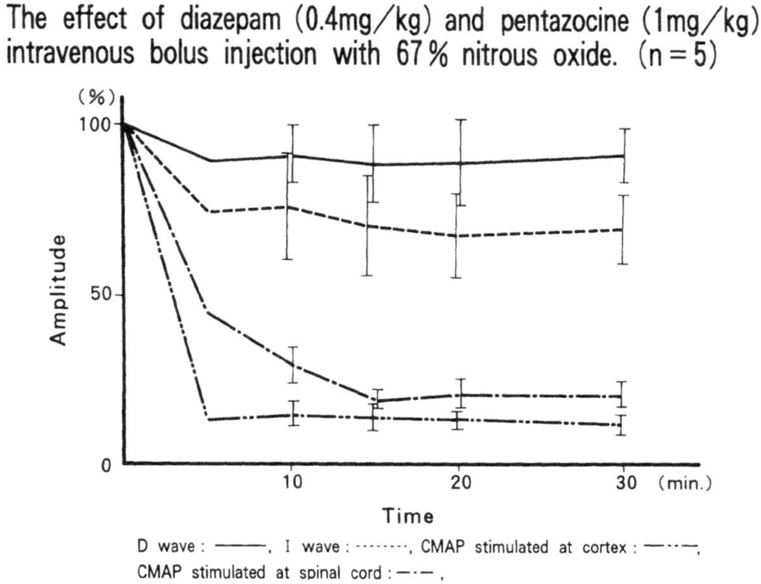

Figure 3 The effect upon MEP and CMAP of inhalational anaesthetics at 1 M.A.C. with 67% nitrons oxide in oxygen. Note the absence of CMAP at 30 minutes

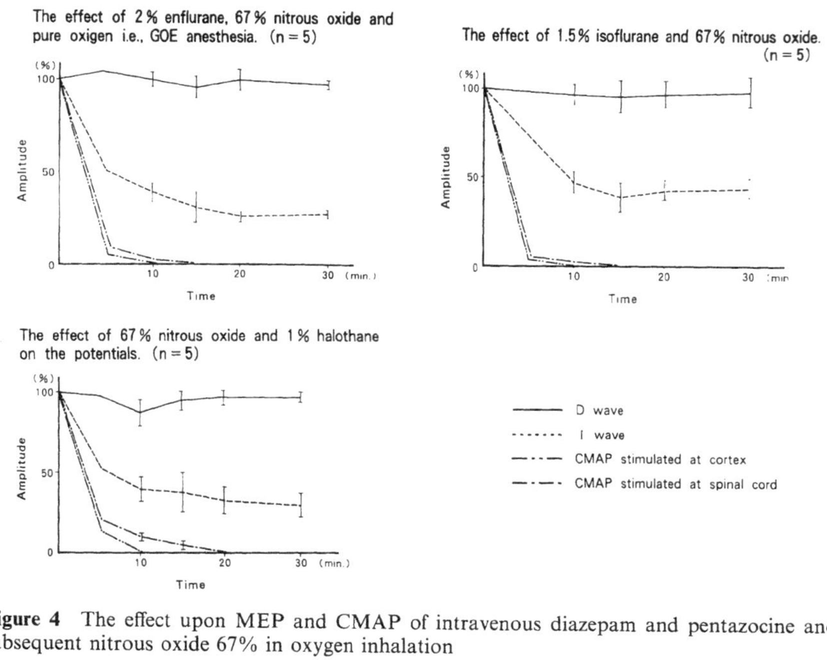

Figure 4 The effect upon MEP and CMAP of intravenous diazepam and pentazocine and subsequent nitrous oxide 67% in oxygen inhalation

DISCUSSION

In conclusion, the inhalational anaesthetic agents enflurane, halothane and isoflurane administered with oxygen significantly reduced the amplitude of each recorded potential except the MEP D-wave. The I-wave was reduced to less than 50% of its original amplitude and compound muscle action potentials to less than 30%, with the action potentials evoked by cortical stimulation being abolished. These effects appeared to be concentration dependent.

Nitrous oxide at 67% caused similar but less pronounced reduction of the potentials and its effect appeared to summate with each inhalational agent.

Thus even in the absence of injury to the spinal cord changes in anaesthesia significantly affected motor evoked potentials[1].

The site and mode of action of most anaesthetic agents is unknown although several writers including Zentoner[2], Levy[3] and de Jong[4] have suggested synaptic blockade as a possibility in the brain and in the ventral horn of the spinal cord. Taylor and Fennelly[5] at the Royal National Orthopaedic Hospital, London have shown similar effects from isoflurane and nitrous oxide on the motor potential in man during spinal surgery in this symposium.

The most stable potential in these experiments was the MEP D-wave and this may be the most appropriate monitor of motor pathway function during anaesthesia with nitrous oxide and inhalational agents as used in many clinical facilities.

The similarity of the responses recorded during administration of 67% nitrous oxide alone or with diazepam and pentazocine neuroleptanalgesia suggests that this may be a suitable technique to allow cord monitoring by I-wave or compound muscle action potential recording in future.

References

1. Toda Y. The effect of anesthetic agents on descending spinal evoked potentials and the compound muscle action potentials elicited by stimulation of the cerebral motor cortex and the spinal cord. Journal of the Japanese Orthopaedic Association. 1992;66:279–290.
2. Zentner J, Ebner A. Nitrous oxide suppresses the electromyographic response evoked by electrical stimulation of the motor cortex. Neurosurgery. 1989;24(1):60–62.
3. Levy WJ, McCaffrey M, York DH, Tauzer F. Motor evoked potentials from transcranial stimulation of the motor cortex in cats. Neurosurgery. 1984;15(2):214–227.
4. de Jong RH, Robles R, Heavner JE. Suppression of impulse transmission in the cats doral horn by inhalational anesthetics. Anesthesiology. 1970;32:440–445.
5. Taylor BA, Fennelly ME, Farrell J. Intraoperative spinal cord monitoring with combined motor and sensory spinal evoked potentials. Journal of Bone and Joint Sugery (Br). 1992;74B Supp. 1:83–84.

60
Acute motor evoked potential changes following spinal cord ischaemia in the rat

W. WANG, R. A. ZAPPULLA and J. NIEVES

MATERIAL AND METHODS

Twenty-two female Sprague-Dawley rats, 250–300 g, were used in the study. All rats were allowed free access to food and water the night before surgery. The experimental rats (n = 15) underwent a right craniectomy and laminectomy at the T13 spinal level under anaesthesia (ketamine 80mg/kg, i.p. and Rompun 20 mg/kg, i.m.). The animals were intubated, mechanically ventilated and chemically paralysed. The body temperature was constantly monitored by a rectal probe and maintained at 37–38°C with a thermal pad. The blood gases were tested randomly among the experimental and control groups and no significant difference was found between the two.

The animals were placed in a prone position and an incision was made over the left upper thorax exposing the descending aorta at the level where the azygos vein crosses the aorta. A ligature was passed around the aorta at this position. Prior to closure of the ligature baseline MEPs were collected and verified by three consecutive recordings collected three minutes apart. Immediately following aortic occlusion by ligature closure the MEPs were recorded and repeated at 2.5 minute intervals for up to 30 minutes following occlusion. Transcortical monopolar stimulation (10.1 Hz, 15 mA, 0.05 ms) was delivered between an insulated stainless steel electrode placed on the pial matter over the sensorimotor cortex (anode) and a needle electrode placed in the hard palate (cathode).

MEPs were recorded epidurally from the spinal cord through the laminectomy and referenced to a needle electrode in the paraspinal muscles. The analog signals were amplified, filtered (150–2800 Hz), and averaged on a QSI 9000 system (Quantified Signal Imaging Inc., Toronto, Ontario, Canada). Each MEP consisted of the average of 2000 single sweep epochs. The control group (n = 7) received the same surgical procedure as the experimental group except for the aortic occlusion.

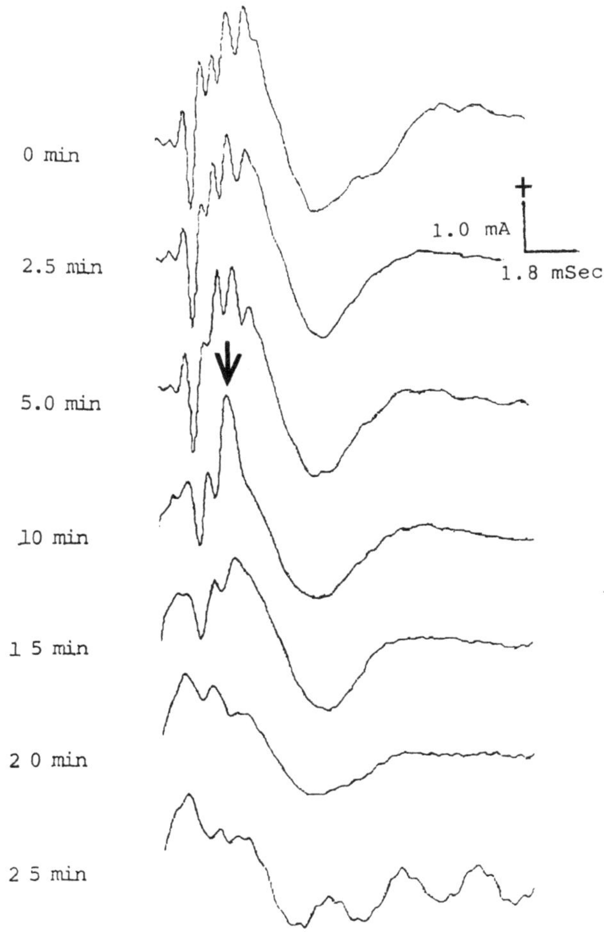

Figure 1 MEP changes following aortic occlusion. This figure represents the MEP changes in five of the 15 experimental animals following spinal cord ischaemia. In this figure the amplitude enhancement (phase 1) is shown at 5 minutes followed by a positive, slow injury potential (phase 2) at 10 minutes (arrow). At 20 minutes after ischaemia the MEP is lost completely (phase 3)

Five peaks were observed for each MEP recording and each peak was analysed separately. The group effects of aortic occlusion on the MEP were evaluated by the difference in peak latency and amplitude between recordings obtained immediately prior to and following aortic occlusion compared with the control group.

RESULTS

Figures 1 and 2 are representatives of the MEP changes recorded from all animals following aortic occlusion. Changes in peak amplitude following

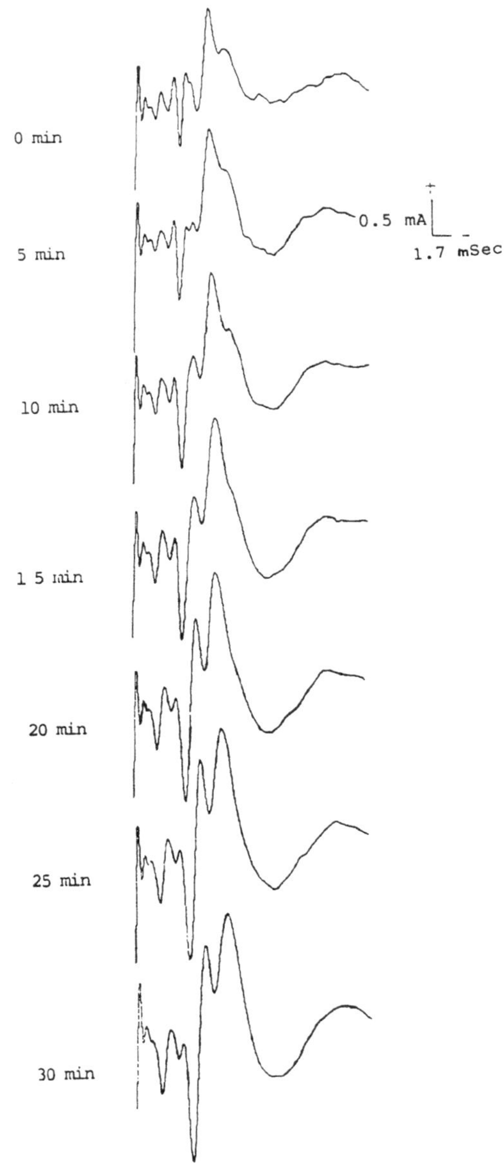

Figure 2 MEP changes following aortic occlusion. This figure represents the MEP changes in 10 of the 15 experimental animals. The amplitude increase is obvious from 15 to 30 minutes. In addition most of these peaks become slower and partially merged. However, the MEP persists throughout our observation period. The difference between this and another pattern shown in Figure 1 may reflect the difference in the collateral circulation to the spinal cord

aortic occlusion had three distinct phases. An initial increase in peak amplitude (phase 1) occurred immediately following aortic occlusion, followed by a replacement of the later peaks by a large voltage, slow response (phase 2) which heralded the loss of all peaks in five animals (Figure 1). In the remaining ten animals, following the initial peak amplitude enhancement, the later peaks became slower, less distinctive and partially merged but persisted throughout the observation period (Figure 2), although in some animals the peak amplitudes were below baseline levels.

The increase in peak amplitude observed in phase 1 occurred for at least one peak of the responses in all experimental animals. Analysis of group data demonstrated a significant ($p < 0.05$) amplitude enhancement as compared to the control group for peaks 1 to 4 (Figure 3A).

There was a progressive increase in peak latency following aortic occlusion. At the conclusion of the observation period (30 minutes) there was a significant increase ($p < 0.05$) in the latency of peaks 2 to 5 when compared to the control group (Figure 3B).

DISCUSSION

A previous study employing a similar experimental model has demonstrated the effectiveness of aortic occlusion in the rat as a technique for inducing spinal cord ischaemia[1]. The difference between animals that lost their MEPs and those animals whose MEPs persisted may reflect a difference in the collateral circulation to the spinal cord. Consequently, the effects of collateral circulation on spinal cord viability must be accounted for in experimental spinal cord ischaemic models.

The early enhancement of peak amplitude (phase 1) reported here has been previously described in studies of global cerebral ischaemia in the dog[2]. The fact that the cerebral circulation in our study was not compromised argues against a cortical mechanism for the peak enhancement observed. It is more likely that it reflects altered activity of spinal pathways and/or grey matter. This is supported by recent findings from our laboratory that demonstrated similar MEP amplitude enhancement following compression injury of the rat spinal cord[3]. It is possible that compressive injury results in vascular compromise of the cord and that the peak enhancement subsequent to cord compression shares a common mechanism as that operating during spinal ischaemia produced by aortic occlusion. While the exact mechanism for the peak enhancement is unknown it is possible to speculate that ischaemia leads to a hyper-excitability of spinal neurons due to a selective loss of inhibition or increased activity or excitatory neural elements. Previous studies have demonstrated the involvement of excitatory amino acids following ischaemia to the central nervous system[4]. Alternatively, the increase in peak amplitude may result from a decrease in inhibitory mechanisms operating at the spinal level. The spinal grey matter has a high concentration of glycine which is responsible for inhibitory activity of the spinal interneuronal system. Strychnine, a glycine blocker, has been shown to enhance the MEP[5].

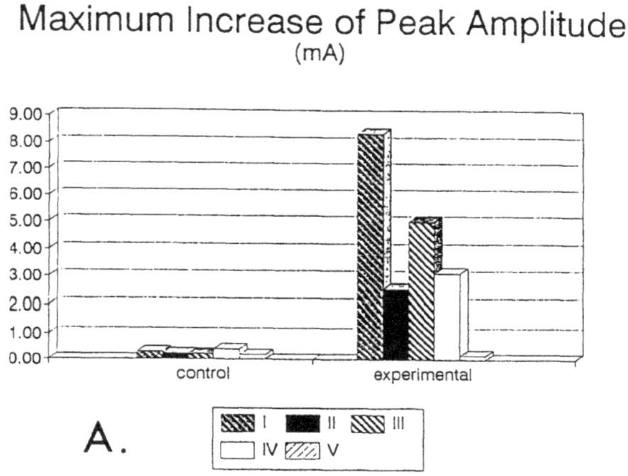

A.

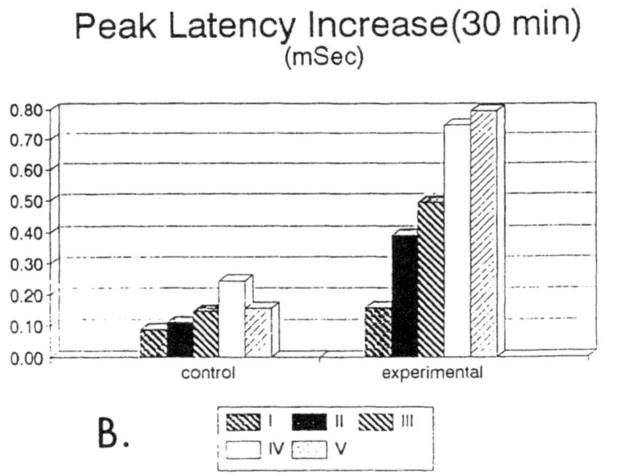

B.

Figure 3 The maximum peak increase during the 30 minute spinal ischaemia (A) and peak latencies at the end of the 30 minute (B). This is a group comparison between experimental and ischaemic animals. The amplitude difference is significant for peak 1 to 4. The latency difference is significant for peak 2 to 5

The injury potential recorded in all animals that went on to lose their MEP may represent a progressive change related to calcium influxes secondary to injury and to changes in the ionic milieu[6,7]. The occurrence of this potential heralds the more serious injury as represented by the loss of the MEP.

In conclusion this study demonstrates that there are three distinct patterns of change in the MEP following aortic occlusion. All experimental animals demonstrated an early enhancement of MEP amplitude. However only those animals that developed an injury potential ultimately lost their MEP. While the exact mechanism for these changes is uncertain, the data suggest that specific alterations of the MEP may be a predictor of spinal cord injury.

References

1. LeMay D, Neal S, Neal S, Zelenock G, D'Alecy L. Paraplegia in the rat induced by aortic cross-clamping: Model characterization and glucose exacerbation of neurologic deficit. J Vasc Surg. 1987;5:383–390.
2. Konrad P, Tacker W, Levy W, Reedy P, Cook J, Geddes L. Motor evoked potentials in the dog: effects of global ischemia on spinal cord and peripheral nerve signals. Neurosurg. 1987;20:117–124.
3. Shiau J, Zappulla R, Nieves J. The effect of graded spinal cord injury on the extrapyramidal and pyramidal motor evoked potentials of the rat. Neurosurgery. 1992;30:76–84.
4. Wieloch T. Neurochemical correlates to selective neuronal vulnerability. Progress in Brain Research. 1985;63:69–85.
5. Gennuso R, Zappulla R, Nieves J. The effect of spinal gray activation by strychnine on the motor evoked potential in the rat. Neurosurgery. 1991;29:563–567.
6. Goodman R, Wachs K, Keller S, Black P. Spontaneous spinal cord 'injury potential' in the rat. Neurosurgery. 1985;17:757–759.
7. Borgens R. Voltage gradients and ionic currents in injured and regenerating axons. Advances in Neurology. 1988;47:52–66.

Index of Authors by Chapter Number

Subject Index

www.ingramcontent.com/pod-product-compliance
Lightning Source LLC
LaVergne TN
LVHW011901200926
848484LV00002B/496
9780792388333